Y. K. Lin · I. Elishakoff (Eds.)

Stochastic Structural Dynamics 1

New Theoretical Developments

Second International Conference
on Stochastic Structural Dynamics
May 9–11, 1990, Boca Raton, Florida, USA

Springer-Verlag

Berlin Heidelberg NewYork
London Paris Tokyo
Hong Kong Barcelona Budapest

Prof. Y. K. Lin
Center for Applied Stochastics Research
Florida Atlantic University
Boca Raton, FL 33431-0991
USA

Prof. I. Elishakoff
Center for Applied Stochastic Research and
Department of Mechanical Engineering
Florida Atlantic University
Boca Raton, FL 33431-0991
USA

ISBN-13:978-3-642-84533-8 e-ISBN-13:978-3-642-84531-4
DOI: 10.1007/978-3-642-84531-4

Library of Congress Cataloging-in-Publication Data
International Conference on Stochastic Structural Dynamics (2nd : 1990 : Boca Raton, Fla.)
Stochastic structural dynamics /Y. K. Lin, I. Elishakoff.
Contents: v. 1. New theoretical developments.

1. Structural dynamics. 2. Stochastic processes.
I. Lin, Y. K. (Yu-Kweng). II. Elishakoff, Isaac. III. Title.
TA654.I567 1990
624.1'7--dc 20 91-18956

Softcover reprint of the hardcover 1st edition 1991

Typesetting: Camera ready by authors

61/3020-543210 – Printed on acid-free paper

Dedicated
to the 25th Anniversary
of Florida Atlantic University

Y. K. L., I. E.

Preface

This volume contains eighteen selected papers presented at the Second International Conference on Stochastic Structural Dynamics, which are related to new theoretical developments in the field. This and a companion volume, related to new practical applications, constitute the proceedings of the conference, and reflect the state of the art of the rapidly developing subject.

The conference was held in Boca Raton, Florida during May 9-11, 1990 hosted by the Center for Applied Stochastics Research of Florida Atlantic University. A total of 20 technical sessions were organized, and attended by eighty participants from 12 countries.

Special emphases of the conference were placed on two areas: applications to earthquake engineering and stochastic stability of nonlinear systems. Two sessions were dedicated to the memory of late Professor Frank Kozin, one of the founders and most active contributors to the stochastic stability theory.

We are indebted to the National Center for Earthquake Engineering Research (NCEER) for financial support. Most credit belongs to each of the authors whose contributions were the very basis for the undoubted success of the conference. We are grateful to the reviewers who carefully refereed the contributions for these two volumes. Our special thanks are due to Mrs. Christine Mikulski, who carried out all the necessary secretarial tasks associated with the conference with dedication.

Y.K. Lin, I. Elishakoff
Editors
Boca Raton, February 1991

List of Participants

(The chairmen of the sessions are identified by asterisks)

* Ang, A.H-S., Department of Civil Engineering, University of California, Irvine, CA 92717, USA

* Ariaratnam, S.T., Solid Mechanics Division, University of Waterloo, Waterloo, Ontario, CANADA N2L 3G1

* Benaroya, H., Department of Mechanical and Aerospace Engineering, Rutgers University, Piscataway, NJ 08855-0909, USA

Billah, K. Y. R., Department of Civil Engineering and Operations Research, Princeton University, E-209 Engineering Quadrangle, Princeton, NJ 08544, USA

Cai, G., Center for Applied Stochastics Research, Florida Atlantic University, Boca Raton, FL 33431-0991, USA

* Casciati, F., Dipartamento di Meccanica Strutturale, Dell'Universita Di Pavia, Via Abbiategrasso 211, 27100 Pavia, ITALY

Choi, S. T., Institute of Aeronautics and Astronautics, National Cheng Kung University, Tainan, TAIWAN 70101

Davoori, H., Department of Mechanical Engineering, University of Puerto Rico, Mayaguez, PUERTO RICO 00708

Debchaudhury, A., Rockwell International, Rocketdyn, Department 545, 6633 Canoga Ave., Canoga Park, California 91303, USA

Di Paola, M., Dipartimento di Ingegneria Strutturale e Geotecnica, Universita' degli Studi di Palermo, Viale delle Scienze, I 90128, Palermo, ITALY

Donley, M. G., Structural Dynamics Research Corporation, 2000 Eastman Drive, Milford, OH 45150-2789, USA

Elishakoff, I., Center for Applied Stochastics Research and Department of Mechanical Engineering, Florida Atlantic University, Boca Raton, FL 33431-0991, USA

Falsone, G., Dipartimento di Ingegneria Strutturale e Geotecnica, Universita' degli Studi di Palermo, Viale delle Scienze, I 90128, Palermo, ITALY

Faravelli, L., Dipartamento di Meccanica Strutturale, Dell'Universita Di Pavia, Via Abbiategrasso 211, 27100 Pavia, ITALY

Gaonkar, G.H., Department of Mechanical Engineering, Florida Atlantic University, Boca Raton, FL, 33431-0991, USA

* Hilton, H. H., Department of Aeronautical and Astronautical Engineering, University of Illinois at Urbana-Champaign, 104 S. Mathews, Urbana, IL 61801-2997, USA

* Hoshiya, M., Department of Civil Engineering, Musashi Institue of Technology, 1-28-1 Tamazutsumi Setagaya-ku, Tokyo 157, JAPAN

Hou, Z.K., California Institute of Technology, 104-44, Pasadena, CA 91125, USA

* Irschik, H., Department of Civil Engineering, Institut für Allgemeine Mechanik, Technische Universität Wien, Karlsplatz 13/201 A-1040 Wien AUSTRIA

Kareem, A., Department of Civil Engineering, University of Notre Dame, Notre Dame, IN 46556-0767 USA

Larsen, C. E., Mail Code ES42, Johnson Space Center, NASA, Houston, TX 77058 USA

Leng, G., Department of Aeronautical and Astronautical Engineering, University of Illinois at Urbana-Champaign, 104 South Mathews, Urbana, IL 61801, USA

Li, Z., Department of Civil, Mechanical and Environmental Engineering, The George Washington University, Washington, D.C. 20052, USA

Lin, Y. K., Center for Applied Stochastics Research, College of Engineering, Florida Atlantic University, Boca Raton, Fl, 33431-0991, USA

Liu, W. K., Department of Mechanical Engineering, Northwestern University, Evanston, IL 60201, USA

* Lutes, L. D., Department of Civil Engineering, Texas A&M University, College Station, TX 77843, USA

Maymon, G., Ministry of Defence, RAFAEL - Div. 46, P.O. Box 2250, Stop 46 Haifa, 31 021 ISRAEL

* Minai, R., Disaster Prevention Research Institute, Kyoto University, Gokasho, Uji, Kyoto 611, JAPAN

Muscolino, G., Dipartimento di Ingegneria Strutturale e Geotecnica, Universita' degli Studi di Palermo, Viale delle Science, Palermo, ITALY

* Namachchivaya, N. Sri, Department of Aeronautical and Astronautical Engineering, University of Illinois at Urbana-Champaign, 104 South Mathews, Urbana, IL 61801, USA

Noori, M., Department of Mechanical Engineering, Worcester Polytechnic Institute, 100 Inst. Rd., Worcester, MA 01609, USA

Norwood, M., 5670 Spring Garden Road, Halifax, CANADA B3J 1H6

Orisamolu, I. R., Martec Limited, 5670 Spring Garden Rd., Halifax, Nova Scotia, CANADA B3J 1H6

Palo, P., Code L44, Naval Civil Engineering Laboratories, Ft. Hueneme, CA 93043, USA

Pradlwarter, H.J., Institut für Mechanik, Universität Innsbruck, Technikerstrasse 13, A-6020, Innsbruck, AUSTRIA

Quek, S.T., Department of Civil Engineering, National University of Singapore, 10 Kentridge Crescent, 0511, SINGAPORE

Rahman, M.S., Department of Civil Engineering, North Carolina State University, Box 7908, Raleigh, N.C. 27695-7908, USA

Roy, R.V., R. L. Spencer Laboratories, Department of Mechanical Engineering, University of Delaware, Newark, DE 19716, USA

* Scanlan, R. H., Department of Civil Engineering, The Johns Hopkins University, Baltimore, MD 21218, USA

* Schuëller, G.I., Institut für Mechanik, Universität Innsbruck, Technikerstrasse 13, A-6020, Innsbruck, AUSTRIA

Shiao, M. C., Sverdrup Technology, Inc., NASA Lewis Research Center Group, 2001 Aerospace Parkway, Brook Park, OH 44142, USA

* Shinozuka, M., Department of Civil Engineering & Operations Research, Princeton University, E-209 Engineering Quadrangle, Princeton, NJ 08544, USA

* Singh, M. P., Dept. of Engineering Science and Mechanics, Virginia Polytechnic Institute and State University, Blacksburg, VA 24061-0219, USA

Socha, L., Department of Civil Engineering, 212 Ketter Hall, State University of NY, Buffalo, NY 14260, USA

* Spanos, P.D., Department of Mechanical Engineering, Rice University, P.O. Box 1892, Houston, TX 77251, USA

Sternberg, A., RAPHAEL, Kiriat Mozkin, P.O.B. 2250, Haifa, 31021, ISRAEL

Su, H., Department of Mechanical Engineering, S-B 201, Concordia University, 1455 de Maisonneuve Blvd. W., Montreal, Quebec, CANADA H3G 1M8

Su, T-C., Center for Applied Stochastics Research and Department of Ocean Engineering, College of Engineering, Florida Atlantic University, Boca Raton, FL, 33431-0991, USA

* Timashev, S., Science and Engineering Center, USSR Academy of Sciences, Ural Dept., 91 Pervomaiskaya St., 620219, Sverdlovsk, GSP-207, U.S.S.R

* To, C.W.S., Department of Mechanical Engineering, University of Western Ontario, London, Ontario, CANADA N6A 5B9

* Wedig, W.V., Institut für Technische Mechanik, Universität of Karlsruhe, Kaiserstrasse 12, D-7500 Karlsruhe 1, FRG

Xu, K., Dept. of Civil Engineering, Northwestern University, 2145 Sheridan Rd., Evanston, IL 60208, USA

* Yang, C.Y., Department of Civil Engineering, 137 DuPont Hall, University of Delaware, Newark, DE 19716, USA

Yim, S., Department of Civil Engineering, Apperson Hall 206, Oregon State University, Corvallis, OR 97331, USA

Yong, Y., Center for Applied Stochastics Research and Department of Ocean Engineering, Florida Atlantic University, Boca Raton, FL 33431-0991, USA

Zerva, A., Department of Civil Engineering, Drexel University, Philadelphia, PA 19104, USA

Zhang, R., Center for Applied Stochastics Research, Florida Atlantic University, Boca Raton, FL 33431-0991, USA

Zhang, X.T., Department of Engineering Mechanics, Tongji University, Shanghai, 200092 PEOPLE'S REPUBLIC OF CHINA

Zhang, Z.Y., Department of Electrical Engineering and Computer Science, Polytechnic University, 333 Jay Street, Brooklyn, NY 11201, USA

Zhu, W.Q., Department of Mechanics, Zhejiang University, Hangzhou, PEOPLE'S REPUBLIC OF CHINA

Contents

Lyapunov Exponents of Two-Degrees-of-Freedom Linear Stochastic Systems

S. T. Ariaratnam, D.S.F. Tam and Wei-Chau Xie,
Solid Mechanics Division,
Faculty of Engineering, University of Waterloo,
Waterloo, Ontario, Canada, N2L 3G1.

Summary

The almost-sure asymptotic stability of a class of two degrees-of-freedom linear systems subjected to wide-band random parametric excitation of small intensity is investigated. By combined use of the method of stochastic averaging and a well-known procedure due to Khas'minskii, asymptotic expressions for the largest Lyapunov exponent of the system for various values of the system parameters are derived, from which conditions for stochastic almost-sure asymptotic stability are obtained.

Introduction

Lyapunov exponents play an important role in the modern theory of nonlinear dynamical systems. They characterize the rate of exponential changes of the states or responses of the systems. For continuous stochastic dynamical systems, the almost sure stability of the trivial solution is determined by the sign of the largest Lyapunov exponent of the linearized system. If the largest Lyapunov exponent is negative, the trivial solution is stable with probability 1 (w.p.1) while, if the largest Lyapunov exponent is positive, the trivial solution is unstable w.p.1. Hence, the vanishing of the largest Lyapunov exponent gives the boundary of almost-sure stability.

A method for the exact evaluation of the Lyapunov exponent of linear systems described by stochastic differential equations of the Itô type was given by Khas'minskii [1,2] in 1967. This method has been fruitfully employed to evaluate numerically Lyapunoy exponents and stochastic stability conditions for certain two-dimensional systems by several investigators, notably Mitchell and Kozin [3], and Nishioka [4]. Asymptotic expressions for Lyapunov exponents for such systems subjected to weak excitations have been obtained by Arnold et al. [5] using essentially the method of Khas'minskii, and by Stratonovich and Romanovskii [6] using the method of stochastic averaging. The direct use of Khas'minskii's method to higher dimensional systems has not met with much success because of the difficulty of discussing diffusion processes occurring on surfaces of unit hyperspheres in higher dimensional spaces.

† This paper is dedicated to the memory to the late Professor Frank Kozin

Y. K. Lin · I. Elishakoff (Eds.)
Stochastic Structural Dynamics 1
New Theoretical Developments
2nd International Conference on Stochastic Structural Dynamics
May 9–11, 1990, Boca Raton, Florida

In this paper the stability of a class of coupled two-degrees-of-freedom systems subjected to parametric excitation by a wide band stochastic process of small intensity is considered. The study is motivated by problems in the dynamic stability of elastic systems subjected to stochastically fluctuating loads. The stochastic moment stability of such systems was examined previously by Ariaratnam and Srikantaiah [7] using the method of stochastic averaging. In the present study, the almost-sure stability of the same class of problems is studied using a combination of the method of averaging and the technique of Khas'minskii [1]. Explicit asymptotic expressions for the largest Lyapunov exponent for various values of the system parameters are obtained. The result for single degree-of-freedom systems is obtained in the special case when the coupling parameters are set equal to zero.

Formulation

The systems considered are described by stochastic differential equations of the form:

$$\begin{aligned} \ddot{q}_1+2\beta_1\dot{q}_1+\omega_1^2 q_1+\omega_1(k_{11}q_1+k_{12}q_2)\xi(t) &= 0, \\ \ddot{q}_2+2\beta_2\dot{q}_2+\omega_2^2 q_2+\omega_2(k_{21}q_1+k_{22}q_2)\xi(t) &= 0, \end{aligned} \tag{1}$$

where q_1, q_2 are generalized coordinates, β_1, β_2 are damping constants, ω_1, ω_2 are natural frequencies, and k_{ij}, i, $j=1, 2$ are constants. The excitation is represented by $\xi(t)$, which is taken to be a wide-band, stationary, stochastic process with zero mean value. It is assumed that $\xi(t)$ has a nearly constant spectral density S over a wide range of frequencies which includes ω_1 and ω_2, so that it may be approximated by a white noise process, and that S is of the same order of smallness as the damping coefficients β_1, β_2. The spectral density S is defined by the relation

$$S = 2\int_0^\infty E[\xi(t)\xi(t+\tau)]\cos\omega\tau\, d\tau,$$

where $E[\ ]$ denotes the expectation.

Transforming to polar coordinates using the relations

$$q_i = a_i\cos\Theta_i, \quad \dot{q}_i = -a_i\omega_i\sin\Theta_i, \quad \Theta_i = \omega_i t+\theta_i, \quad i=1, 2, \tag{2}$$

and then applying stochastic averaging to the resulting equations (see Ariaratnam and Srikantaiah[7]) lead to the following Itô equations for the averaged amplitudes a_1, a_2, which are a uniformly valid first approximation to the exact values:

$$da_1 = m_1 dt + \sigma_{11} dW_1 + \sigma_{12} dW_2,$$
$$da_2 = m_2 dt + \sigma_{21} dW_1 + \sigma_{22} dW_2, \tag{3}$$

where $W_1(t)$, $W_2(t)$ are independent Wiener processes of unit intensity and the drift coefficients m_i and the diffusion coefficients σ_{ij}, $i,j=1,2$ are given by

$$m_1 = -\beta_1 a_1 + \frac{S}{16}\Big(3k_{11}^2 a_1 + 2k_{12}^2 \frac{a_2^2}{a_1}\Big),$$

$$m_2 = -\beta_2 a_2 + \frac{S}{16}\Big(3k_{22}^2 a_2 + 2k_{21}^2 \frac{a_1^2}{a_2}\Big),$$

$$[\sigma\sigma^T]_{11} = \frac{S}{8}(k_{11}^2 a_1^2 + 2k_{12}^2 a_2^2),$$

$$[\sigma\sigma^T]_{22} = \frac{S}{8}(k_{22}^2 a_2^2 + 2k_{21}^2 a_1^2),$$

$$[\sigma\sigma^T]_{12} = [\sigma\sigma^T]_{21} = 0.$$

It may be noted that the averaged amplitude vector (a_1, a_2) is a two-dimensional diffusion process and that the coefficients of the right hand side terms of equations (3) are homogeneous in a_1, a_2 of degree one. Hence, the procedure of Khas'minskii may be employed to derive an expression for the largest Lyapunov exponent of the amplitude process. To this end a further polar transformation is applied:

$$\rho = \frac{1}{2}\log(a_1^2 + a_2^2), \quad \phi = \mathrm{Tan}^{-1}(a_2/a_1), \quad 0 \le \phi \le \frac{1}{2}\pi.$$

By the use of Itô's differential rule, or otherwise, the following pair of Itô equations governing ρ, ϕ may be obtained:

$$d\rho = Q(\phi)dt + \Sigma(\phi)dW,$$
$$d\phi = \Phi(\phi)dt + \Psi(\phi)dW, \tag{4}$$

where $W(t)$ is a Wiener process of unit intensity and

$$Q(\phi) = \lambda_1 \cos^2\phi + \lambda_2 \sin^2\phi + \Psi^2(\phi),$$

$$\Phi(\phi) = \frac{1}{2}(\lambda_2 - \lambda_1)\sin 2\phi + \frac{S}{64}[(k_{11}^2 + k_{22}^2)\sin 4\phi + 16(k_{12}^2 \sin^4\phi + k_{21}^2 \cos^4\phi)\cot 2\phi],$$

$$\Sigma^2(\phi) = \frac{S}{16}[2(k_{11}^2 \cos^4\phi + k_{22}^2 \sin^4\phi) + (k_{12}^2 + k_{21}^2)\sin^2 2\phi], \tag{5}$$

$$\Psi^2(\phi) = \frac{S}{32}[(k_{11}^2 + k_{22}^2)\sin^2 2\phi + 8(k_{12}^2 \sin^4\phi + k_{21}^2 \cos^4\phi)].$$

The constants λ_1, λ_2 are defined by

$$\lambda_1 = -\beta_1+\frac{S}{8}k_{11}^2, \quad \lambda_2 = -\beta_2+\frac{S}{8}k_{22}^2,$$

and are, as will be seen later, the Lyapunov exponents of the two uncoupled single degree-of-freedom systems that result when the coupling coefficients k_{12}, k_{21} are set equal to zero.

From the second of equation (4) it is clear that the ϕ—process is a diffusion on the first quadrant of the unit circle. Since $\Psi^2(\phi)$ does not vanish in $0\leq\phi\leq\pi/2$, the diffusion is non-singular, the density $\mu(\phi)$ of the invariant measure being governed by the Fokker-Planck equation:

$$\frac{1}{2}[\Psi^2(\phi)\mu(\phi)]'' - [\Phi(\phi)\mu(\phi)]' = 0, \tag{6}$$

where a prime signifies differentiation with respect to ϕ. The general solution of equation (6) is

$$\mu(\phi) = \frac{2C}{\Psi^2(\phi)U(\phi)} - \frac{2G}{\Psi^2(\phi)U(\phi)}\int^{\phi} U(\theta)d\theta, \tag{7}$$

where C, G are integration constants and

$$U(\phi) = \exp[-2\int^{\phi} \{\Phi(\theta)/\Psi^2(\theta)\}d\theta]$$
$$= 2\operatorname{cosec}2\phi \exp\Big[(\lambda_2-\lambda_1)\int^{\sin^2\phi} \frac{d\theta}{a\theta^2+b\theta+c}\Big], \tag{8}$$

the constants a, b, c being given by

$$a = \frac{S}{8}[2(k_{12}^2+k_{21}^2)-(k_{11}^2+k_{22}^2)],$$
$$b = \frac{S}{8}(k_{11}^2+k_{22}^2-4k_{21}^2), \tag{9}$$
$$c = \frac{S}{4}k_{21}^2.$$

It can be shown (Appendix) that the boundaries $\phi=0$, $\phi=\pi/2$ are both entrance points in the sense of Feller, and hence the stationary probability flux represented by G is zero. Thus, there is no accumulation of probability mass at the boundaries, and the ϕ—process is ergodic throughout the interval $0<\phi<\pi/2$. The invariant density $\mu(\phi)$ is therefore given by

$$\mu(\phi) = \frac{2C}{\Psi^2(\phi)U(\phi)}, \tag{10}$$

C being the normalizing constant. The form of the integral in equation (8) depends on the sign of the discriminant Δ, where

$$\Delta = b^2 - 4ac$$
$$= \frac{S^2}{64}[(k_{11}^2 + k_{22}^2)^2 - 16k_{12}^2 k_{21}^2].$$

For $\Delta > 0$, the invariant density $\mu(\phi)$ is of the form

$$\mu(\phi) = \frac{C\sin 2\phi}{\Psi^2(\phi)}\exp\left[-\frac{2(\lambda_2-\lambda_1)}{\Delta^{1/2}}\tanh^{-1}\frac{b+2a\sin^2\phi}{\Delta^{1/2}}\right], \quad (\Delta > 0), \tag{11a}$$

where C is determined from the condition

$$\int_0^{\pi/2} \mu(\phi)d\phi = 1.$$

For $\Delta < 0$, the hyperbolic term in (11) is to be replaced appropriately by its trigonometric equivalent, while, for $\Delta = 0$, the right hand side of (11) is replaced by its limit as $\Delta \to 0$. Explicitly these expressions are

$$\mu(\phi) = \frac{C\sin 2\phi}{\Psi^2(\phi)}\exp\left[\frac{2(\lambda_2-\lambda_1)}{(-\Delta)^{1/2}}\tan^{-1}\frac{b+2a\sin^2\phi}{(-\Delta)^{1/2}}\right], \quad (\Delta < 0), \tag{11b}$$

and

$$\mu(\phi) = \frac{C\sin 2\phi}{\Psi^2(\phi)}\exp\left[-\frac{2(\lambda_2-\lambda_1)}{b+2a\sin^2\phi}\right], \quad (\Delta = 0). \tag{11c}$$

Lyapunov Exponent

The Lyapunov exponent of the system (1) is defined by

$$\lambda = \lim_{t\to\infty} \frac{1}{t}\log[a_1^2(t)+a_2^2(t)]^{1/2}$$
$$= \lim_{t\to\infty} \frac{1}{t}\rho(t).$$

Integrating the first of equations (4),

$$\rho(t) = \rho(0) + \int_0^t Q(\phi)dt + \int_0^t \Sigma(\phi)dW,$$

so that

$$\frac{1}{t}\rho(t) = \frac{1}{t}\rho(0) + \frac{1}{t}\int_0^t Q(\phi)dt + \frac{1}{t}\int_0^t \Sigma(\phi)dW.$$

Since the ϕ—process is ergodic, and $E[dW(t)] = 0$,

$$\frac{1}{t}\int_0^t Q(\phi)dt \;\rightarrow\; E[Q(\phi)],$$

and

$$\frac{1}{t}\int_0^t \Sigma(\phi)dW \;\rightarrow\; 0,$$

as $t \rightarrow \infty$, with probability one. Hence λ is given by

$$\lambda = E[Q(\phi)] = \int_0^{\pi/2} Q(\phi)\mu(\phi)d\phi. \tag{12}$$

Substituting from equations (5) and (11) in equation (12) and performing the indicated integration yields the following expression for the Lyapunov exponent:

$$\lambda = \frac{1}{2}\Big[(\lambda_1+\lambda_2) + (\lambda_1-\lambda_2)\coth(\frac{\lambda_1-\lambda_2}{\Delta^{1/2}}\alpha)\Big], \quad (\Delta>0), \tag{13a}$$

where α is given by

$$\tanh\alpha = \Big[1-(\frac{4k_{12}k_{21}}{k_{11}^2+k_{22}^2})^2\Big]^{1/2}. \tag{14a}$$

Again, when $\Delta<0$, the corresponding trigonometric forms are substituted in the right hand sides of equations (13) and (14) to give

$$\lambda = \frac{1}{2}\Big[(\lambda_1+\lambda_2) + (\lambda_1-\lambda_2)\coth(\frac{\lambda_1-\lambda_2}{(-\Delta)^{1/2}}\alpha)\Big], \quad (\Delta<0), \tag{13b}$$

where α is given by

$$\tan\alpha = \Big[(\frac{4k_{12}k_{21}}{k_{11}^2+k_{22}^2})^2-1\Big]^{1/2}. \tag{14b}$$

In the exceptional case when $\Delta=0$, the limiting form of (13) is

$$\lambda = \frac{1}{2}\Big[(\lambda_1+\lambda_2) + (\lambda_1-\lambda_2)\coth\frac{2(\lambda_1-\lambda_2)}{|k_{12}k_{21}|S}\Big]. \tag{15}$$

The Lyapunov exponent for a single degree-of-freedom system may be recovered from equations (13) and (14). Thus, setting the coupling coefficients k_{12}, k_{21} to zero, it is evident from equation (14) that $\alpha=0$, and equation (13) then gives $\lambda=\lambda_1$ if $\lambda_1>\lambda_2$ and $\lambda=\lambda_2$ if $\lambda_2>\lambda_1$, confirming that the expression (13) is, in fact, the largest Lyapunov exponent of the system.

The system is asymptotically stable with probability one (w.p.1) if λ is negative and unstable w.p.1 if λ is positive.

Conclusions

A method of calculating the Lyapunov exponent of a class of two degrees-of-freedom systems subjected to wide-band, parametric, random excitation having a uniform spectral density has been presented. An explicit expression for the largest Lyapunov exponent, valid in the first approximation, has been obtained. The result obtained is applicable to a wide class of mechanical systems whose stiffness is subjected to random fluctuations of small intensity.

The method has been extended also to the case where the excitation spectrum is non-uniform and to certain multi-degrees-of-freedom linear systems. These results will be reported in a forthcoming publication.

Acknowledgement

The research reported in this paper was supported by the Natural Sciences and Engineering Research Council of Canada, Grant No. A-1815.

References

1. Khas'minskii, R.Z. Necessary and sufficient conditions for the asymptotic stability of linear stochastic systems, *Theory of Probability and Its Applications* (English translation) 1967, **12**(1), 144.
2. Khas'minskii, R.Z. *Stochastic Stability of Differential Equations*, Sijthoff and Noordhoff, Alphen an den Rijn, 1980 (Translation of the Russian edition, Nauka, Moscow, 1969).
3. Mitchell, R.R. and Kozin, F. Sample stability of second order differential equations with wide band noise coefficients, *SIAM Journal on Applied Mathematics* 1974, **27**, 571.
4. Nishioka, K. On the stability of two-dimensional linear stochastic systems, *Kodai Mathematics Seminar Reports* 1976, **27**, 211.
5. Arnold, L., Papanicolaou, G. and Wihstutz, V. Asymptotic analysis of the Lyapunov exponent and rotation number of the random oscillator and applications, *SIAM Journal on Applied Mathematics* 1986, **46**, 427.
6. Stratonovich, R.L. and Romanovskii, Yu.M. Parametric effect of a random force on linear and non-linear oscillatory systems, In *Non-Linear Transformations of Stochastic Processes*, Edited by Kuznetsov, P.I., Stratonovich, R.L. and Tikhonov, V., Pergamon Press, Oxford, 1965, 322. (English translation edited by Wise, J. and Cooper, D.C)
7. Ariaratnam, S.T. and Srikantaiah, T.K. Parametric instabilities in elastic structures under stochastic loading, *J. Struct. Mech.* 1978, **6**, 349.
8. Karlin, S. and Taylor, H.M. *A Second Course in Stochastic Processes*, Academic Press, 1981.
9. Gikhman, I.I. and Skorokhod, A.V. *Stochastic Differential Equations*, Springer-Verlag, Berlin, 1972.

Appendix

To establish the nature of the boundaries $\phi=0$, $\phi=\pi/2$ of the diffusion process $\phi(t)$, a procedure described by Karlin and Taylor [8] will be followed.

The scale measure $S[\phi_1, \phi_2]$ of the ϕ—process is defined by

$$S[\phi_1, \phi_2] = \int_{\phi_1}^{\phi_2} U(\theta)d\theta, \tag{A.1}$$

where the scale density $U(\phi)$ is as given by equation (8). It can be seen that since k_{12}, k_{21} are not zero, and by scaling of variables one can make $|k_{12}| \neq |k_{21}|$, there exist positive constants K_1, K_2 such that

$$K_1 \operatorname{cosec} 2\phi \leq U(\phi) \leq K_2 \operatorname{cosec} 2\phi.$$

Hence, if $0<\phi<\pi/2$,

$$S(0,\phi] = \int_0^{\phi} U(\theta)d\theta \geq \int_0^{\phi} K_1 \operatorname{cosec} 2\theta \, d\theta = \infty,$$

and

$$S[\phi,\pi/2) = \int_{\phi}^{\pi/2} U(\theta)d\theta \geq \int_{\phi}^{\pi/2} K_1 \operatorname{cosec} 2\theta \, d\theta = \infty,$$

implying that $U(\phi)$ is not integrable in the vicinity of both the boundaries $\phi=0$ and $\phi=\pi/2$. These boundaries are therefore *natural* under the Gikhman-Skorokhod [9] classification.

To further determine whether they are also *entrance* boundaries as defined by Feller, it is necessary to consider the speed measure defined by

$$M[\phi_1, \phi_2] = \int_{\phi_1}^{\phi_2} \frac{1}{\Psi^2(\phi)U(\phi)} d\phi, \tag{A.2}$$

and the integrals

$$N(0) = \int_0^{\phi} U(\theta)M(0, \theta]d\theta, \tag{A.3}$$

and

$$N(\pi/2) = \int_{\phi}^{\pi/2} U(\theta)M[\theta,\pi/2)d\theta. \tag{A.4}$$

$N(0)$ and $N(\pi/2)$ approximately measure the times it takes to reach an interior point ϕ, $0<\phi<\pi/2$, starting at the boundary points $\phi=0$ and $\phi=\pi/2$, respectively. If these times are finite, then the boundaries are classified as entrance boundaries. By integrating the expressions in (11), it is found that

$$M[\phi_1,\phi_2] = \begin{cases} 2(\lambda_2-\lambda_1)\exp\left[-\dfrac{2(\lambda_2-\lambda_1)}{\Delta^{1/2}}\tanh^{-1}\dfrac{v}{\Delta^{1/2}}\right]_{\phi_1}^{\phi_2}, & \Delta>0, \\ 2(\lambda_2-\lambda_1)\exp\left[-\dfrac{2(\lambda_2-\lambda_1)}{(-\Delta)^{1/2}}\tan^{-1}\dfrac{v}{(-\Delta)^{1/2}}\right]_{\phi_1}^{\phi_2}, & \Delta<0, \\ 2(\lambda_2-\lambda_1)\exp\left[-\dfrac{2(\lambda_2-\lambda_1)}{v}\right]_{\phi_1}^{\phi_2}, & \Delta=0, \end{cases}$$

where $v(\phi)=b+2a\sin^2\phi$. Considering the case $\Delta>0$, and taking $\phi_1=0$, with α denoting the constant $2(\lambda_2-\lambda_1)/\Delta^{1/2}$,

$$\begin{aligned} N(0) &= \int_0^{\phi} U(\theta)M(0,\theta]d\theta \\ &\le \int_0^{\phi} 2|\lambda_2-\lambda_1|\,\left|\exp(-\alpha\tanh^{-1}\frac{v(\theta)}{\Delta^{1/2}}) - \exp(-\alpha\tanh^{-1}\frac{v(0)}{\Delta^{1/2}})\right| K_2 \operatorname{cosec} 2\theta\, d\theta \\ &< \infty, \end{aligned}$$

since the integrand has a finite limit as $\theta \to 0^+$. Similar arguments apply to $N(\pi/2)$ and for the cases $\Delta<0$, $\Delta=0$.

Random Eigenvalues and Structural Dynamic Models

Haym Benaroya

Mechanical & Aerospace Engineering
RUTGERS University
P.O. Box 909
Piscataway, New Jersey 08855–0909

Abstract

A review of the literature is provided of the study of structural vibration problems with random parameters. There have been several approaches to this problem. These have been encapsulated in this paper for the benefit of those who need to assess such possibilities.

In particular, the algebraic theory of random variables is delineated here with the application in mind being the determination or estimation of the statistics of the eigenvalues of linear dynamical systems. Various transformation techniques are summarized and discussed in addition to simple application.

Background

There is no single more important descriptor of a (linear) dynamic system than its eigenstructure: eigenvalues and eigenvectors. System characteristics, properties, and thus behavior, are all embedded in system eigenvalues and eigenvectors. The effects of mass, damping, and stiffness distributions determine the magnitude and distribution of the eigenvalues, as do structural imperfections, concentrations, and constraints.

Inherent inaccuracies in manufacturing, measurements, uncertainties of geometry and material, and other factors make engineering systems uncertain in the sense

[1]This work is dedicated to the memory of Frank Kozin

Y. K. Lin · I. Elishakoff (Eds.)
Stochastic Structural Dynamics 1
New Theoretical Developments
2nd International Conference on Stochastic Structural Dynamics
May 9–11, 1990, Boca Raton, Florida

that their parameter values cannot be specified exactly. The values can only be determined within a prescribed range in terms of a probability distribution. This requires the analyst and designer to work with uncertainties and to be able to estimate the sensitivities of system performance to such uncertainties.

The study of such systems, those with random valued parameters, is known under several names: disordered structural systems, random systems, random eigenvalue problems. We choose the latter as being the most concise and of broadest implications. Random eigenvalue problems may be viewed as an extension of the classical branch of applied mathematics which deals with deterministic linear homogeneous eigenvalue problems for ordinary differential operators. Here, random eigenvalue problems are considered in the physical context of vibration engineering.

Before proceeding, we refer the reader to the wonderful review by Ibrahim [1987].

Introduction

In general, one studies $2n$ order differential equations of the form

$$Lw(x) = \lambda M w(x) \tag{1}$$

with $2n$ associated boundary conditions $U_i(w) = 0,\ i = 1, ..., 2n$ at $x = a, x = b$. The equivalent integral equation formulation is:

$$w(x) = \lambda \int_a^b G(x, y) M w(x) dy, \tag{2}$$

where $G(x, y)$ is the Green's function associated with the differential operator L subject to the above initial conditions.

In problems of physical origin, coefficients in operators L and M, and boundary conditions U_i are usually only approximate. Therefore, it may more suitable to regard these coefficients as random variables with probability distributions. The eigenvalues and eigenfunctions are then to be interpreted as random variables and random functions, respectively. It is then of interest to address questions such as: *What is the probability that one or more eigenvalue lies in a given interval or is less than a given value?* Such information, if derivable, permits one to proceed with numerous applied studies.

The study of random eigenvalues [v.Scheidt and Purkert 1983] is intimately involved with the theory of random algebraic polynomials [Bharucha–Reid 1986], which has wide applicability. A random algebraic polynomial will arise (i) if the coefficients of an algebraic polynomial are subject to random error, (ii) in the study of difference and differential equations with random coefficients, (iii) in the study of random matrices, (iv) in the study of approximate solutions of operator equations, (v) in regression models of data, (vi) in the analysis of filters and random differential equations. It is possible to see from this list that an understanding of random algebraic polynomials will permit application to an extensive body of science and engineering.

The spectral theory of random matrices has been developed [Bharucha–Reid 1972] primarily by mathematical statisticians and physicists. In quantum mechanics, the energy levels of a system are described by the eigenvalues of a Hamiltonian operator which is linear symmetric (Hermitian). In order to determine the eigenvalues and eigenfunctions of the system, the following eigenvalue equation must be solved:

$$(H - \lambda I)\psi = 0. \tag{3}$$

Due to the complex nature of the Hamiltonian, H is not known. Thus, probabilistic hypotheses on H are introduced and H is represented by a random matrix with the following assumption made: The statistical behavior of the energy levels is identical with the behavior of the eigenvalues of the random matrix.

Problems concerning the distribution of natural frequencies of an elastic continuum were first formulated in connection with the statistical theory of the heat capacity of solids. Early interest in structural elements has been primarily in the distribution of natural frequencies in thin and thin–walled systems. If one is able to assume that the frequencies are discrete, it is possible to introduce the concept of a function of the distribution of natural frequencies. This function is defined as the number of natural frequencies smaller than a particular frequency value. The derivative of this function is, of course, the density of natural frequencies. These functions are characteristics of the given elastic body. Simple examples have been worked by Bolotin [1984] for one and two dimensional problems of rods and shells. Thus, it is in general of interest to estimate the distribution of eigenvalues $N(\Lambda)$ of $\mathbf{K}(\omega)$:

$$N(\Lambda) = \sum_{\Lambda_k < \Lambda} H(\Lambda - \Lambda_k), \tag{4}$$

where H is the Heaviside function, Λ is a realization of a natural frequency, and the density of natural frequencies $\nu(\Lambda)$ is given by the derivative of $N(\Lambda)$

$$\nu(\Lambda) = \sum_{\Lambda_k < \Lambda} \delta(\Lambda - \Lambda_k), \tag{5}$$

where δ is the Dirac delta function.

Consider the discretized model of a structural system:

$$\mathbf{M}\ddot{\mathbf{y}} + \mathbf{C}\dot{\mathbf{y}} + \mathbf{K}\mathbf{y} = \mathbf{f}(t), \tag{6}$$

where in this initial value problem, $\mathbf{f}(t),\ t > 0$ is a stochastic vector function of time. In addition, elements of the matrices $\mathbf{M}, \mathbf{C}, \mathbf{K}$ will contain random variables. This equation may be written using operator notation as

$$\mathbf{Ly} = \mathbf{f}(t), \ t > 0, \tag{7}$$

where $\mathbf{L}$ is a linear homogeneous ordinary differential vector operator. If instead, the formulation is a boundary value problem, then

$$\mathbf{Ly} = \mathbf{f}(x), \ 0 < x < 1, \tag{8}$$

where now the stochastic nonhomogeneous term on the right hand side represents determinant timewise behavior but statistical spacewise behavior.

If randomness (or complexity) exists in the spatial properties of the system parameters, then not only is $\mathbf{f}(x)$ a random field, but so is $\mathbf{L} = \mathbf{L}(x)$, in effect resulting in the random Green's function of the equivalent discretized formulation

$$\lambda^{-1}\mathbf{y}(x_i) = \int_0^1 \mathbf{G}(x_i, x_{i+1})\mathbf{h}(x_{i+1})\mathbf{y}_{i+1}dx_{i+1}, \tag{9}$$

where $\mathbf{f}(x) \equiv \lambda\mathbf{h}(x)\mathbf{y}$, and a recursive relation has been implied. The eigenvalues λ of this system are to be interpreted as random variables.

One can extend the formulation to random matrices, for example, randomly distributed stiffness properties, by assuming the form

$$\begin{aligned} \mathbf{K}(\omega) &= \mathbf{K}_0 + \mathbf{B}(\omega) \\ \mathbf{K}(\omega) &= [k_{ij}(\omega)]_{1\leq(i,j)\leq n} \end{aligned} \tag{10}$$

where the mean value of the random matrix can generally taken equal to $\mathbf{K}_0$. (The notation (ω) denoting random function is adopted here.) Considering a two degree-of-freedom system results in the 2x2 matrix (where no mass coupling exists and zero damping is assumed)

$$\mathbf{K}(\omega) = \begin{bmatrix} k_{11}(\omega) & k_{12}(\omega) \\ k_{21}(\omega) & k_{22}(\omega) \end{bmatrix} \tag{11}$$

with associated equation for characteristic values

$$|\mathbf{K}(\omega) - \lambda I| = 0. \tag{12}$$

$\lambda =$ the square of the modal frequencies, resulting in the random quadratic polynomial (characteristic equation)

$$a_2(\omega)\lambda^2 + a_1(\omega)\lambda + a_0(\omega) = 0, \tag{13}$$

with

$$\begin{aligned} a_2(\omega) &= 1 \quad \text{almost surely.} \\ a_1(\omega) &= -[k_{11}(\omega) + k_{22}(\omega)] \\ a_0(\omega) &= k_{11}(\omega)k_{22}(\omega) - k_{12}(\omega)k_{21}(\omega). \end{aligned} \tag{14}$$

It has been assumed that all mass elements have numerical value 1. As with the deterministic case, $\lambda_1(\omega)$ and $\lambda_2(\omega)$ are the two solutions of the above characteristic equation. In the probabilistic case, however, these are random variables. Their distributions provide significant insight to properties of the system which they represent. It is their joint density, as a function of the joint density of the matrix elements $k_{ij}(\omega)$, which is of interest. Note that $a_0(\omega)$ is a function of the coupling properties $k_{12}(\omega)$ of the two degree-of-freedom system. Three possible solution sets must be considered:

1. $\lambda_1(\omega)$, $\lambda_2(\omega)$ are real and unequal,
2. $\lambda_1(\omega)$, $\lambda_2(\omega)$ are real and equal,
3. $\lambda_1(\omega)$, $\lambda_2(\omega)$ are complex conjugates.

Of particular interest is the effect of various types of coupling, k_{12}, k_{21} ($k_{12} = k_{21}$ almost surely), on the distributions $N(\Lambda), \nu(\Lambda)$. It is possible to consider more general systems with inertial as well as elastic coupling. The system characteristic values are then defined by

$$\begin{vmatrix} k_{11}(\omega) - \lambda m_{11}(\omega) & k_{12}(\omega) - \lambda m_{12}(\omega) \\ k_{21}(\omega) - \lambda m_{21}(\omega) & k_{22}(\omega) - \lambda m_{22}(\omega) \end{vmatrix} = 0. \tag{15}$$

Early work by Hamblen [1955, 1956] considered random algebraic equations of the form

$$\eta^2 - \xi_1 \eta + \xi_2 = 0, \tag{16}$$

which has roots

$$\eta_{1,2} = \frac{\xi_1}{2} \pm \sqrt{\frac{\xi_1^2}{2} - \xi_2}, \tag{17}$$

or $\xi_1 = \eta_1 + \eta_2$, $\xi_2 = \eta_1 \eta_2$. The roots are either both real or complex conjugates. The events $P(R)$ and $P(R)$, the probability of real or complex conjugate roots are then defined by

$$\begin{aligned} P(R) &= P(\xi_2 \leq \frac{\xi_1^2}{4}) \\ P(C) &= P(\xi_2 > \frac{\xi_1^2}{4}). \end{aligned} \tag{18}$$

Given the joint density function $f(\xi_1, \xi_2)$ of coefficients ξ_1 and ξ_2 can be estimated. Therefore, one can evaluate (24) by integrating $f(\xi_1, \xi_2)$ over the appropriate domains. Given $P(R)$ and $P(C)$, the conditional densities of real and complex conjugate roots can be evaluated:

$$\begin{aligned} f(\xi_1, \xi_2 | C) &= f(\xi_1, \xi_2)/P(C), \quad \xi_2 > \xi_1^2/4 \\ f(\xi_1, \xi_2 | R) &= f(\xi_1, \xi_2)/P(R), \quad \xi_2 \leq \xi_1^2/4. \end{aligned} \tag{19}$$

For example, given a *gamma* joint distribution $f(\xi_1, \xi_2) = exp\{-\xi_1 - \xi_2\}$, for coefficients ξ_1, ξ_2 one finds that $P(R) = 0.24$ approximately, and the marginal densities for the roots are found to be

$$g_1(v_1|R) = \frac{1}{0.24}\left[\frac{v_1^2 + v_1 - 1}{(1+v_1)^2} exp\{-v_1\} + (1+v_1)^{-2} exp\{-(v_1^2 + 2v_1)\}\right], \; 0 \le v_1 < \infty$$

$$g_2(v_2|R) = \frac{1}{0.24}\left[(1+v_2)^{-2} exp\{-(v_2^2 + 2v_2)\}\right], \; 0 \le v_2 < \infty, \tag{20}$$

where v_1, v_2 represent the real roots and g_1, g_2 their marginal densities. See Figure 1. Expressions are derived also for the joint densities of the complex conjugate roots $\eta_{1,2} = \alpha \pm i\beta$. One can then derive $g_1(v_1|R), g_2(v_2|R)$.

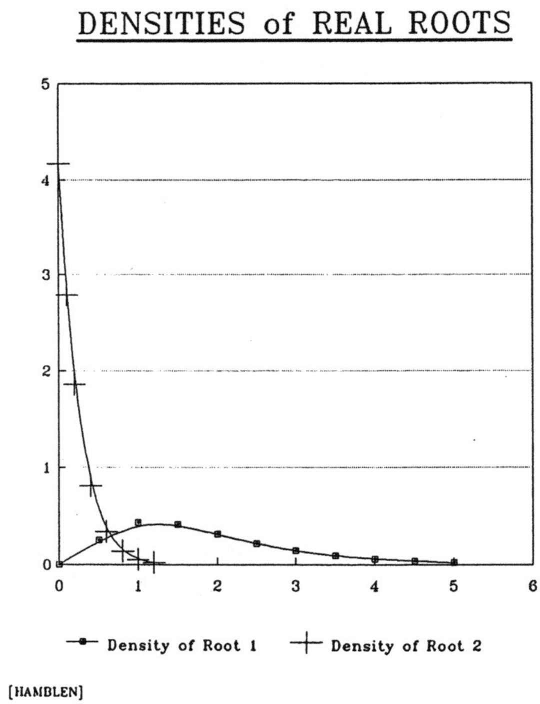

Figure 1.

Algebraic Transform Techniques

Transform techniques have also been used to study random variable systems [Giffin 1975, Springer 1979]. The idea is to utilize transforms such as the Mellin [Epstein 1948]

$$M(s) = \int_0^\infty f(x) x^{s-1} dx \tag{21}$$

in order to determine the densities of functions of random variables. This approach intends to benefit from the advantages of the operational calculus. In fact, characteristic functions (Fourier transform) and moment generating functions (exponential and Laplace transforms) already are widely utilized.

Pursuing the Mellin transform, if $f(x)$ above is interpreted to be a probability density function, then

$$M(s) = E(x^{s-1}), \tag{22}$$

the mathematical expectation. Thus,

$$\begin{aligned} E(x^0) &= M(1) \\ E(x) &= M(2) \\ \mathrm{var}(x) &= M(3) - [M(2)]^2 \end{aligned} \tag{23}$$

a second moment analysis is straightforward. For example, the probability density function of the product $Y = XY$ of two nonnegative independent random variables with probability density functions $f_1(x_1)$ and $f_2(x_2)$ is expressible as a Mellin convolution

$$f_y(y) = \int_0^\infty \frac{1}{x_2} f_1(\frac{y}{x_2}) f_2(x_2) dx_2 = \int_0^\infty \frac{1}{x_1} f_2(\frac{y}{x_1}) f_1(x_1) dx_1. \tag{24}$$

The greatest utility of these techniques is for independent random variables that appear as sums, products, and quotients. However, dependent random variables are also considered. For example, the two-dimensional random variable (U, V) with joint density function $f(u, v)$ that is positive in the first quadrant and zero everywhere else has the Mellin transform

$$M(s_1, s_2) = \int_0^\infty \int_0^\infty u^{s_1 - 1} v^{s_2 - 1} f(u, v) du dv \tag{25}$$

with the inverse

$$f(u, v) = \frac{1}{(2\pi i)^2} \int_{h-i\infty}^{h+i\infty} \int_{h-i\infty}^{h+i\infty} u^{-s_1} v^{-s_2} M(s_1, s_2) ds_1 ds_2, \tag{26}$$

where the inversion line of integration is known as the Bromwich path.

In the following, we do not consider work in statistical energy analysis [Lyon, Zeeman and Bogdanoff 1969].

Structures

When working with probabilistic problems, it is possible to classify analytical methods according to Keller [1960] as "honest" and "dishonest". In essence, the honest methods apply analytical tools to mathematical models and then utilize the probabilistic nature of parameters to obtain the moments of the solution. Dishonest methods, on the other hand, average the mathematical models first, then apply the analytical tools. In the latter, one works with averaged equations. In the former, one works with averages of solutions. In principle, honest methods are preferred since the two approaches do not usually result in identical solutions and the "honest" approach works with the physics before dealing with the uncertainty. The primary honest method is the perturbation method [Bellman 1964, Rellich 1969]. However, it may be necessary in some instances to approach problems by averaging equation first, as will be observed in the discussion below. The presentation that follows is not chronological, but loosely grouped by topic and method.

A series of papers [Boyce and Goodwin 1964, Goodwin and Boyce 1964, Boyce 1966, 1967, 1978, Boyce and Xia 1983, 1985] have considered how random system properties will affect the eigenvalues. The small random transverse vibration of elastic beams [1964] of random cross section will lead to a governing equation of the form

$$\{[1+\epsilon R(x)]^4 u(x)''\}'' - \lambda[1+\epsilon R(x)]^2 u(x) = 0, \tag{27}$$

with associated boundary conditions, where $\epsilon R(x)$ is a small stationary perturbation of zero mean, and ϵ is the expansion parameter used in a perturbation analysis. The eigenvalues and eigenvectors are expanded in terms of this small parameter. Asymptotic analysis is also performed when the eigenvalue is large. The cases of random support mechanisms are also studied.

In another study [1964], the method of integral equations is used to model the vibrations of a random elastic string. The boundary value problem governing the free transverse vibrations of a tightly stretched elastic string having both ends fixed can be written in the form

$$U(x)'' + \lambda(1 + A(x))U(x) = 0, \tag{28}$$

where U is the dimensionless transverse displacement, $A(x)$ is proportional to the random deviation of the linear density from its mean value at the point x and $[1 + A(x)] > 0$. This equation is transformed into a Fredholm integral equation:

$$\begin{aligned} u(x) - \lambda \int_0^1 K(x,y)u(y)dy = 0, \quad 0 \le x \le 1 \\ K(x,y) = T(x,y)[1+A(x)]^{1/2}[1+A(y)]^{1/2}, \end{aligned} \tag{29}$$

and $T(x,y)$ is the Green's function for the integral equation. The eigenvalues λ_h are the same as for the original differential equation. Upper and lower bounds for the statistical moments of the frequencies are given in terms of corresponding moments and appropriate correlation functions for the random linear density. This analysis is based on the equation

$$\sum_{h=1}^{\infty} \lambda_h^{-m} = \int_0^1 K_m(x,x)dx, \quad m = 1,2,\ldots \tag{30}$$

where an iterated scheme for the kernels has been established.

Subsequently [1966], nonhomogeneous boundary value problems of the Sturm–Liouville type having random forcing functions are considered. Estimates for the statistical moments of the response are found for the case that the forcing function is stationary and weakly correlated. Specifically, the equation considered is

$$L[y] = f(x), \quad 0 < x < 1, \tag{31}$$

where $L[y] = -(py')' + qy$ is self–adjoint, with linear homogeneous boundary conditions. The solution is given by

$$y(x_0) = \int_0^1 G(x_0, x_1) f(x_1) dx_1, \tag{32}$$

where Green's function $G(x_0, x_1)$ is symmetric. The statistical consequences of this solution are examined in significant detail.

As mentioned previously, one may approach a random system "dishonestly", meaning that one will work with averaged equations. Boyce [1967] does this with the linear self–adjoint random eigenvalue problem

$$Ly = \lambda h(x) y, \quad 0 \le x \le 1, \tag{33}$$

where $h(x)$ is a stochastic function and therefore λ are random variables. The differential equation can be written as the integral equation

$$\lambda^{-1} y(x_0) = \int_0^1 G(x_0, x_1)[1 + \eta(x_1)] y(x_1) dx_1, \tag{34}$$

which is first averaged. To solve the averaged equation, *a priori* assumptions must be made about $< \lambda^{-1} y(x_0) >$, $< \eta(x_1) y(x_1) >$, where the notation $< \cdot >$ denotes mathematical expectation. As Boyce points out, these necessary assumptions cannot be justified at this stage, but only by examining the adequacy of the resulting predictions. An increasing number of assumptions must be made at each step in the iterative solution of the integral equation. This approach is compared with a perturbation approach for three examples: a fixed elastic string, a fixed–free elastic string, and a simply supported elastic beam. Two conclusions are drawn. Since the iterative "dishonest" method makes no assumptions regarding the magnitude of the random parameters, it may be applicable over a wider range of uncertainties than an honest method which depends on some "smallness" criteria. (It should be noted that implicit in the iterative approach is a statement that the series will formally converge. Also, if one wishes to use only a few terms in the iterative series solution, it will be necessary that subsequent terms be smaller in some sense.) The second conclusion is that the perturbation result will be more plausible for higher eigenvalues.

As mentioned above, the term ***disordered*** has been used to denote large degree-of-freedom systems which have random variable parameters. An old problem is the consideration of how one can best represent such complex (linear) systems by ones which are simpler, yet which respond to the class of inputs in a manner which is representative of the given systems. An earlier review of disordered structural systems is provided by Soong and Cozzarelli [1976].

To the best of our knowledge, the earliest paper on disordered systems is by Chenea and Bogdanoff [1958]. The mean and variances of the impedances F_0/x and of one of the frequency responses x/δ_{st} of damped linear one degree-of-freedom systems are investigated when spring constant, mass, and damping constant are considered as random variables. Specifically, the system parameters are defined as $k = \bar{k}L$, $m = \bar{m}M$, $c = \bar{c}N$, where dimensionless random variables L, M, N have unity mean values and are assumed to be independent. For lightly damped systems, considerable dispersion as well as deviation between the mean and deterministic value are encountered in the neighborhood of resonance. In certain heavily damped systems, on the other hand, highly accurate components are not necessary to ensure that the system performance is close in the mean sense with little dispersion to the performance of the mean or design system. Furthermore [Bogdanoff and Chenea 1961] conclude that, in the presence of disorder, the highest natural frequencies may deviate considerably from the values predicted using deterministic or designed parameter values, and therefore, very little confidence can be placed in the calculated values of the top few natural frequencies. However, the lowest frequencies can be estimated with considerable accuracy. More generally, considerable dispersion in frequency response, impedance and impulsive admittance is encountered with relatively small dispersion in parameters.

Soong and Bogdanoff [1963, 1964] continued this study with an N degree-of-freedom disordered linear chain, one end fixed, the other end free, governed by

$$\begin{aligned} m_p\ddot{u}_p &= -k_p(u_p - u_{p-1}) - k_{p+1}(u_p - u_{p+1}), \; p = 1, 2, \ldots, N-1 \\ m_N\ddot{u}_N &= Fe^{i\omega t} - k_N(u_N - u_{N-1}) \end{aligned} \tag{35}$$

with $u_0 = 0$. New variables are introduced: $w_p = u_{p+1} - u_p$, and the solution is assumed of the form

$$
\begin{aligned}
u_p &= \phi_p e^{i\omega t} \\
w_p &= \psi_p e^{i\omega t}.
\end{aligned}
\tag{36}
$$

The equations of motion are re–written in a vector–matrix form transforming the displacement vector of the pth mass $\{\phi_p \; \psi_p\}^T$ into that of the next mass $\{\phi_{p+1} \; \psi_{p+1}\}^T$.

This study is continued for the case where only the mass is assumed random $M_j = m(1 + X_j)$, with deterministic damping introduced between the masses. The masses are assumed to have small variances, allowing an expansion of the natural frequencies in powers of X_j.

In a series of papers [Bogdanoff 1965, Schiff and Bogdanoff 1972, Kayser and Bogdanoff 1975, Bogdanoff and Kozin 1985], it is assumed that the frequency response of the system can be approximated on a frequency band of interest by a linear function of the inputs with constant coefficients. A mean–square error function integrated over the frequency band is constructed. The constant coefficients which make this error function a minimum provide the approximate values to the frequency response. The approximate complex system is created by imposing constraints on the given system which do not obscure structural motion of interest. The sequence of approximants $\{S_{x_0}, S_{x_1}, \ldots\}$ having an increasing number of degrees–of–freedom can then be constructed by successively relaxing the constraints introduced to produce $\{S_{x_0}\}$. At no stage in the procedure is it necessary to compute the response of the original complex system for purposes of judging accuracy. The choice of approximate system will be made based on the character of the excitation. For example, high frequency input will require a model which will predict local motion, while low frequency input requires gross behavior modeling. As a measure of the quality of successive approximants, the mean–square difference between approximants of the frequency response is used. This difference is maximized since in relaxing constraints the resonance peaks in the frequency response cannot increase and can never be to the left of the original system.

Several works [Fox and Kapoor 1968, Collins and Thomson 1969, Kiefling 1970] have discussed the eigenvalue problem for systems with statistical structural properties with the approach based on sensitivity or perturbation of the eigenvalues and eigenvectors. Such equations are now well known. An example of such a perturbation approach is provided in an appendix. Fox and Kapoor derive the rate of change (with respect to any system parameter) in terms of only the eigenvalue and eigenvector under consideration. Thus a complete solution of the eigenproblem is not needed. This is not the case in Collins and Thomson. Kiefling suggests an approach to circumvent the numerical difficulties experienced by Fox and Kapoor. There exists much interest and activity in estimating eigenvalue sensitivities in the aerospace communities, since the eigenvalue structure is directly related to system behavioral characteristics and system stability. This literature is not included in this paper, but may be approached beginning with Haug et al. [1986].

Hasselman and Hart [1972] study the modal analysis of random structural systems utilizing Taylor series representations for the eigenvalues and eigenvectors. Modal characteristics of structures may be treated as random variables which are linear combinations of the assumed independent random variables associated with specific structural properties. Therefore, it is suggested, by the central limit theorem, the modal characteristics will tend to have Gaussian distributions regardless of how the basic random variables are distributed. The effects of modal truncation on accuracy of mode statistics are examined, and reasonably accurate results may be obtained for the lower modes. Hart [1973] continued these studies.

Shinozuka and Astill [1972] have studied the statistical properties of the eigenvalues of a spring supported beam–column. The spring supports and axial force are treated as random variables. The distributions of material and geometric properties are considered to be correlated homogeneous random functions. Simulation and perturbation methods are used and accuracy compared in the calculation of the variance of the nth vibration and for static buckling eigenvalues. The perturbation method predicts values for the variance of the first 4 eigenvalues which are less accurate in the buckling case than in the vibration case. The perturbation method can either overestimate or underestimate the variance of the fundamental vibration eigenvalue.

Elishakoff [1975] has studied the deterministic distribution of the natural frequencies in certain structural elements. The outcome of an analysis is an estimate of the number of modes not exceeding a given cutoff level. The case of a spherical panel undergoing random vibration is studied.

Work to derive the probability densities of the output of random systems has met with less success than methods for first and second order statistics due to the difficulty of such problems. Kozin [1961] has obtained the probability densities of the output of some random systems. In particular, for the system

$$\ddot{x}(t) + \omega^2 x(t) = 0, \tag{37}$$

where $x(t_0) = x_0$, $\ddot{x}(t_0) = u_0$, and x_0, u_0, ω^2 may only be known probabilistically, $p_0(x, u, \omega^2)$. It is shown that the equation governing the density of the output is given by

$$\frac{\partial p}{\partial t} = -u\frac{\partial p}{\partial x} + \omega^2 x \frac{\partial p}{\partial u}, \tag{38}$$

with $p(x, u, \omega; t_0) = p_0(x, u, \omega^2)$. Two examples are considered: (i) ω^2 is a random variable, x, u, ω are independent, $x_0 = 0, u_0 = 1$, and (ii) x_0 is Gaussian with zero mean and variance 1, u_0 is Gaussian with mean and variance equal to 1, and the rest of the specifications as per (i). These results are generalized [1963] and approximate solutions are obtained for more complex systems.

Iyengar and Manohar [1989] derive the probability distribution of the eigenvalues of the random string equation, a second order stochastic boundary value problem,

$$\frac{d}{dx}[(1+\delta g(x))\frac{dy}{dx}]+\lambda^2[1+\epsilon f(x)]y=0, \tag{39}$$

$y(0)=y(1)=0$. f and g are known bounded random processes, δ and ϵ are parameters such that $(1+\delta g)>0$ and $(1+\epsilon f)>0$. Techniques are based on assuming parameters filtered through white–noise processes and thus the eigenvalue problem can be characterized indirectly through an associated Fokker–Planck equation.

A Two Degree–of–Freedom Model

Consider the two degree–of–freedom system with no damping of Figure 2. The equations of free vibration are

$$\begin{bmatrix} m_1 & 0 \\ 0 & m_2 \end{bmatrix}\left\{\begin{matrix}\ddot{x}_1\\ \ddot{x}_2\end{matrix}\right\}+\begin{bmatrix} k_1+k_2 & -k_2 \\ -k_2 & k_2+k_3 \end{bmatrix}\left\{\begin{matrix}x_1\\ x_2\end{matrix}\right\}=\left\{\begin{matrix}0\\ 0\end{matrix}\right\}. \tag{40}$$

Letting $m_1 = m_2 = 1$ for ease of presentation, the characteristic equation for this coupled system is

$$\begin{aligned} \eta^2-\zeta_1\eta+\zeta_2&=0\\ \eta&=\omega^2\\ \zeta_1&=k_1+k_3+2k_2>0\\ \zeta_2&=k_1k_3+k_2(k_1+k_3)>0 \end{aligned} \tag{41}$$

where ω^2 represents the squares of the system natural frequencies. Note that in the characteristic equation, there will always be sums and products of random variables.

In this development, it is assumed that the coupling spring constant k_2 is randomly distributed according to $f_{k_2}(k_2)$, and that k_1 is also randomly distributed according to $f_{k_1}(k_1)$. k_3 is taken to be a constant.

It is possible to proceed to derive the densities of the transformed variables ζ_1, ζ_2 in terms of the densities of the spring constants k_1, k_2, but it will be advantageous to proceed directly to derive the joint density $f_{\zeta_1\zeta_2}(\zeta_1,\zeta_2)$ as functions of $f_{k_1k_2}(k_1,k_2)$. Some details on the Jacobian of the transformation is provided in the appendix. In order to transform between the given and the derived densities, it is necessary to

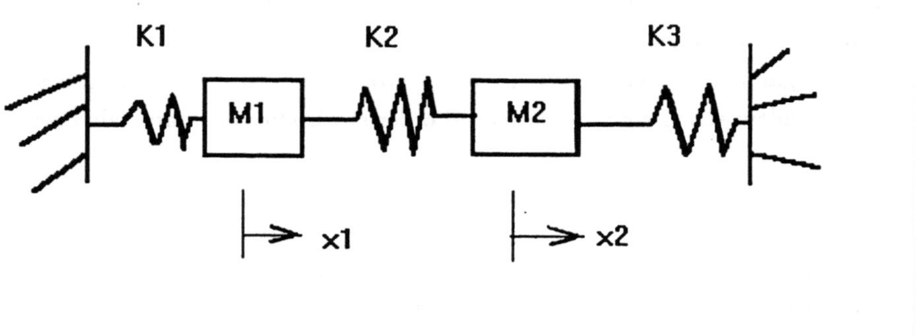

Figure 2. Simple Spring–Mass System

solve for the given functions k_1, k_2 in terms of the derived functions ζ_1, ζ_2. There are two such solution pairs here:

$$\begin{aligned} k_{11} &= \zeta_1 - k_3 - 2k_{21} \\ k_{12} &= \zeta_1 - k_3 - 2k_{22} \\ k_{21} &= -\frac{a}{2} + \frac{1}{2}\sqrt{a^2 - 4b} \\ k_{22} &= -\frac{a}{2} - \frac{1}{2}\sqrt{a^2 - 4b} \\ a &= -\frac{1}{2}(-\zeta_1 + 2k_3) \\ b &= -\frac{1}{2}(\zeta_2 - k_3\zeta_1 + k_3^2) \end{aligned} \tag{42}$$

The derived density is then

$$f_{\zeta_1\zeta_2}(\zeta_1, \zeta_2) = \frac{f_{k_1k_2}(k_{11}, k_{21})}{|J(k_{11}, k_{21})|} + \frac{f_{k_1k_2}(k_{12}, k_{22})}{|J(k_{12}, k_{22})|}, \tag{43}$$

where the Jacobians are, respectively,

$$\begin{aligned} |J(k_{11}, k_{21})| &= |k_{11} - k_3 - 2k_{21}| \\ |J(k_{12}, k_{22})| &= |k_{12} - k_3 - 2k_{22}|. \end{aligned} \tag{44}$$

Marginal densities f_{ζ_1} and f_{ζ_2} may be obtained, if necessary, by integrating out the other variable.

Given the joint density function $f_{\zeta_1\zeta_2}(\zeta_1,\zeta_2)$, the next task is to derive the joint density function of the roots

$$\eta_{1,2} = \frac{\zeta_1}{2} \pm \sqrt{\frac{\zeta_1^2}{4} - \zeta_2} \tag{45}$$

of the above characteristic equation, which may also be written as

$$\begin{aligned} \zeta_1 &= \eta_1 + \eta_2 \\ \zeta_2 &= \eta_1\eta_2. \end{aligned} \tag{46}$$

The solution of a quadratic equation admits solutions which are real or complex conjugates. The line between these two classes of solutions is given by the parabola $\zeta_2 = \zeta_1^2/4$. Using the notation $f_{\zeta_1\zeta_2}(\zeta_1,\zeta_2) = f(x,y)$, the following expressions define the probability of real and complex roots.

$$\begin{aligned} P(R) = P(\zeta_2 \leq \zeta_1^2/4) &= \int\int_{y \leq x^2/4} f(x,y)dydx \\ P(C) = P(\zeta_2 > \zeta_1^2/4) &= \int\int_{y > x^2/4} f(x,y)dydx, \end{aligned} \tag{47}$$

where R and C are the real and complex conjugate events. The density functions of these events are then

$$\begin{aligned} f(x,y|R) &= \frac{f(x,y)}{P(R)}, \quad y \leq x^2/4 \\ f(x,y|C) &= \frac{f(x,y)}{P(C)}, \quad y > x^2/4 \end{aligned} \tag{48}$$

For the case where the roots are real, let $\eta_{1,2} \equiv v_{1,2}$. Then,

$$\begin{aligned} f_{v_1v_2}(v_1,v_2|R) &= f_{\zeta_1\zeta_2}(v_1+v_2, v_1v_2|R)|J(v_1,v_2)| \\ &= f_{\zeta_1\zeta_2}(v_1+v_2, v_1v_2)|J(v_1,v_2)|/P(R) \end{aligned} \tag{49}$$

where

$$|J(\eta_1,\eta_2)| = \det\begin{bmatrix} \frac{\partial\zeta_1}{\partial\eta_1} & \frac{\partial\zeta_1}{\partial\eta_2} \\ \frac{\partial\zeta_2}{\partial\eta_1} & \frac{\partial\zeta_2}{\partial\eta_2} \end{bmatrix} = |\eta_1 - \eta_2|. \tag{50}$$

The marginal densities of v_1, v_2 can be obtained

$$\begin{aligned} f_{v_1}(v_1|R) &= \int_{-\infty}^{v_1} f_{v_1v_2}(v_1,v_2|R)dv_2 \\ f_{v_2}(v_2|R) &= \int_{v_2}^{\infty} f_{v_1v_2}(v_1,v_2|R)dv_1. \end{aligned} \tag{51}$$

For the case where the roots are complex conjugates, i.e., $\eta_{1,2} = \alpha \pm \beta i$, it is necessary to derive the joint density of (α, β). Thus,

$$\alpha = \frac{\zeta_1}{2}$$
$$\beta = \pm\sqrt{\zeta_2 - \frac{\zeta_1^2}{4}}$$

or

$$\zeta_1 = 2\alpha$$
$$\zeta_2 = \alpha^2 + \beta^2. \tag{52}$$

The joint probability density of (α, β) is given by

$$\begin{aligned} f_{\alpha\beta}(\alpha, \beta|C) &= f_{\zeta_1\zeta_2}(2\alpha, \alpha^2 + \beta^2|C)|J(\alpha, \beta)| \\ &= f_{\zeta_1\zeta_2}(2\alpha, \alpha^2 + \beta^2)|J(\alpha, \beta)|/P(C), \end{aligned} \tag{53}$$

where

$$J(\alpha, \beta) = \det \begin{bmatrix} \frac{\partial\zeta_1}{\partial\alpha} & \frac{\partial\zeta_1}{\partial\beta} \\ \frac{\partial\zeta_2}{\partial\alpha} & \frac{\partial\zeta_2}{\partial\beta} \end{bmatrix} = 4\beta \tag{54}$$

for $\beta > 0$. For $\beta < 0$, the same solution applies, except that $J(\alpha, \beta) = -4\beta$.

Complex Stiffness

The question of damping is difficult one, even for deterministic and simple systems [Caughey and O'Kelly 1965]. For the classes of problems considered here, damping may be modeled according to techniques which have proven promising elsewhere. Possibilities include mean square proportional damping, or models for structural damping. This type of damping is due to the dissipation of energy generated through internal friction. It is a hysteresis phenomenon associated with cyclic stress in elastic materials. It can be shown [Meirovitch 1986] that the equation of motion will be of the complex form:

$$m\ddot{y} + k(1 + i\gamma)y = 0. \tag{55}$$

The term $k(1 + i\gamma)$ is called either complex stiffness or complex damping, and the formulation is valid for harmonic excitation. The characteristic equation for this equation is

$$\lambda^n - a_1(\omega)\lambda^{n-1} + \cdots + (-1)^n a_n(\omega) = 0, \tag{56}$$

with roots $\xi_1(\omega) \ldots \xi_n(\omega)$ where $a_k(\omega) = \alpha_k(\omega) + i\beta_k(\omega), \quad k = 1, 2, \ldots, n$ are complex variables.

The following transformations exist between the coefficients of the characteristic equation and the roots:

$$a_1(\omega) = \sum_{k=1}^{n} \xi_k(\omega)$$
$$a_2(\omega) = \sum_{i<j} \xi_i(\omega)\xi_j(\omega)$$
$$\vdots$$
$$a_n(\omega) = \Pi_{k=1}^{n} \xi_k(\omega). \tag{57}$$

Assume the random variables $\alpha_k(\omega), \beta_k(\omega)$ to be independent, normal with zero mean, $N(0,\sigma)$, and with joint density

$$\left(\frac{1}{(2\pi)^{1/2}\sigma}\right)^{2n} exp\left[-\frac{1}{2\sigma^2}\sum_{k=1}^{n} a_k a_k^*\right], \tag{58}$$

then it can be shown [Girshick 1942] that the joint densities of the real and imaginary parts of the zeros are

$$\left(\frac{1}{(2\pi)^{1/2}\sigma}\right)^{2n} exp\left(-\frac{1}{2c^2}\left[\sum_{k=1}^{n}\xi_k\sum_{k=1}^{n}\xi_k^* + \cdots + \xi_1\xi_1^*\cdots\xi_n\xi_n^*\right]\right)\sum_{i=1}^{n}\sum_{j=i+1}^{n}|\xi_i - \xi_j|^2, \tag{59}$$

where the Jacobian of the transformations is

$$D = |d^2 = \sum_{i=1}^{n}\sum_{j=i+1}^{n}|\xi_i - \xi_j|^2. \tag{60}$$

Concluding Remarks

Much can be learned by considering the behavior of simple dynamical systems with parameter uncertainties using the techniques discussed and brought together above.

Perturbation expansions [Bellman 1964], techniques based on these [Benaroya and Rehak 1988, Ettouney et al. 1989], and Monte Carlo simulations [Shinozuka 1972] may, however, be the only recourse for considering systems where the number of degrees of freedom will number in the thousands.

Acknowledgements

The interest of the Federal Aviation Administration and program manager Larry Neri in this work is appreciated. We also thank Harry Kemp of the FAA Technical

Center for graciously performing numerous on–line literature searches for us, saving us many weeks of effort.

References & Bibliography

Baldock, G.R., Bridgeman, T., **The Mathematical Theory of Wave Motion**, John Wiley & Sons, 1981

Bellman, R., **Perturbation Techniques in Mathematics, Physics, and Engineering**, Holt, Reinhart and Winston, New York 1964

Benaroya, H., Rehak, M., "Parametric Random Excitation I: Exponentially Correlated Parameters", Journal of Engineering Mechanics, 113(6) June 1987 861–874

Benaroya, H., Rehak, M., "Parametric Random Excitation II: White–Noise Parameters", Journal of Engineering Mechanics, 113(6) June 1987 875–884

Benaroya, H., Rehak, M., *Finite Element Methods and Probabilistic Structural Analysis — A Selective Review*, Applied Mechanics Reviews, 41(5) May 1988 201–213

Benaroya, H., Rehak, M., "Response and Stability of a Random Differential Equation – Part I: Moment Equation Method", Journal of Applied Mechanics, (56) 1989 192–195

Benaroya, H., Rehak, M., "Response and Stability of a Random Differential Equation – Part II: Iterative Method", Journal of Applied Mechanics, (56) 1989 196–201

Bharucha–Reid, A.T., Sambandham, M., **Random Polynomials**, Academic Press, 1986

Bharucha–Reid, A.T., **Random Integral Equations**, Academic Press, 1972

Bogdanoff, J.L., Chenea, P.F., "Dynamics of Some Disordered Linear Systems", Int. J., Mech. Sci., 1961, vol. 3, pp 157–169

Bogdanoff, J.L., "Mean–Square Approximate Systems and Their Application in Estimating Response in Complex Disordered Linear Systems", J. of the Acoustical Society, Vol. 38, No. 2, August 1965

Bogdanoff, J.L., Kozin, F., "Simple Approximants for Complex Linear Systems", AIAA 26th Structures, Structural Dynamics and Materials Conference, pp. 218–223, Orlando, FL, 15–17 April 1985

Bolotin, V.V., **Random Vibration of Elastic Systems**, Martinus Nijhoff Publishers, 1984

Boyce, W.E., Goodwin, B.E., "Random Transverse Vibrations of Elastic Beams", J. SIAM, Vol. 12, No. 3, Sept. 1964

Boyce, W.E., "A 'Dishonest' Approach to Certain Stochastic Eigenvalue Problems", SIAM J. Appl. Math., Vol. 15, No. 1, Jan. 1967

Boyce, W.E., "Random Eigenvalue Problems", pp. 1–73, in **Probabilistic Methods in Applied Mathematics**, Ed. A.T. Bharucha–Reid, Academic Press, 1968

Boyce, W.E., Xia, N.-M., "Upper Bounds for the Means of Eigenvalues of Random Boundary Value Problems with Weakly Correlated Coefficients", Quarterly of Applied Mathematics, January 1985

Caughey, T.K., O'Kelly, M.E.J., "Classical Normal Modes in Damped Linear Dynamic Systems", J. of Applied Mechanics, pp. 583–588, Sept. 1965

Chenea, P.F., Bogdanoff, J.L., "Impedance of Some Disordered Systems", Coll. on Mech. Impedance Methods for Mech. Vibr., ASME Annual Mtg. 1958

Collins, J.D., Thomson, W.T., "The Eigenvalue Problem for Structural Systems with Statistical Properties", AIAA Journal, Vol. 7, No. 4, April 1969

Edelman, A., "Eigenvalues and Condition Numbers of Random Matrices", SIAM J. Matrix Anal. Appl., Vol. 9, No. 4, October 1988

Elishakoff, I., "Distribution of Natural Frequencies in Certain Structural Elements", J. of the Acoustical Society, Vol. 57, No. 2, February 1975

Epstein, B., "Some Applications of the Mellin Transform in Statistics", Ann. Math. Stat. Vol. 19 (1948) 370–379

Ettouney, M., Benaroya, H., Wright, J., "Boundary Element Methods in Probabilistic Structural Analysis", Applied Mathematical Modelling, (13) July 1989 432–441

Ettouney, M., Benaroya, H., Wright, J., "Probabilistic Boundary Element Methods," **Computational Mechanics of Probabilistic and Reliability Analysis**, W. K. Liu, T. Belytschko, Eds, Elmpress International, 1989

Fox, R.L., Kapoor, M.P., "Rates of Change of Eigenvalues and Eigenvectors", AIAA Journal, Vol. 6, No., 12, December 1968

Giffin, W.C., **Transform Techniques for Probability Modeling,** Academic Press, New York, 1975

Girshick, M.A., "Note on the Distribution of Roots of a Polynomial with Random Complex Coefficients", Ann. Math. Stat., Vol. 13, 235–238 (1942)

Goodwin, B.E., Boyce, W.E., "The Vibrations of a Random Elastic String: The Method of Integral Equations", Q. of Appl. Math., Vol. 22, No. 3, Oct. 1964

Hamblen, J.W., "Distributions of Roots of Algebraic Equations with Variable Coefficients", Doctoral Dissertation, Purdue University, June 1955

Hamblen, J.W., "Distributions of Roots of Quadratic Equations with Random Coefficients", Ann. Math. Stat. Vol. 27, 1136–1143 (1956)

Hart, G.C., "Eigenvalue Uncertainty in Stressed Structures", J. Engineering Mechanics, June 1973

Hasselman, T.K., Hart, G.C., "Modal Analysis of Random Structural Systems", J. Engineering Mechanics, June 1972

Haug, E.J., Choi, K.K., Komkov, V., **Design Sensitivity Analysis of Structural Systems**, Academic Press, New York, 1986

Ibrahim, R.A., "Structural Dynamics with Parameter Uncertainties", Applied Mechanics Reviews, vol. 40, no.3, Mar. 1987, 309–328

Iyengar, R.N., Manohar, C.S., "Probability Distributions of the Eigenvalues of the Random String Equation", J. of Applied Mechanics, Vol. 56, Mar. 1989, 202–207

Kayser, K.W., Bogdanoff, J.L., "A New Method for Predicting Response in Complex Linear Systems", J. of Sound and Vibration, Vol. 38, No. 3, 1975

Keller, J.B., "Wave Propagation in Random Media", Proc. Sympos. Appl. Math. Vol. 13, Amer. Math. Soc., Providence, RI, 1960

Kiefling, L.A, "Comment on 'The Eigenvalue Problem for Structural Systems with Statistical Properties' ", AIAA Journal, Vol. 8, No. 7, July 1970

Kozin, F., "On the Probability Densities of the Output of Some Random Systems", J. of Applied Mechanics, vol. 28, 161–164 (1961)

Kozin, F., "On approximations to the densities and moments of a class of stochastic systems", Proc. Camb. Phil. Soc., (1963), vol. 59, 463

Leong, Y.K., "Stochastic Approach to Matrices", SIAM J. Appl. Math., Vol. 47, No. 5, October 1987

Lumley, J.L., **Stochastic Tools in Turbulence**, Academic Press, 1970

Lyon, R.H., **Statistical Energy Analysis**, MIT Press, 1975

Meirovitch, L., **Elements of Vibration Analysis**, 2nd Ed., McGraw–Hill, 1986

Rellich, F., **Perturbation Theory of Eigenvalue Problems**, Gordon and Breach, New York, 1969

vom Scheidt, J., Purkert, W., **Random Eigenvalue Problems**, North Holland, 1983

Schiff, A.J., Bogdanoff, J.L., "An Estimator for the Standard Deviation of a Natural Frequency - Part 1", J. of Applied Mechanics, June 1972

Schiff, A.J., Bogdanoff, J.L., "An Estimator for the Standard Deviation of a Natural Frequency - Part 2", J. of Applied Mechanics, June 1972

Shahruz, S.M., Ma, F., "Approximate Decoupling of the Equations of Motion of Linear Undamped Systems", J. of Applied Mechanics, September 1988

Shinozuka, M., Astill, J.J., "Random Eigenvalue Problems in Structural Analysis", AIAA Journal, Vol. 10, No. 4, April 1972

Shinozuka, M., "Monte Carlo Solution of Structural Dynamics", Int. J. Computers Structures, Vol. 2, 1972, pp 855–874

Soong, T.T., Bogdanoff, J.L., "On the Impulsive Admittance and Frequency Response of a Disordered Linear Chain of N Degrees of Freedom", Int. J. Mech. Sci., 1963, vol. 5, pp. 237

Soong, T.T., Bogdanoff, J.L., "On the Impulsive Admittance and Frequency Response of a Disordered Linear Chain of N Degrees of Freedom", Int. J. Mech. Sci., 1964, vol. 6, pp. 225–237

Soong, T.T., Cozzarelli, F.A., "Vibration of Disordered Structural Systems", Shock and Vibration Digest, Vol. 8, No. 5, May 1976

Springer, M.D., **The Algebra of Random Variables**, John Wiley & Sons, New York, 1979

Zeman, J.L., Bogdanoff, J.L., "A Comment on Complex Structural Response to Random Vibrations", AIAA Journal, Vol. 7, No. 7, July 1969

Zeman, J.L., Bogdanoff, J.L., "A Statistical Approach to Complex Random Vibration", J. of the Acoustical Society, Vol. 50, No. 3, 1971

Wave Attenuation in Disordered Periodic Structures

G. Q. CAI and Y. K. LIN

Center for Applied Stochastics Research
College of Engineering
Florida Atlantic University
Boca Raton, FL

Abstract

The ability of a periodic structure to transmit wave motions from one location to another is reduced due to disorder, namely random variations of material and geometric properties. The phenomenon is known as localization, and the average exponential decay rate of wave transmission with respect to the wave propagation distance is called the localization factor. Two systematic procedures are presented in this paper for calculating the localization factor of a disordered periodic structure on the basis of probability theory. One employs the concept of invariant probability of phase difference between two waves traveling in opposite directions; another takes into account reflections from nearby disordered periodic units successively. In both methods, the random variation of each physical parameter in a disordered unit is represented by a random variable, and a sequence of such random variables corresponding to different units is assumed to be an ergodic sequence; therefore, the sequential averages may be replaced by ensemble averages. Application of the method is illustrated by an example and the results are compared with Monte Carlo simulations.

Introduction

An ideal periodic structure is one which is composed of an array of identically constructed units. It possesses some interesting dynamic properties, one of which is the existence of alternate wave-passage frequency-band and wave-stoppage frequency-band. If a periodic structure is infinitely long and is undamped, then a disturbance can propagate indefinitely without attenuation if it is at a frequency within a wave-passage frequency-band, or it decays in a short distance if it is at a frequency within a wave-stoppage frequency band. Extensive investigations of periodic structures are available in the literature, for example, Brillouin [1], Miles [2], Lin [3-5], Lin and McDaniel [6], Mead [7-10].

Y. K. Lin · I. Elishakoff (Eds.)
Stochastic Structural Dynamics 1
New Theoretical Developments
2nd International Conference on Stochastic Structural Dynamics
May 9–11, 1990, Boca Raton, Florida

Because of manufacturing and material irregularities, perfectly periodic structures do not exist in reality. Departure from perfect periodicity is known as disorder, which may cause attenuation of wave propagation in wave-passage frequency bands, even if such a structure is undamped. This is known as the localization effect, first shown by Anderson [11].

The localization effect of disordered periodic structures has aroused considerable interest in the engineering research community. Deterministic analyses, conducted by Valero and Bendiksen [12], Cornwell and Bendiksen [13], and Pierre and his co-workers [14-16], have shown that the normal modes which would be periodic along the length of a perfectly periodic structure, are localized in a small region when periodicity is disrupted. It is the modal localization that prevents a disturbance to propagate far away from the excitation point.

The average exponential rate at which a structural wave decays with respect to the wave propagation distance in a disordered infinite periodic structure is known as the localization factor. If a structure were exactly periodic; namely, any two adjoining substructures (cell units) were identical, then a structural wave would pass through the interface of every pair of adjoining units. However, in a disordered system, two adjoining units are generally different; therefore, some reflection generally takes place at the interfaces, and a total transmission is no longer possible.

The localization factors due to disorder in periodic structures have been evaluated by several investigators, using various perturbation schemes [17-19]. In particular, Hodgs and Woodhouse [17] calculated the localization factor for a chain of coupled pendula. Two perturbation schemes were devised to treat the cases of weak and strong localizations, depending on the relative measures of disorder and internal (namely, unit-to-unit) coupling. Pierre [18] also calculated the localization factor for both weak and strong localization, using both the wave propagation and normal mode formulations and obtained essentially the same results. He devised two perturbation schemes to treat the two cases when the ratio of the disorder measure to the internal coupling measure is either large or small. As expected, his results were not very accurate when this ratio is of order one.

The first analysis of a generic disordered periodic system was given by Kissel [19], making use of the concept of wave transmission and reflection, and a limit theorem on the product of random matrices due to Furstenberg [20]. However, Kissel's analysis requires that the reflection be small; thus, his results are valid only for the case of weak localization. Unlike previous investigators, Kissel has also explored the case of multi-coupling; namely, the internal coupling which permits the transmission of several types of waves. An outline was provided of an approach based on the multiplicative ergodic theorem of Oseledets [21]. Unfortunately, the theorem only predicts the existence of a localization factor (analogous to the Lyapunov exponent in the time domain, of interest to Oseledets), without providing the necessary clues as how it can be calculated.

The ability to treat a generic system is clearly very desirable, since the analysis is applicable to a general class of problems, not restricted to a specific type of governing equations. In this paper, two systematic procedures are presented for calculating the localization factor for a generic disordered periodic structure. The first one, called the method of invariant probability, is applicable to undamped disordered periodic structures. In such a structure the probability distribution of the phase difference between two waves of the same type traveling in opposite directions is identical at every cell-to-cell interface. With the knowledge of this probability distribution, the localization factor can be obtained simply. The second procedure, called the method of multiple reflections, permits successive improvement of the accuracy. This is done by taking into account reflections from additional nearby cells [22]. Account is also taken of structural damping; thus, the results are more useful to practicing engineers. The key to the successful development of both methods under rather general settings is the substitution of ensemble averages for sequential averages of certain statistical properties of the structure on the basis of assumed spatial ergodicity. The simplicity of these methods is illustrated by applying them to a multiply supported Euler-Bernoulli beam with an additional torsional spring at each support. The accuracy of the results is substantiated by Monte-Carlo simulations.

Wave Transmission and Reflection Matrices

Consider two typical cell-units in a disordered periodic structure, denoted as cells n and n+1 and shown in Fig. 1a. The relation between the state vector at the left end of cell-unit n and that at the left end of cell-unit n+1 may be written as follows

$$\begin{Bmatrix} w(n+1_\ell) \\ f(n+1_\ell) \end{Bmatrix} = [T(n)] \begin{Bmatrix} w(n_\ell) \\ f(n_\ell) \end{Bmatrix} \tag{1}$$

where each **w** is a p-dimensional vector of generalized displacements, each **f** is a p-dimensional vector of generalized forces, and [T(n)] is a 2px2p transfer matrix associated with the nth cell. It is implied in Eq.(1) that the motion is timewise sinusoidal, and the displacements and forces are interpreted as their complex-valued amplitudes. Because of disorder, [T(n)] is generally different from the ideal transfer matrix [T] associated with the ideal design condition of exact periodicity. Nevertheless, it is always possible to write

$$[T(n)] = [P(n)][T] \tag{2}$$

which means that the effect of disorder in the nth cell can be lumped at its right end and represented by a "point" transfer matrix [P(n)]. In other words, the original system shown in Fig. 1a can be replaced by an equivalent system shown in Fig.1b. In typical transfer matrix analysis [23], a point

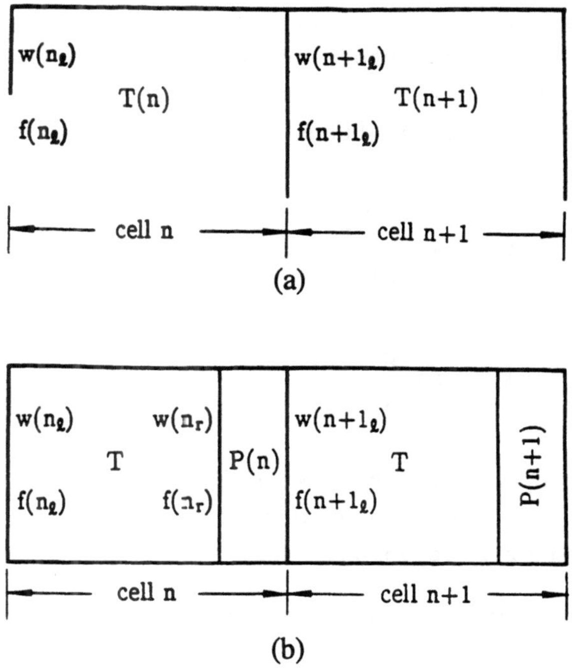

Fig. 1 State vector representation
(a) Original system
(b) Equivalent system

transfer matrix is used to relate state vectors on the two sides of a structural discontinuity which may arise from a concentrated mass, spring, damper, etc. However, [P(n)] in Eq.(2) is merely a mathematical device, and it may even correspond to, e.g., a negative mass. With this device, Eq.(1) is replaced by

$$\begin{Bmatrix} w(n_r) \\ f(n_r) \end{Bmatrix} = [T] \begin{Bmatrix} w(n_\ell) \\ f(n_\ell) \end{Bmatrix} \tag{3a}$$

$$\begin{Bmatrix} w(n+1_\ell) \\ f(n+1_\ell) \end{Bmatrix} = [P(n)] \begin{Bmatrix} w(n_r) \\ f(n_r) \end{Bmatrix} \tag{3b}$$

One well-known property shared by every transfer matrix is that its eigenvalues are reciprocal pairs. Thus, we can denote the eigenvalues of the transfer matrix [T] as λ_1, λ_2,

..., λ_p and λ_1^{-1}, λ_2^{-1}, ..., λ_p^{-1}, where $|\lambda_1| \leq |\lambda_2| \leq \ldots \leq |\lambda_p| \leq 1$. The eigenvectors corresponding to these eigenvalues constitute a transformation matrix [D], with which the state vectors in Eq.(3a) are transformed to wave vectors as follows [24]

$$\begin{Bmatrix} w(n_\ell) \\ f(n_\ell) \end{Bmatrix} = [D]\begin{Bmatrix} \mu^r(n_\ell) \\ \mu^\ell(n_\ell) \end{Bmatrix} \tag{4a}$$

$$\begin{Bmatrix} w(n_r) \\ f(n_r) \end{Bmatrix} = [D]\begin{Bmatrix} \mu^r(n_r) \\ \mu^\ell(n_r) \end{Bmatrix} \tag{4b}$$

where the superscript r or ℓ associated with a wave vector indicates the direction of wave propagation, to be either right-going or left-going.

Substituting Eqs.(4a) and (4b) into Eq.(3a), one obtains

$$\begin{Bmatrix} \mu^r(n_r) \\ \mu^\ell(n_r) \end{Bmatrix} = [D]^{-1}[T][D]\begin{Bmatrix} \mu^r(n_\ell) \\ \mu^\ell(n_\ell) \end{Bmatrix} = [Q]\begin{Bmatrix} \mu^r(n_\ell) \\ \mu^\ell(n_\ell) \end{Bmatrix} \tag{5}$$

where [Q] is also a transfer matrix, but it transfers wave vectors, instead of state vectors. This wave transfer matrix is clearly diagonal. In fact,

$$[Q] = [D]^{-1}[T][D] = \begin{bmatrix} \Lambda & 0 \\ 0 & \Lambda^{-1} \end{bmatrix} \tag{6}$$

where

$$\Lambda = \begin{bmatrix} \lambda_1 & & 0 \\ & \ddots & \\ 0 & & \lambda_p \end{bmatrix} \tag{7}$$

Eq.(5) can be rewritten in another form

$$\begin{Bmatrix} \mu^\ell(n_\ell) \\ \mu^r(n_r) \end{Bmatrix} = [S]\begin{Bmatrix} \mu^r(n_\ell) \\ \mu^\ell(n_r) \end{Bmatrix} \tag{8}$$

with

$$[S] = \begin{bmatrix} 0 & \Lambda \\ \Lambda & 0 \end{bmatrix} \tag{9}$$

Eq.(8) has been written to suggest that $\mu^r(n_\ell)$ and $\mu^\ell(n_r)$ are the input waves and $\mu^\ell(n_\ell)$ and $\mu^r(n_r)$ are output waves, as viewed by the structural unit which is characterized by [S].

Similar transformation of the state vectors in Eq.(3b) to wave vectors results in

$$\begin{Bmatrix} \mu^r(n+1_\ell) \\ \mu^\ell(n+1_\ell) \end{Bmatrix} = [Q(n)]\begin{Bmatrix} \mu^r(n_r) \\ \mu^\ell(n_r) \end{Bmatrix} \tag{10}$$

where

$$[Q(n)] = [D]^{-1}[P(n)][D] = [D]^{-1}[T(n)][D]\begin{bmatrix} \Lambda^{-1} & 0 \\ 0 & \Lambda \end{bmatrix} \tag{11}$$

Eq.(10) can be re-cast in the form of Eq.(8); namely,

$$\begin{Bmatrix} \mu^\ell(n_r) \\ \mu^r(n+1_\ell) \end{Bmatrix} = [S(n)]\begin{Bmatrix} \mu^r(n_r) \\ \mu^\ell(n+1_\ell) \end{Bmatrix} \tag{12}$$

The elements in [S(n)] can be obtained from those of [Q(n)] as follows

$$s_{11} = -\, q_{22}^{-1}\, q_{21} \tag{13a}$$

$$s_{12} = q_{22}^{-1} \tag{13b}$$

$$s_{21} = q_{11} - q_{12}\, q_{22}^{-1}\, q_{21} \tag{13c}$$

$$s_{22} = q_{12}\, q_{22}^{-1} \tag{13d}$$

In Eqs. (13a-13d) s_{ij} and q_{ij} are pxp submatrices of [S(n)] and [Q(n)], respectively, and the argument (n) is omitted in denoting the submatrices for simplicity.

It is physically meaningful to re-name the submatrices of [S(n)] as $r^r(n)=s_{11}$, $t^\ell(n)=s_{12}$, $t^r(n)=s_{21}$, and $r^\ell(n)=s_{22}$, and re-write Eq.(12) as

$$\begin{Bmatrix} \mu^\ell(n_r) \\ \mu^r(n+1_\ell) \end{Bmatrix} = \begin{bmatrix} r^r(n) & t^\ell(n) \\ t^r(n) & r^\ell(n) \end{bmatrix} \begin{Bmatrix} \mu^r(n_r) \\ \mu^\ell(n+1_\ell) \end{Bmatrix} \tag{14}$$

The submatrices t and r are called the transmission and reflection matrices, respectively. The superscript for each of these submatrices signifies the direction of an incoming wave group to be transmitted or reflected. For example, $t^r(n)$ is a transmission matrix for an incoming wave group traveling in the right direction.

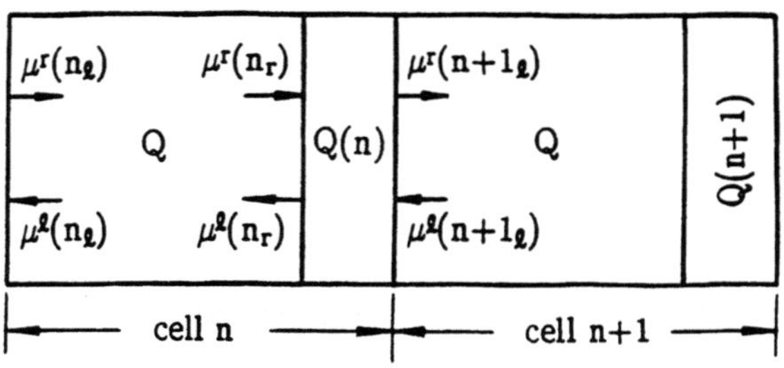

Fig. 2 Wave vector representation, equivalent system.

Matrix [S(n)], which contains the transmission and reflection submatrices, is known as a scattering matrix. This particular scattering matrix characterizes the effects of disorder in cell n which are lumped at the right end of cell n, as shown schematically in Fig.2. In wave mechanics, the concept of a scattering matrix is quite general; it characterizes the behavior of any identifiable structural element. In fact, matrix [S] in Eq.(8) is also a scattering matrix, in which the transmission and reflection matrices are $t^r=t^\ell=\Lambda$ and $r^r=r^\ell=0$. These submatrices of [S] are also related to those of a matrix [Q] according to Eqs.(13a-d). It is of interest to note that all scattering matrices are symmetric, a consequence of the reciprocity theorem for linear acoustic systems.

We shall now consider the case of mono-coupling in more detail, which means that only one type of wave motion, (either right-going, or left-going, or both) is permitted at every cell-to-cell interface. In this case, the wave transfer matrices Q and the scattering matrices S become 2x2, and the transmission and reflection matrices reduce to the transmission and reflection coefficients. Moreover, $t^r(n) = t^\ell(n)$, as can be proven from Eqs.(13b) and (13c) and from the fact $|Q(n)|=1$.

The exponential decay rate of wave amplitude due to passage through cell n is given by

$$\gamma(n) = -\ell n \left| \frac{\mu^r(n+1_\ell)}{\mu^r(n_\ell)} \right| \tag{15}$$

The localization factor can then be obtained as

$$\gamma = \lim_{N\to\infty} \frac{1}{N} \sum_{n=1}^{N} \gamma(n) \tag{16}$$

Method of Invariant Probability

Since disorder in a periodic structure is expected to have a similar effect as damping on wave propagation, it is reasonable to neglect damping if the objective of an analysis is to estimate the effect due to disorder alone [17-19]. Consider, therefore, an undamped disordered chain, which consists of cells n through N. External excitations, if any, are assumed to be on the left of cell n; thus, they are not located on the chain. Energy conservation requires that

$$|\mu^r(n_\ell)|^2 + |\mu^\ell(N_\ell)|^2 = |\mu^\ell(n_\ell)|^2 + |\mu^r(N_\ell)|^2 \tag{17}$$

The left and right hand sides of Eq.(17) represent, respectively, the average energy in-flow and energy out-flow of the chain within one period of time $2\pi/\omega$. As $N \to \infty$, the waves $\mu^r(N_\ell)$ and $\mu^\ell(N_\ell)$ vanish due to localization; namely, $|\mu^r(N_\ell)|\to 0$ and $|\mu^\ell(N_\ell)|\to 0$. It follows $|\mu^r(n_\ell)|^2=|\mu^\ell(n_\ell)|^2$ and

$$|\eta(n)| = \frac{|\mu^r(n_\ell)|}{|\mu^\ell(n_\ell)|} = 1 \qquad (18)$$

where $\eta(n)$ is the ratio of the right-going and left-going waves. Eq.(18) implies that $\eta(n)$ can be expressed as

$$\eta(n) = e^{i\theta(n)} \qquad (19)$$

where $\theta(n)$ is purely real; it represents the phase difference between the two waves $\mu^r(n_\ell)$ and $\mu^\ell(n_\ell)$.

It can be shown from Eqs.(5) and (10) that $\theta(n)$ and $\theta(n+1)$ are related as follows

$$e^{i\theta(n)} = \frac{q_{12}(n) - e^{i\theta(n+1)}\, q_{22}(n)\, \lambda^2}{e^{i\theta(n+1)}\, q_{21}(n)\lambda^2 - q_{11}(n)}$$

$$n = N - 1, \ldots, 2, 1 \qquad (20)$$

where λ may be taken to be either of the two eigenvalues of transfer matrix T. However, when $N \to \infty$, the probability density, $p[\theta(n)]$, of $\theta(n)$ must be independent of n, for any finite n. In other words, the probability distribution for the phase difference between the right-going and left-going waves becomes invariant in a infinite disordered chain. The existence of such an invariant probability density (also referred to as invariant measure) has been proved by Furstenberg [20] and is known in the mathematical literature as Furstenberg's theorem. Due to the functional dependence of the random variables $\theta(n)$ and $\theta(n+1)$ shown in Eq.(20), the invariant probability density $p(\theta)$ must satisfy the relation

$$\{p[\theta(n)]\}_{\theta(n)=\theta}$$

$$= \{\int p[\theta(n+1)]p[\mathbf{x}(n)] \left|\frac{\partial\theta(n+1)}{\partial\theta(n)}\right| d\mathbf{x}(n)\}_{\theta(n)=\theta}$$

$$= \{p[\theta(n+1)]\}_{\theta(n+1)=\theta} \tag{21}$$

where $\mathbf{x}(n) = \{x_1(n), x_2(n), \ldots x_k(n)\}$ is a random vector, and each $x_j(n)$ represents a disordered parameter in cell n. Since the total number of cells is infinite, it is reasonable to assume that, for different n, each $x_j(n)$ represents an ergodic sequence of random variables. In obtaining Eq.(21), use has been made of the independence of the two random variables $\theta(n+1)$ and $\mathbf{x}(n)$; namely,

$$p[\theta(n+1), \mathbf{x}(n)] = p[\theta(n+1)]\ p[\mathbf{x}(n)] \tag{22}$$

due to the fact that $\theta(n+1)$ is only related to $\theta(n+2)$ and random vectors $\mathbf{x}(n+1)$, as can be seen from the recursive relationship, Eq.(20).

An explicit analytical expression for the invariant measure $P(\theta)$ is generally difficult to obtain; however, for specific cases, it may be calculated approximately from Eq.(21) by using a perturbation approach, as shown later in an example. With the knowledge of the probability density $p(\theta)$, the localization factor can be obtained as follows.

We begin by rewriting Eq.(10) as follows

$$\begin{Bmatrix} \mu^r(n_r) \\ \mu^\ell(n_r) \end{Bmatrix} = [V(n)]\begin{Bmatrix} \mu^r(n+1_\ell) \\ \mu^\ell(n+1_\ell) \end{Bmatrix} \tag{23}$$

where $[V(n)] = [Q(n)]^{-1}$. The first row of Eq.(23) reads

$$\mu^r(n_r) = v_{11}(n)\ \mu^r(n+1_\ell) + v_{12}(n)\ \mu^\ell(n+1_\ell) \tag{24}$$

where $v_{ij}(n)$ are elements of $[V(n)]$. Comparing Eqs.(5) and (24), we obtain

$$\mu^r(n_\ell) = v_{11}(n)\lambda^{-1}\mu^r(n+1_\ell) + v_{12}(n)\lambda^{-1}\mu^\ell(n+1_\ell) \tag{25}$$

Dividing both sides of Eq.(25) by $\mu^r(n+1_\ell)$ and expressing the wave ratio $\eta(n+1)$ as $e^{i\theta(n+1)}$, we obtain

$$\frac{\mu^r(n_\ell)}{\mu^r(n+1_\ell)} = v_{11}(n)\lambda^{-1} + v_{12}(n)\lambda^{-1}e^{-i\theta(n+1)} \tag{26}$$

It is known that if the frequency of the wave motion is within a wave-passage band, then $|\lambda|=1$ and $\lambda^{-1}= \lambda^*$ for an undamped periodic structure. Thus, the exponential decay rate of wave amplitude due to passage through cell n is given by

$$\gamma(n) = \ell n \left| \frac{\mu^r(n_\ell)}{\mu^r(n+1_\ell)} \right| = \frac{1}{2}\, \ell n\{| v_{11}(n)|^2 + | v_{12}(n)|^2$$

$$+ 2\mathrm{Re}[v_{11}(n)v_{12}^*(n)e^{i\theta(n+1)}]\} \tag{27}$$

where Re[ ] denotes the real part of a complex quantity, and an asterisk denotes the complex conjugate. Being a function of the disordered parameters, each $v_{ij}(n)$ in Eq.(27) may be denoted by

$$v_{ij}(n) = v_{ij}[\mathbf{x}(n)] \tag{28}$$

As shown in Eq.(22), the random variables $\mathbf{x}(n)$ and $\theta(n+1)$ are independent. Hence, $p[\mathbf{x}(n),\theta(n+1)]=p[\mathbf{x}(n)]p[\theta]$. The localization factor can then be obtained as

$$\gamma = \lim_{N\to\infty} \frac{1}{N} \sum_{n=1}^{N} \gamma(n) = E[\gamma(n)]$$

$$= \frac{1}{2} \int\int \ell n\{| v_{11}(\mathbf{x})|^2 + | v_{12}(\mathbf{x})|^2 + 2\mathrm{Re}[v_{11}(\mathbf{x})v_{12}^*(\mathbf{x})e^{i\theta}]\}$$

$$p(x)p(\theta)\ dx\ d\theta. \qquad (29)$$

in which spatial averaging has been replaced by ensemble averaging on the basis of the ergodicity assumption for the random sequence $x(n)$, $(n=1,2,\ldots)$.

Eq.(29)is an exact expression, and is applicable to any randomly disordered mono-coupling chain, whether localization is weak, moderate or strong. Given the probability distribution of the disorder, the accuracy of the computed localization factor depends only on the accuracy of the computed invariant measure $p(\theta)$.

Method of Multiple Reflections

As indicated earlier, wave reflection takes place at every interface between two non-identical cell units and contributes to the wave decay along its path. Therefore, the localization factor can be computed on the basis of such reflections. Several specific cases will now be considered in the order of increasing complexity:

(1) Small reflections --- If all the reflection coefficients due to disorder are small, such that the effect of multiple reflections is negligible, then from Eqs.(8) and (14),

$$\mu^r(n+1_\ell) \approx t^r(n)\lambda\ \mu^r(n_\ell) \qquad (30)$$

Moreover, since matrix $[Q(n)]$ is 2x2 with a unit determinant, $t^r(n)=t^\ell(n)=t(n)=q_{22}^{-1}(n)$. Whence

$$\gamma = -\ \ell n\ |\ \lambda\ | + \lim_{N\to\infty} \frac{1}{N} \sum_{n=1}^{N} \ell n\ |\ q_{22}(n)\ | \qquad (31)$$

The sequential average in Eq.(31) may be replaced by an ensemble average; thus,

$$\gamma = -\ \ell n\ |\ \lambda\ |\ +\int \ell n\ |\ q_{22}(x)\ |\ p(x)\ dx \qquad (32)$$

Given the forms of $q_{22}[x(n)]$ and the probability distribution of $x(n)$, the integral can be evaluated at least numerically. In simple cases, it can often be carried out in closed form, as shown later in an example.

If the system disorder is also small, then Eq.(32) may be further approximated by

$$\gamma = -\ln|\lambda| + \frac{\sigma_i^2}{2}\left\{\frac{\partial^2}{\partial x_i^2}\ln|q_{22}(x)|\right\}_{x=0} \tag{33}$$

where σ_i^2 are the variances of $x_i(n)$. If, in addition, the structure is undamped and γ is evaluated at a wave-passage frequency, then $|\lambda|=1$, and Eq.(33) reduces to

$$\gamma = \frac{\sigma_i^2}{2}\left\{\frac{\partial^2}{\partial x_i^2}\ln|q_{22}(x)|\right\}_{x=0} \tag{34}$$

Eq.(34) has been obtained previously by Kissel [19].

It should be emphasized that Eq.(32) is valid only for small reflection coefficients which imply that localization is weak. Eq.(33) requires both small reflection coefficients and small disorder, and Eq.(34) requires additionally a zero damping and a wave-passage frequency.

(2) Moderate reflections -- If the reflection coefficients due to disorder are not sufficiently small, the approximation Eq.(32) may not be adequate. For improved accuracy, certain multiple reflections will be considered. As shown in Fig. 3, the right-going wave $\mu^r(n_r)$ gives rise to a right-going transmitted wave $t(n)\mu^r(n_r)$ and a left-going reflected wave $r^r(n)\mu^r(n_r)$. The reflected wave travels through cell n and is itself partially reflected back at the interface between cells n and n-1. This back-reflected wave travels also through cell n, whereupon it splits again into reflected and the transmitted parts. The latter is clearly $t(n)\lambda^2 r^\ell(n-1)r^r(n)\mu^r(n_r)$, and the former travels back through

cell n, etc. Summing up all the right-going waves which pass through the interface between cells n and n+1, we obtain

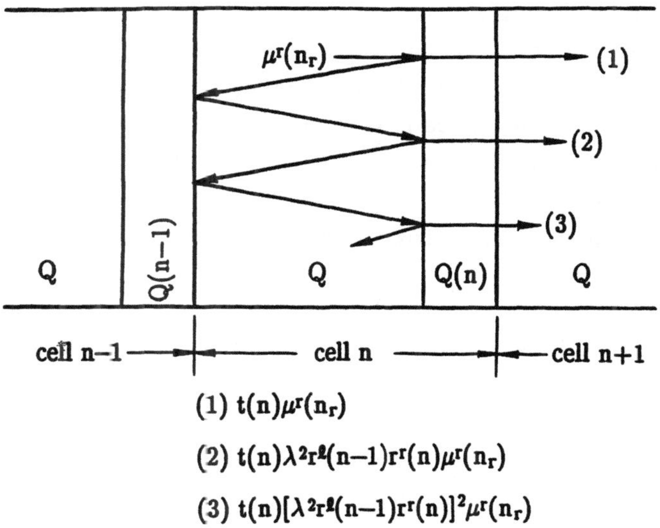

Fig. 3 Transmitted waves through the interface between cells n and n + 1

$$\mu^r(n+1_\ell)=t(n)\{1+\lambda^2 r^\ell(n-1)r^r(n)+[\lambda^2 r^\ell(n-1)r^r(n)]^2+\ldots\}\mu^r(n_r)$$

$$= \frac{t(n)}{1-\lambda^2\, r^\ell(n-1)\, r^r(n)}\mu^r(n_r) \qquad (35)$$

Thus

$$\eta(n) = \frac{t(n)\lambda}{1-\lambda^2\, r^\ell(n-1)\, r^r(n)} \qquad (36)$$

It follows from Eqs.(13a-d) and the ergodicity assumption,

$$\gamma = -\ln|\lambda| + \int \ln|q_{22}(x)|\,p(x)dx + \iint \ln\left|1+\frac{\lambda^2\, q_{12}(x_2)q_{21}(x_1)}{q_{22}(x_2)\, q_{22}(x_1)}\right| p(x_1)p(x_2)\,dx_1dx_2 \qquad (37)$$

The third term is seen to provide added accuracy over Eq.(32). Eq.(37) is valid for the case of moderate localization.

(3) Strong reflection -- Eq.(37) may still be inadequate if the reflection coefficients are relatively large (although each cannot have a magnitude greater than one). Higher accuracies, however, can always be obtained by including more reflection terms in the formulation. It can be shown that if the effect of reflection at the interface between cells n-1 and n-2 is also taken into consideration, then

$$\mu^r(n+1_\ell) = t(n)\left[\frac{1}{1-a} + \frac{c}{(1-a)(1-b)(1-c)}\right]\mu^r(n_r) \tag{38}$$

where

$$a = \lambda^2 \, r^\ell(n-1)r^r(n) \tag{39a}$$

$$b = \lambda^2 \, r^\ell(n-2)r^r(n-1) \tag{39b}$$

$$c = t^2(n-1)\lambda^4 \, r^\ell(n-2)r^r(n) \tag{39c}$$

Again, using Eqs.(13a-d) and invoking the ergodicity assumption,

$$\gamma = -\ell n|\lambda| + \int \ell n|q_{22}(x)|p(x)\,dx$$

$$-\iiint \ell n\left|\frac{1}{1+f_1(x_1,x_2)} - \frac{f_2(x_1,x_2,x_3)}{[1+f_1(x_1,x_2)][1+f_1(x_2,x_3)][1+f_2(x_1,x_2,x_3)]}\right|$$

$$p(x_1)p(x_2)p(x_3)\,dx_1dx_2dx_3 \tag{40}$$

where

$$f_1(x_1,x_2) = \frac{\lambda^2 \; q_{12}(x_2)q_{21}(x_1)}{q_{22}(x_1)q_{22}(x_2)} \tag{41a}$$

$$f_2(x_1,x_2,x_3) = \frac{\lambda^4 \; q_{12}(x_3)q_{21}(x_1)}{q^2_{22}(x_2)q_{22}(x_3)q_{22}(x_1)} \tag{41b}$$

Formulas of still higher accuracies can be derived in an analogous manner.

<u>An Example</u>

For illustration, the above theory is applied to an Euler-Bernoulli beam on evenly spaced hinge supports and with an additional torsional spring at each support, as shown in Fig. 4. In order to focus our attention on certain key issues in the theory without being obscured by unnecessary complexities, it is assumed that the beam is undamped and that only the torsional spring stiffnesses are random and are described by

$$k_n = k_0[1+ x(n)] \qquad n = 1, 2, \ldots \tag{42}$$

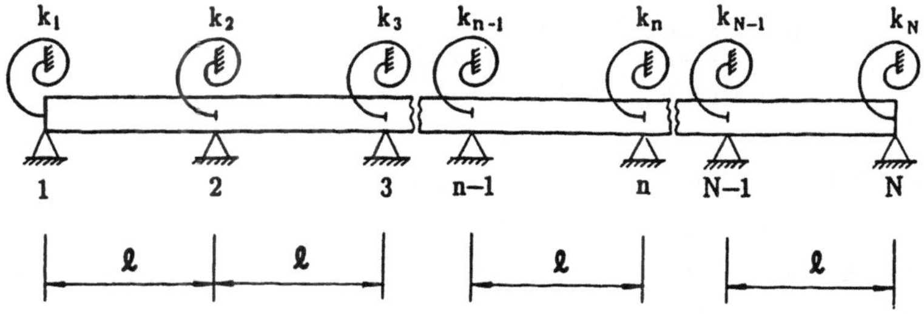

Fig. 4 A beam on hinge supports with additional torsional spring at each support

where k_0 is the average k_n and $x(n)$ are random variables with zero means. The other physical parameters are assumed to be deterministic, including the distances between neighboring supports ℓ, the bending rigidity of the beam EI, and the mass of the beam per unit length m. The theory is, of course, applicable when several or all the parameters are random.

Several choices can be made for a typical cell-unit in the analysis. The one selected for the following discussion is a typical beam element between two neighboring supports plus the entire right spring. The entire left spring is treated as belonging to the preceding cell. Then the transfer matrix for the nth cell is given by [6].

$$[T(n)] = \begin{bmatrix} \beta & \dfrac{\alpha}{\nu} \\ -\dfrac{\nu(1-\beta^2)}{\alpha} + \dfrac{\beta}{\delta}[1+x(n)] & \beta + \dfrac{\alpha}{\delta\nu}[1+x(n)] \end{bmatrix} \tag{43}$$

where $\delta = \dfrac{EI}{\ell k_0}$, $\nu = \ell(\dfrac{\omega^2 m}{EI})^{1/4}$, ω = frequency

$$\alpha = \frac{\cosh\nu \cos\nu - 1}{\sinh\nu - \sin\nu} \tag{44a}$$

and

$$\beta = \frac{\sinh\nu \cos\nu - \cosh\nu \sin\nu}{\sinh \nu - \sin \nu} \tag{44b}$$

The non-dimensional δ is called the internal (cell-to-cell) coupling parameter, which is a measure of relative resistance against a rotation at a support between the beam element and an average torsional spring. If k_0 is very large such that δ is very small, then there is little cell-to-cell coupling, and strong localization is expected to occur [18].

The transfer matrix [T] for the ideal cell-unit without disorder and the point transfer matrix [P(n)], lumping the effect of disorder, are obtained as follows

$$[T] = \begin{bmatrix} \beta & \dfrac{\alpha}{\nu} \\ -\dfrac{\nu(1-\beta^2)}{\alpha} + \dfrac{\beta}{\delta} & \beta + \dfrac{\alpha}{\delta\nu} \end{bmatrix} \tag{45a}$$

$$[P(n)] = \begin{bmatrix} 1 & 0 \\ \dfrac{x(n)}{\delta} & 1 \end{bmatrix} \tag{45b}$$

The eigenvalues of transfer matrix [T] are a reciprocal pair which may be denoted conveniently by $\lambda=e^{i\psi}$ and $\lambda^{-1}=e^{-i\psi}$. It can be shown that ψ can be found from [6]

$$\cos\psi = \beta + \frac{\alpha}{2\delta\nu} \tag{46}$$

and the wave-passage frequency bands are determined by the inequality

$$-1 \leq \beta + \frac{\alpha}{2\delta\nu} \leq 1 \tag{47}$$

The wave transfer matrix [Q(n)] is obtained as

$$[Q(n)] = \begin{bmatrix} 1-iAx(n) & -iAx(n) \\ iAx(n) & 1+iAx(n) \end{bmatrix} \tag{48}$$

where $A = \alpha/(2\delta\nu\sin\psi)$.

The following physical properties have been used in the numerical calculations: ℓ = 0.1651m, m = 1.8043 kg/m, and EI = 0.3143 N-m^2. In addition, x(n) are assumed to be identically and uniformly distributed between $-\sqrt{3}\sigma$ and $\sqrt{3}\sigma$ where σ is the standard deviation of x(n).

The invariant probability density of the wave-phase difference for a infinite disordered chain must satisfy Eq.(21). For the present multi-span beam, Eq.(21) has the following specific form

$$[p(\theta')]_{\theta'=\theta} = \{\frac{1}{2\sqrt{3}\sigma}\int_{-\sqrt{3}\sigma}^{\sqrt{3}\sigma} \left|\frac{\partial\theta}{\partial\theta'}\right| p[\theta(\theta',x)]dx\}_{\theta'=\theta} = p(\theta) \tag{49}$$

where

$$\left|\frac{\partial\theta}{\partial\theta'}\right| = G(\theta',x) = \{1-2A\, x\sin(\theta'+2\psi) + 2A^2x^2[1+\cos(\theta'+2\psi)]\}^{-1} \tag{50}$$

and $\theta(\theta',x)$ represents an implicit relation given by

$$\sin\theta = \{\sin(\theta'+2\psi)-2Ax[1+\cos(\theta'+2\psi)]\}G(\theta',x) \tag{51a}$$

and

$$\cos\theta = \{\cos(\theta'+2\psi)+2Ax\sin(\theta'+2\psi) - 2A^2x^2[1+\cos(\theta'+2\psi)]\}G(\theta',x) \tag{51b}$$

Since $|x|$ is bounded by $\sqrt{3}\,\sigma$, a perturbation procedure, applicable when $|A\sigma|$ is small, can be devised as follows. Let

$$p(\theta) = \frac{1}{2\pi} + \sum_{j=1}^{\infty} (C_j \sin j\theta + D_j \cos j\theta) \tag{52}$$

where

$$C_j = C_{j0} + C_{j1}A\sigma + C_{j2}(A\sigma)^2 + \ldots \tag{53a}$$

$$D_j = D_{j0} + D_{j1}A\sigma + D_{j2}(A\sigma)^2 + \ldots \tag{53b}$$

In Eq.(52), $p(\theta)$ is expanded into a Fourier series because it is a periodic function with a period 2π. The constant term in this expression is known to be $1/2\pi$; otherwise, the normalization condition for $p(\theta)$ will not be satisfied.

Substituting Eqs.(52) and (53) into Eq.(49) and equating terms of the same power in $A\sigma$, the coefficients C_{jk} and D_{jk} can be obtained. The final expressions for $p(\theta)$ are given below:

$$p(\theta) = \frac{1}{2\pi}\{1 + A^2\sigma^2 \frac{\sin(\theta+\psi)}{\sin\psi}[1 + \frac{\cos(\theta+\psi)}{\cos\psi}]\} + O(A\sigma)^4;$$

$$\cos 2\eta\psi \neq 1, \quad -\pi \leq \theta \leq \pi \tag{54}$$

The above perturbation scheme is not applicable when $\cos 2n\psi = 1$.

Fig. 5 depicts the computed approximate invariant measure $p(\theta)$ for the cases σ = 0.01, 01 and 0.2. The results are in good agreement with those obtained from Monte Carlo simulations, also shown in the figure as circles, triangles and stars. The simulations were carried out according to Eq.(20), starting from an arbitrarily assumed right boundary. The distribution of θ became essentially invariant after 1000 cells, and the simulation results shown in Fig. 5 correspond to the relative frequencies calculated for the subsequent 10^6 cells.

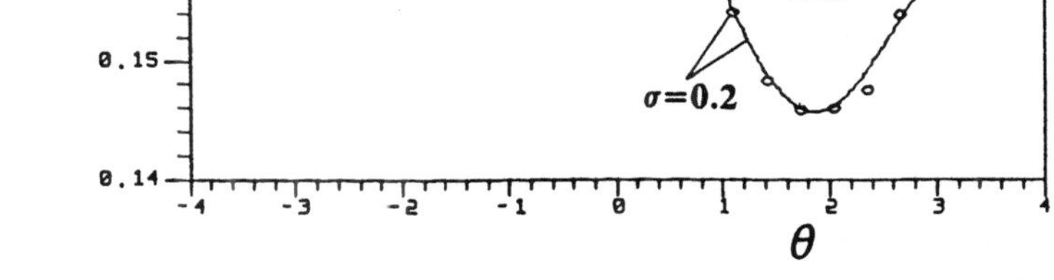

Fig. 5 Invariant probability density of phase difference θ between the right-going and left-going waves, ω = 250 rad/s, δ = 0.1

For the present case, Eq.(29) is reduced to

$$\gamma = \frac{1}{3\sqrt{3}\sigma} \int_{-\pi}^{\pi} P(\theta)d\theta \int_{-\sqrt{3}\sigma}^{\sqrt{3}\sigma} \ell n[1+2Ax\sin\theta + 2A^2x^2(1+\cos\theta)]dx \tag{55}$$

The localization factor γ can be computed numerically from Eq.(55) and Eq.(54). Although Eq.(55) is exact, the computed γ is approximate because Eq.(54) is approximate. We recall that Eq.(54) is valid only for small $| A \sigma |$, which implies a weak localization.

Substituting Eq.(48) into Eq.(32), we obtain an approximate localization factor as follows for the case of wave-passage frequency and weak localization on the basis of multiple reflections,

$$\gamma = \frac{1}{2}\ell n(1+3A^2\sigma^2)-1 + \frac{1}{\sqrt{3}A\sigma} \tan^{-1} (\sqrt{3}A\sigma) \tag{56}$$

where σ^2 is the variance of $x(n)$ which is assumed to be the same for every n. If, in addition, the disorder is small, we obtain from Eq.(34),

$$\gamma = \frac{\alpha^2 \sigma^2}{8\delta^2 v^2 \sin^2\psi} \tag{57}$$

For other cases for which the use of Eq.(32), Eq.(37) or Eq.(40) is more appropriate, numerical integrations are generally required using an assumed probability density for $x(n)$.

The computed localization factors in the first four wave-passage frequency bands are shown in Fig.6(a) and (b) for $\delta=1$, $\sigma=0.1$ and for $\delta=0.1$, $\sigma=0.1$, respectively. Except for frequencies very close to the lower boundary of a wave-passage frequency band (not shown in Fig.6), all the five equations (28), (32), (34), (37) and (40) yield essentially the same results, and in excellent agreement with those obtained from Monte

Carlo simulations, also shown in the figure. That the values of the localization factors are below 0.02 is an indication of weak localization in these cases. At the lower boundary of a

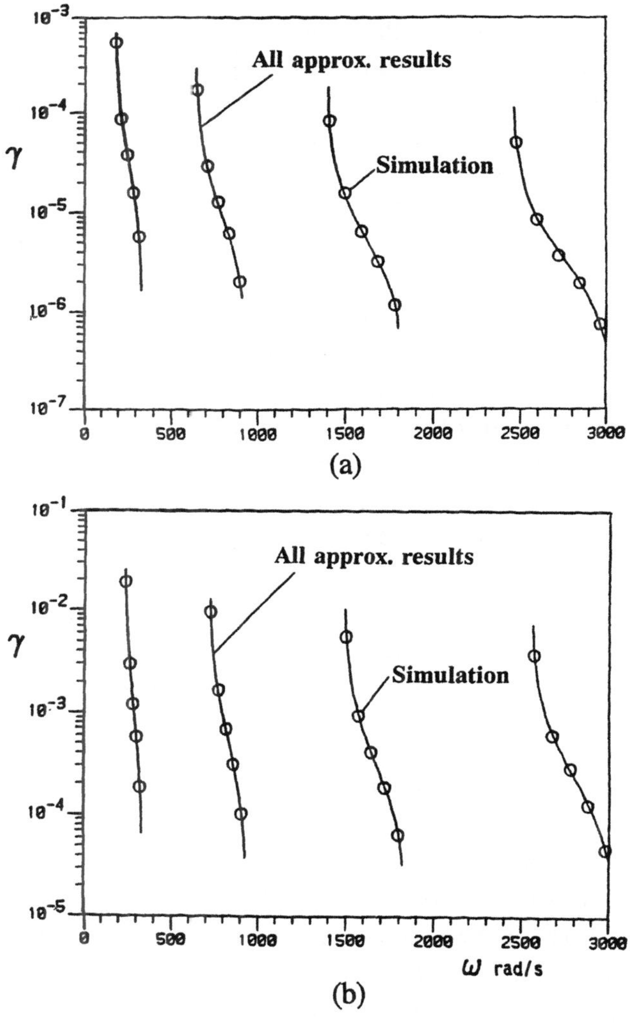

Fig. 6 Localization factors at frequencies in wave-passage frequency bands 1 through 4
(a) $\delta = 1$, $\sigma = 0.1$
(b) $\delta = 0.1$, $\sigma = 0.1$

wave-passage frequency band, the deformations of two neighboring spans tend to be out of phase (exactly out of phase for an ideal periodic structure) and the torsional spring at the supports influence the beam motion to a great extent. Therefore even a small disorder of the torsional springs may give rise to strong localization. On the other hand, the deformations of two neighboring spans are nearly in phase at the upper boundary of a wave-passage frequency and the torsional springs have little effect on the beam motion. Hence, their disorder is relatively unimportant.

The Monte-Carlo simulations referred to above were carried out by multiplying a large number of random wave transfer matrices at a given frequency to obtain one realization of the wave transfer matrix for an overall N-span system and thus one realization of the overall transmission coefficient $t^r(N,1)$. The values of $N^{-1}\ell n\,|\,t^r(N,1)\,|$ for different realizations were averaged over a sufficient number of realizations to obtain an estimate for the localization factor. This procedure is different from the one adopted by Kissel [19] in which an average is taken of a large number of $\ell n|\,t(n)\,|$ where each $t(n)$ is the transmission coefficient of one realization of a single cell. It is felt that our simulation procedure is physically more meaningful. In our simulations, 300 cells were used in each realization, and the average was taken over 2000 realizations.

Fig.7(a) and (b) depict the computed localization factors for systems with $\delta=0.01$, $\sigma=0.25$ and $\delta=0.01$, $\sigma=0.50$, respectively, and with frequencies falling within the lower part of the first wave-passage frequency band. It is seen that the closer a frequency is to the lower boundary of the wave-passage frequency band, the more Eqs.(32) and (34) overestimate the localization factor, because the reflection coefficients become larger and the effect of multiple reflection is no longer negligible. By taking into account the reflections from one additional cell on the immediate left, Eq.(37) yields much more accurate results than those of Eqs.(32) and (34) when compared with simulation results. Fig. 7 shows that for a beam system with fixed δ and σ, localization is strong at a frequency very near to the lower boundary of a wave-passage frequency band, and it becomes weaker as the frequency

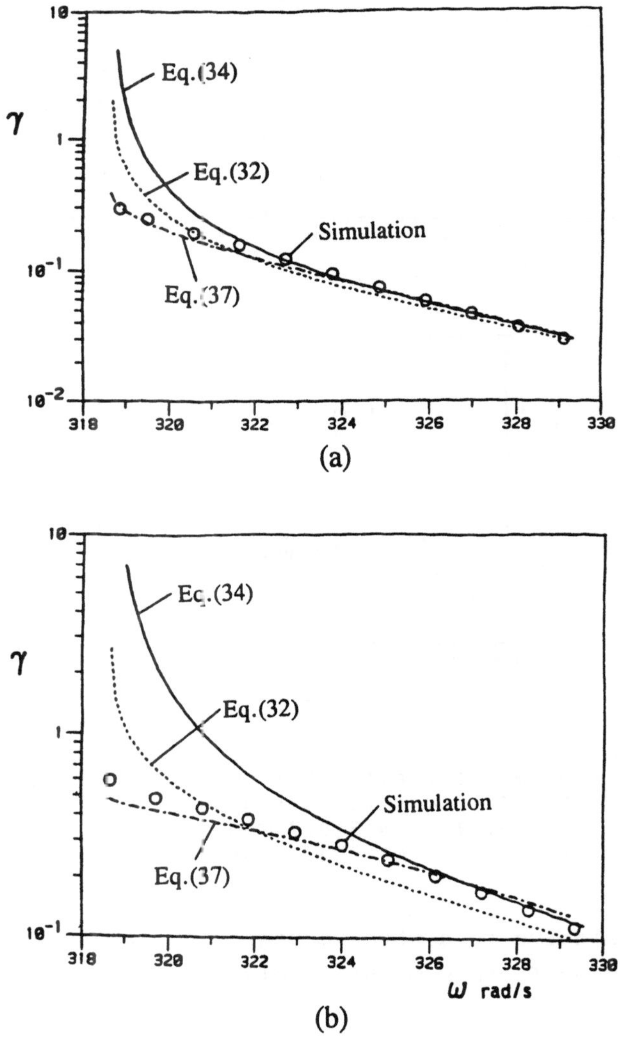

Fig. 7 Localization factor at frequencies in the lower half of the first wave-passage band
(a) $\delta = 0.01$, $\sigma = 0.25$
(b) $\delta = 0.01$, $\sigma = 0.5$

increases. This implies that the disorder to internal coupling ratio (σ/δ) is not the only factor in determining the intensity of localization, as once suggested; the frequency

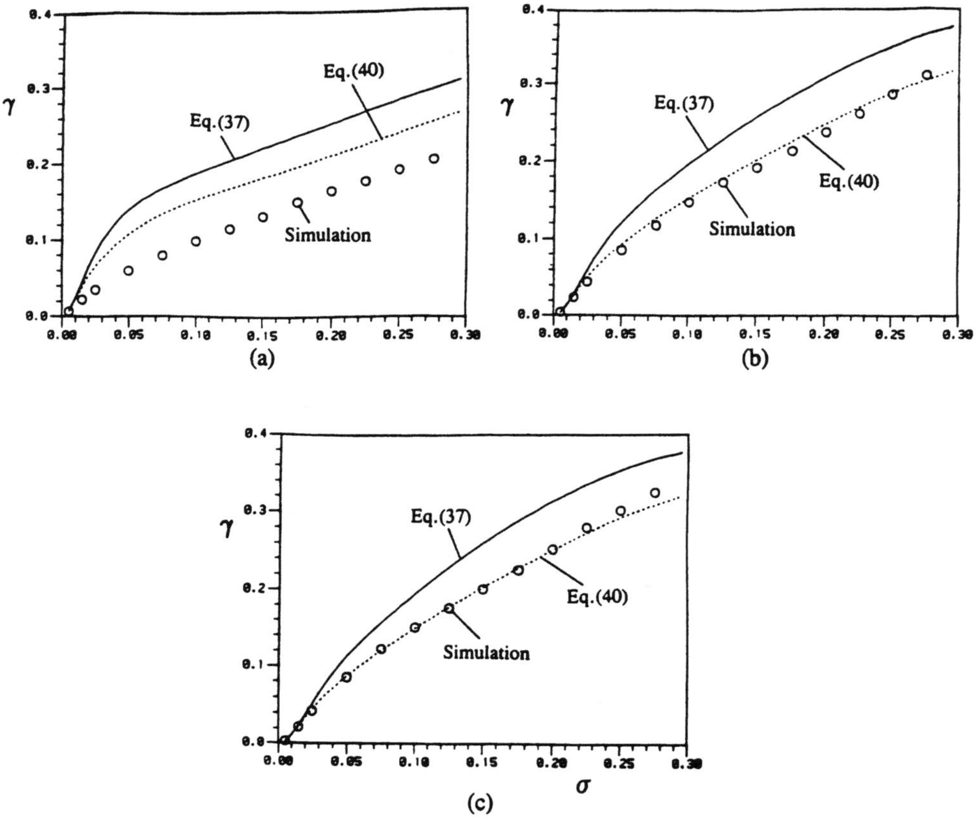

Fig. 8 Localization factors at frequencies near the lower boundaries of the first wave-passage frequency band
(a) $\delta = 0.1$, $\omega = 232.5$ rad/s
(b) $\delta = 0.01$, $\omega = 318.7$ rad/s
(c) $\delta = 0.001$, $\omega = 339.72$ rad/s

location relative to a wave-passage band is at least another factor. Other factors might be also important in more complicated systems.

Results obtained from Eq.(32) are generally more accurate than those obtained from Eq.(34). However, exceptions can occur, as shown in the higher frequency region in Figs.7(a) and 7(b). This is due to the fact that both equations are approximate. Although Eq.(32) essentially retains more terms in the approximation, it does not follow that those additional terms will increase the accuracy consistently.

As pointed out above, localization may be strong at frequencies near the lower boundaries of wave-passage frequency bands, even when the disorder is relatively small. Three such cases: (a) $\delta=0.1$, $\omega=232.5$ rad/s, (b) $\delta=0.01$, $\omega=318.7$ rad/s and (c) $\delta=0.001$, $\omega=339.72$ rad/s have been examined in more detail. The results obtained are shown in Fig. 8(a)-(c), respectively, with varying standard deviation of disorder σ. In all these cases Eq.(37) is clearly inadequate. Eq.(40) leads to satisfactory results for cases (b) and (c). For case (a) in which the disorder is also greater, Eq.(40) is still inadequate although it is much superior to Eq.(37). The use of even higher order approximations is suggested.

Concluding Remarks

Two systematic procedures are presented herein for calculating the localization factor of a randomly disordered periodic structure. The method of invariant probability is only applicable to undamped disordered structure, however, it is exact and the accuracy of the results only depends on the accuracy of the invariant probability. The method of multiple reflections, although being an approximate one, permits successive improvement in the accuracy and can be applied to damped disordered structures. The formulations for both methods are based on a generic disordered periodic structure and can be used to either weak, moderate or strong localization.

Some important characteristics of the localization phenomenon have been observed from the computed results. It is found that those important factors, which affect the degree of localization of a disordered system, include not only the ratio of disorder to internal coupling, but also the frequency location with respect to wave-passage frequency bands. The last factor was not reported in other investigations [17,18].

Acknowledgement

This paper is prepared under grant AFOSR-91-0073 from the Air Force Office of Scientific Research, Air Force Systems Command, USAD, monitored by Dr. Spencer Wu. The U.S. Government is authorized to reproduce and distribute reprints for governmental purposes, notwithstanding any copyright notation thereon.

References

1. Brillouin, L.: Wave propagation in periodic structures. Dover, New York, 1953.

2. Miles, J. W.: Vibration of beams on many supports. Journal, Engineering Mechanics Division, ASCE 82 (1956) 1-9.

3. Lin, Y. K.: Free vibration of continuous skin-stringer panels. Journal of Applied Mechanics 27 (1960) 669-676.

4. Lin, Y. K.: Stresses in continuous skin-stringer panels under random loading. Journal of Aero/Space Sciences 29 (1962) 67-86.

5. Lin, Y. K.: Free vibration of continuous beams on elastic supports. Journal of Mechanical Science 4 (1962) 409-423.

6. Lin, Y. K.; McDaniel, T. J.: Dynamics of beam-type periodic structures. Journal of Engineering for Industry 91 (1969) 1133-1141.

7. Mean, D. J.: Free wave propagation in periodically-supported infinite beam. Journal of Sound and Vibration 11 (1970) 181-197.

8. Mead, D. J.: A general theory of harmonic wave propagation in linear periodic systems with multiple coupling. Journal of Sound and Vibration 27 (1973) 235-260.

9. Mead, D. J.: Wave propagation and natural modes in periodic systems: I. mono-coupled system, II. multi-coupled systems, with and without damping. Journal of Sound and Vibration 40 (1975) 1-39.

10. Mead, D. J.; Markus, S.: Coupled flexural-longitudinal wave motion in periodic beams. Journal of Sound and Vibration 90 (1983) 1-24.

11. Anderson, P. W.: Absence of diffusion in certain random lattices. Physical Review 109 (1958) 1492-1505.

12. Valero, N. A.; Bendiksen, O. O.: Vibration characteristics of mistuned shrouded blade assemblies. Journal of Engineering for Gas Turbines and Power 108 (1986) 293-299.

13. Cornwell, P. J.; Bendiksen, O. O.: Localization of vibrations in large space reflectors. AIAA Paper 87-0949, Proceedings of 28th AIAA/ASME/ASCE/AHS Structures, Structural Dynamics and Materials Conference, April 6-8, 1987, Monterey, CA. 925-935.

14. Pierre, C.; Dowell, E. H.: Localization of vibrations by structural irregularity. Journal of Sound and Vibration 114 (1987) 549-564.

15. Pierre, C.; Tang, D. M.; Dowell E. H.: Localized vibrations of disordered multispan beams: theory and experiment. AIAA Journal 25 (1987) 1249-1257.

16. Pierre, C.; Cha, P. D.: Strong mode localization in nearly periodic disordered structures. AIAA Journal 27 (1989) 227-241.

17. Hodges, C. H.; Woodhouse, J.: Vibration isolation from irregularity in a nearly periodic structures: theory and measurements. Journal of Acoustic Society of America 74 (1983) 894-905.

18. Pierre, C.: Weak and strong vibration localization in disordered structures: a statistical investigation. Journal of Sound and Vibration 139 (1990) 549-564.

19. Kissel, G. J.: Localization in disordered periodic structures. Ph.D. Dissertation, Massachusetts Institute of Technology (1988).

20. Furstenberg, H.: Noncommuting random products. Transactions of the American Mathematical Society 108 (1963) 377-428.

21. Oseledets, V. I.: A multiplicative ergodic theorem. Transactions of the Moscow Mathematical Society 19 (1968).

22. Cai, G. Q.; Lin, Y. K.: Localization of wave propagation in disordered periodic structures. to appear in the AIAA Journal.

23. Pestel, E. C.; Leckie, F. A.: Matrix methods in elastomechanics. McGraw-Hill, New York (1963).

24. Yong, Y.; Lin, Y. K.: Propagation of decaying waves in periodic and piecewise periodic structures of finite length. Journal of Sound and Vibration 129 (1989) 99-118.

Modal Analysis for Random Response of MDOF Systems

M. Di Paola, G. Falsone, G. Muscolino and G. Ricciardi

Dipartimento di Ingegneria Strutturale e Geotecnica, DISEG, Universita' di Palermo, Viale delle Scienze, 90128 Palermo, Italy

Summary

The usefulness of the mode-superposition method of multidegrees of freedom systems excited by stochastic vector processes is here presented. The differential equations of moments of every order are written in compact form by means of the Kronecker algebra; then the method for integration of these equations is presented for both classically and non-classically damped systems, showing that the fundamental operator available for evaluating the response in the deterministic analysis is also useful for evaluating the response in the stochastic analysis.

Key Words: Modal analysis, Itô calculus, Moment equations.

INTRODUCTION

Deterministic dynamic analysis of structural systems discretized by finite elements can be performed through two different approaches: (i) direct integration methods, and (ii) mode superposition methods.

The direct integration methods (that can be classified into explicit[1] and implicit[2-5]) solve the nodal equation of motion directly without any space transformation, so that the size of the approximation operators is related to the number of nodal variables.

In mode superposition methods, by using only the first few mode shapes, the nodal dynamic equilibrium equations are transformed into a reduced system of equations expressed in modal (or generalized) coordinates. These methods show advantages with respect to the previous ones, especially for structural systems with numerous degrees of freedom and if the frequency content of the dynamical loadings is at low frequencies[5-8].

Y. K. Lin · I. Elishakoff (Eds.)
Stochastic Structural Dynamics 1
New Theoretical Developments
2nd International Conference on Stochastic Structural Dynamics
May 9-11, 1990, Boca Raton, Florida

In the stochastic analysis the extension of the direct integration methods to the case of structures subjected to stationary and non-stationary random excitations has only recently been introduced[9–11]. These methods need, at each time step, a large number of matricial products, each of them being comparable to an inversion matrix; so they are very expensive in computer time, especially when the structural system has numerous degrees of freedom. Another problem in using the stochastic direct integration methods is related to the fact that these methods are conditionally stable. Furthermore, other limitations in using these methods are connected to the fact that they can be used only for the evaluation of the second order moments and the extension to higher order moments is almost impossible (i.e. they are applicable only to linear or linearized systems subjected to Gaussian inputs).

The use of mode superposition methods in stochastic analysis has been extensively investigated for both linear[6] and non-linear systems[12–14] and for both white noise[12,13] and filtered[6,14] input processes, also by using higher order closure schemes[14].

In this paper the validity of mode-superposition methods in stochastic analysis is further investigated. In particular, attention is devoted to two fundamental aspects: (i) avoiding the evaluation of the complex eigenvalues for non-classically damped systems, and (ii) showing that these methods become more powerful in stochastic than in deterministic analysis.

The previously described results have been obtained by means of Itô's stochastic calculus[15–17] and Kronecker algebra[18]. The differential equations of any order moments are then written in a particular form that reveals a perfect analogy with deterministic motion equations. On account of the particular form of these equations, the corresponding fundamental matrices are closely related to the fundamental matrix of the deterministic case.

Applications to two multidegrees of freedom systems are presented in order to show the applicability and the versatility of the method.

DETERMINISTIC ANALYSIS

For the sake of clarity, in this section some preliminary concepts of deterministic analysis, which are also useful for stochastic analysis, are summarized.

By using the finite element method, the equation of motion of an n-degrees-of-freedom model of an elastic structural system with viscous damping subjected to a nodal forcing function vector $\mathbf{F}(t)$ is written

$$\begin{aligned} &\mathbf{M}\ddot{\mathbf{x}} + \mathbf{C}\dot{\mathbf{x}} + \mathbf{K}\mathbf{x} = \mathbf{F}(t) \quad ; \\ &\mathbf{x}(t_0) = \mathbf{x}_0 \quad ; \quad \dot{\mathbf{x}}(t_0) = \dot{\mathbf{x}}_0 \end{aligned} \tag{1}$$

where $\mathbf{M}$,$\mathbf{C}$ and $\mathbf{K}$ are the inertia, damping and stiffness matrices respectively, $\ddot{\mathbf{x}}$, $\dot{\mathbf{x}}$ and $\mathbf{x}$ are the nodal acceleration, velocity and displacement vectors rispectively, $\mathbf{x}_0$ and $\dot{\mathbf{x}}_0$ are the vectors of initial conditions. In order to reduce the number of equations, the dynamic response is evaluated in a reduced space by means of the following coordinate transformation

$$\mathbf{x} = \boldsymbol{\Phi}\mathbf{y} \tag{2}$$

where $\boldsymbol{\Phi}$ is the modal matrix of the undamped system, containing the first few eigenvectors of $\mathbf{K}^{-1}\mathbf{M}$ normalized with respect to $\mathbf{M}$; in this way the matrix $\boldsymbol{\Phi}$ is of order $n \times m$, m being the number of modes selected for the analysis ($m \ll n$). The matrix $\boldsymbol{\Phi}$ has the properties

$$\boldsymbol{\Phi}^T\mathbf{M}\boldsymbol{\Phi} = \mathbf{I}_m \quad ; \quad \boldsymbol{\Phi}^T\mathbf{K}\boldsymbol{\Phi} = \boldsymbol{\Omega}^2 \tag{3}$$

where $\mathbf{I}_m$ is the identity matrix of order m and $\boldsymbol{\Omega}^2$ is the diagonal matrix listing the square of the natural radian frequencies $\omega_i^2 \quad (i = 1, 2, \ldots, m)$. Letting $\boldsymbol{\Phi}^T\mathbf{C}\boldsymbol{\Phi} = \boldsymbol{\Xi}$, $\boldsymbol{\Xi}$ be an $m \times m$ symmetric non-negative definite matrix, equation (1) can be written in the reduced space of modal coordinates as follows:

$$\begin{aligned} &\ddot{\mathbf{y}} + \boldsymbol{\Xi}\dot{\mathbf{y}} + \boldsymbol{\Omega}^2\mathbf{y} = \boldsymbol{\Phi}^T\mathbf{F}(t) \quad ; \\ &\mathbf{y}_0 = \boldsymbol{\Phi}^T\mathbf{M}\mathbf{x}_0 \quad ; \quad \dot{\mathbf{y}}_0 = \boldsymbol{\Phi}^T\mathbf{M}\dot{\mathbf{x}}_0 \end{aligned} \tag{4}$$

If $\boldsymbol{\Xi}$ is a diagonal matrix, then the system possesses classical normal modes and equations (4) are given in decoupled form; in the most general case, when $\boldsymbol{\Xi}$ is a full matrix, a $2m = r$ dimension vector approach is commonly used[19]. For this purpose equation (4) can be rewritten in the standard form

$$\dot{\mathbf{z}} = \mathbf{D}\mathbf{z} + \mathbf{V}\mathbf{F}(t) \quad ; \quad \mathbf{z}(t_0) = \mathbf{z}_0 \tag{5}$$

where $\mathbf{z}$ is the $r-$vector of modal state variables defined as

$$\mathbf{z}^T = \left[\mathbf{y}^T \quad \dot{\mathbf{y}}^T\right] \quad ; \quad \mathbf{z}_0^T = \left[\mathbf{y}_0^T \quad \dot{\mathbf{y}}_0^T\right] \tag{6}$$

and the matrices $\mathbf{D}$ and $\mathbf{V}$ are given by

$$\mathbf{D} = \begin{pmatrix} \mathbf{0} & \mathbf{I}_m \\ -\boldsymbol{\Omega}^2 & -\boldsymbol{\Xi} \end{pmatrix} \quad ; \quad \mathbf{V} = \begin{pmatrix} \mathbf{0} \\ \boldsymbol{\Phi}^T \end{pmatrix} \tag{7}$$

Solving equation (5) is generally a very hard task and a numerical solution method is necessary to achieve this goal. Let the time axis be divided into small intervals of equal length Δt, and let $t_0 = 0, t_1, \ldots, t_k, \ldots$ be the division times. We can use the stepwise approach, each step being concerned with the evolution of the system within each interval. Assuming that $\mathbf{F}(t)$ is constant in the time interval $t_k, t_k + \Delta t$, the vector solution $\mathbf{z}(t_k + \Delta t)$ can be evaluated in the unconditionally stable step-by-step procedure[6–8]

$$\mathbf{z}(t_k + \Delta t) = \boldsymbol{\Theta}(\Delta t)\mathbf{z}(t_k) + \mathbf{L}(\Delta t)\mathbf{V}\mathbf{F}(t_k) \tag{8}$$

where $\boldsymbol{\Theta}(t)$ is the so-called fundamental matrix and $\mathbf{L}(t)$ is the load matrix related to $\boldsymbol{\Theta}(t)$ in the form

$$\mathbf{L}(t) = \left[\boldsymbol{\Theta}(t) - \mathbf{I}_r\right]\mathbf{D}^{-1} \tag{9}$$

For classically damped systems the closed form of the matrix $\boldsymbol{\Theta}(t)$ can be easily obtained as

$$\boldsymbol{\Theta}(t) = \begin{pmatrix} -\mathbf{G}(t)\boldsymbol{\Omega}^2 & \mathbf{H}(t) \\ -\mathbf{H}(t)\boldsymbol{\Omega}^2 & \dot{\mathbf{H}}(t) \end{pmatrix} \tag{10}$$

where $\mathbf{H}(t)$ is a diagonal matrix, having the $i-$th diagonal element equal to the impulse response function of the $i-$th oscillator in modal coordinates given by

$$\begin{aligned} h_i(t) &= \frac{1}{\bar{\omega}_i} exp(-\xi_i\omega_i t) sin(\bar{\omega}_i t) \quad ; \quad t \geq 0 \\ h_i(t) &= 0 \quad ; \quad t < 0 \end{aligned} \tag{11}$$

and $\mathbf{G}(t)$ is a diagonal matrix such that $\dot{\mathbf{G}}(t) = \mathbf{H}(t)$. In equation (11) ξ_i and $\bar{\omega}_i = \omega_i\sqrt{1-\xi_i^2}$ can be evaluated taking into account the fact that $2\xi_i\omega_i$ is the $i-$th element of the diagonal matrix $\boldsymbol{\Xi}$.

For non-classically damped systems, the fundamental matrix can be constructed starting from the knowledge of the eigenproperties of $\mathbf{D}$, and explicit expressions can be found in ref. $6-8$

In both cases, for classically and non-classically damped systems, using the fundamental matrices evaluated by means of the eigenproperties of the matrix $\mathbf{D}$, the step-by-step procedure is unconditionally stable and the only source of numerical error is connected with the modelling of the forcing vector that has been assumed constant in each temporal step.

For non-classically damped systems, computation of complex eigenproperties of the matrix $\mathbf{D}$ can be avoided by directly evaluating the fundamental matrix by means of a Taylor expansion [19,20]. Here a finite number N of terms in the expansion for $\boldsymbol{\Theta}$, as well as for the load operator, are retained; that is

$$\begin{aligned} \bar{\boldsymbol{\Theta}}(\Delta t) &= \sum_{j=0}^{N} \frac{1}{j!}(\mathbf{D}\,\Delta t)^j \quad ; \\ \bar{\mathbf{L}}(\Delta t) &= \Delta t \left[\sum_{j=0}^{N-1} \frac{1}{(j+1)!}(\mathbf{D}\,\Delta t)^j \right] \end{aligned} \tag{12}$$

Notice that, replacing these approximated operators in equation (8), the unconditionally stable step-by-step procedure becomes conditionally stable. However, this is not a limitation on the effectiveness of the numerical procedure. Indeed the effects connected with the damping matrix $\boldsymbol{\Xi}$ only produce small changes in the frequencies of the damped structure with respect to the undamped one. Because the modal analysis of the undamped system has been performed and $\mathbf{D}$ has been written in a reduced space, the approximate value of the smallest period $T_m = 2\pi/\omega_m$ is already known, and this is essential for the choice of the time step length to ensure stability of the step-by-step integration method. In order to show this, let us introduce the spectral radius of $\bar{\boldsymbol{\Theta}}(\Delta t)$ in the form

$$\rho\left[\bar{\boldsymbol{\Theta}}(\Delta t)\right] = max_i \left[\left| \sum_{k=0}^{N} \frac{1}{k!}(\lambda_i \Delta t)^k \right| \right] = \left| \sum_{k=0}^{N} \frac{1}{k!}(\lambda_{max}\Delta t)^k \right| \tag{13}$$

where λ_{max} is the maximum among the eigenvalues λ_i of $\mathbf{D}$ (and is very close to the largest radian frequency ω_m of the undamped system). Equation (13) shows that the spectral radii and therefore the stability of the integration method depend only on the step size selected and on the number of terms included in the Taylor expansion. In Figs.1 the spectral radii $\rho(\bar{\Theta})$ of a single degree-of-freedom system with natural period T_0 and damping ratios $\xi = 0$, $\xi = 0.1$ versus the ratio $\Delta t/T_0$ with variatios in the number of terms N are depicted. From these figures it is evident that the numerical procedure is stable using a time step $\Delta t < T_0/4$ if $N = 7$ and for any value of the damping ratios. The numerical results can be extended to non-classically damped systems using a natural period $T_0 = 2\pi/|\lambda_{max}| \simeq 2\pi/\omega_m$.

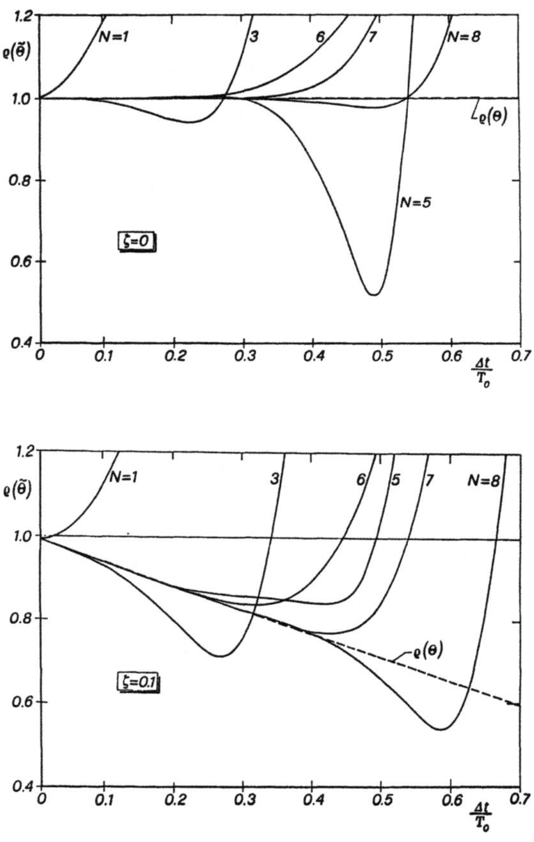

Figs. 1. Solid line: spectral radii of approximate fundamental matrix; dashed line: spectral radii of exact fundamental matrix.

Accuracy tests on an unloaded undamped single oscillator with initial conditions defined by an assigned displacement and zero velocity have been performed. In Table I and in Table II the percentages of period elongations and amplitude decays respectively are reported, varying the ratio $\Delta t/T_0$ and by using $N = 7$, and the results are compared with very common integration schemes, showing greater accuracy of the proposed method over the other ones.

TABLE I - PERCENTAGE PERIOD ELONGATIONS

METHOD \ $\frac{\Delta t}{T_0}$	$\frac{1}{50}$	$\frac{1}{25}$	$\frac{1}{10}$	$\frac{1}{8}$	$\frac{1}{6}$
HOUBOLT	0.7	2.2	9	13	20
WILSON ($\theta = 1.4$)	0.3	1	5.5	8	12.5
NEWMARK ($\delta = \frac{1}{2}$, $\alpha = \frac{1}{4}$)	0.1	0.5	3	5	8.5
N = 7	$< 10^{-3}$	$< 10^{-3}$	$< 10^{-3}$	$< 10^{-3}$	0.0025

TABLE II - PERCENTAGE AMPLITUDE DECAYS

METHOD \ $\frac{\Delta t}{T_0}$	$\frac{1}{50}$	$\frac{1}{25}$	$\frac{1}{10}$	$\frac{1}{8}$	$\frac{1}{6}$
HOUBOLT	0.5	3	13	17	22
WILSON ($\theta = 1.4$)	0.1	0.7	6.5	10	16
NEWMARK ($\delta = \frac{1}{2}$, $\alpha = \frac{1}{4}$)	0	0	0	0	0
N = 7	$< 10^{-8}$	$< 10^{-6}$	$0.5\ 10^{-3}$	$0.2\ 10^{-2}$	$0.12\ 10^{-1}$

DIFFERENTIAL MOMENT EQUATIONS

Let $\mathbf{F}(t)$ be a stochastic vector process, then $\mathbf{z}(t)$ is a stochastic vector process too and it has to be characterized in a probabilistic sense. In particular, if $\mathbf{F}(t)$ is a Gaussian white noise vector process, then equation (5) can be converted into an Itô type stochastic differential equation in the form

$$d\mathbf{z} = \mathbf{D}\mathbf{z}\,dt + \mathbf{V}d\mathbf{b} \tag{14}$$

where $\mathbf{b}$ is an r-vector of Wiener process such that the first two moments of $d\mathbf{b}$ are

$$E[d\mathbf{b}] = \mathbf{0} \quad ; \quad E[d\mathbf{b} \otimes d\mathbf{b}] = \mathbf{q}_2(t)dt \tag{15}$$

while higher order moments are infinitesimals of order greater than dt. In equation (15) $E[\cdot]$ means stochastic average, $\mathbf{q}_2(t)$ is the strength of the white noise vector process, and the symbol $\otimes$ means Kronecker product[18], so that the vector $E[d\mathbf{b} \otimes d\mathbf{b}]$, of order r^2, contains all possible moments of second order of the vector $d\mathbf{b}$.

In order to characterize the vector solution $\mathbf{z}$ in a probabilistic sense, the Itô rule[17] extended[12] to a vector $\boldsymbol{\phi}(\mathbf{z},t)$ of scalar real valued functions of $\mathbf{z}$, continuously differentiable on t and twice differentiable on $\mathbf{z}$, has to be applied in the form

$$\Delta\boldsymbol{\phi}(\mathbf{z},t) = \frac{\partial}{\partial t}\boldsymbol{\phi}(\mathbf{z},t) + \left[\nabla_{\mathbf{z}}^T \otimes \boldsymbol{\phi}(\mathbf{z},t)\right] d\mathbf{z} + \frac{1}{2!}\left[\nabla_{\mathbf{z}}^{[2]T} \otimes \boldsymbol{\phi}(\mathbf{z},t)\right] (d\mathbf{z})^{[2]} \tag{16}$$

where $\nabla_{\mathbf{z}}$ is the differential vector operator, collecting all the derivatives with respect to the elements of $\mathbf{z}$, i.e. $\nabla_{\mathbf{z}}^T = [\partial/\partial z_1 \quad \partial/\partial z_2 \ldots \partial/\partial z_r]$ and the exponents in square brackets mean Kronecker power (i.e. $(d\mathbf{z})^{[2]} = d\mathbf{z} \otimes d\mathbf{z}$). Setting $\boldsymbol{\phi}(\mathbf{z},t) = \mathbf{z}^{[j]}$ and evaluating the stochastic average, we obtain in a compact form the differential equations of all moments $\mathbf{m}_j$ of order j of $\mathbf{z}$, as

$$\dot{\mathbf{m}}_j = \mathbf{D}_j\mathbf{m}_j + \mathbf{V}_j\mathbf{q}_2(t) \quad ; \quad \mathbf{m}_j = E[\mathbf{z}^{[j]}] \tag{17}$$

where $\mathbf{D}_j$ and $\mathbf{V}_j$ are defined as follows

$$\mathbf{D}_j = \mathbf{D} \otimes \mathbf{I}_r^{[j-1]} + \mathbf{I}_r \otimes \mathbf{D} \otimes \mathbf{I}_r^{[j-2]} + \ldots + \mathbf{I}_r^{[j-1]} \otimes \mathbf{D} = \mathbf{D} \otimes \mathbf{I}_r^{[j-1]} + \mathbf{I}_r \otimes \mathbf{D}_{j-1} \tag{18}$$

$$\mathbf{V}_j = \frac{1}{2}\mathbf{Q}_j\left(\mathbf{Q}_{j-1} \otimes \mathbf{I}_r\right)\left(\mathbf{m}_{j-2} \otimes \mathbf{I}_r^{[2]}\right)\mathbf{V}^{[2]} \quad ; \quad \mathbf{V}_1 = \mathbf{0} \tag{19}$$

In equation (19) the matrices $\mathbf{Q}_j$ are defined in the form

$$\mathbf{Q}_j = \sum_{k=0}^{j-1} \mathbf{E}_{r^{j-k},r^k} \tag{20}$$

where $\mathbf{E}_{q,p}$ denotes a permutation matrix[18] of order $(qp)\times(qp)$ consisting of $q\times p$ arrays of $p\times q$ dimensional elementary submatrices $\mathbf{E}^{ij}$. The $p\times q$ dimensional elementary submatrix $\mathbf{E}^{ij}$ is one at the (i,j)-th position and zero in all the other positions.

Comparing equations (5) and (17) the very similar form of the mathematical structure governing the evolution of the state variable $\mathbf{z}$ and the moments $\mathbf{m}_j$ can be seen.

NUMERICAL TECHNIQUE TO SOLVE THE MOMENT EQUATIONS

In order to solve equation (17), the fundamental matrix of $\mathbf{D}_j$ has to be evaluated. In order to do this, the eigenvalues and eigenvectors can easily be computed starting from the eigenvalues and eigenvectors of $\mathbf{D}$; the following relationships hold[13]

$$\mathbf{D}_j\boldsymbol{\Psi}^{[j]} = \boldsymbol{\Psi}^{[j]}\boldsymbol{\Lambda}_j \quad ; \quad \boldsymbol{\Lambda}_j = \boldsymbol{\Lambda}\otimes\mathbf{I}_r^{[j-1]} + \mathbf{I}_r\otimes\boldsymbol{\Lambda}_{j-1} \tag{21}$$

where $\boldsymbol{\Psi}$ is the matrix of eigenvectors of $\mathbf{D}$ and $\boldsymbol{\Lambda}$ is the diagonal matrix listing the eigenvalues of $\mathbf{D}$. It follows that if m modes are taken into account in the deterministic analysis, m^j modes are taken into account in the stochastic analysis for the evaluation of the moments of order j. These modes are the combination of the first m modes of the deterministic analysis. The fundamental matrices $\boldsymbol{\Theta}_j(\Delta t)$ of $\mathbf{D}_j$ are related to the fundamental matrix of $\mathbf{D}$ by means of the following relationship

$$\boldsymbol{\Theta}_j(\Delta t) = \boldsymbol{\Theta}^{[j]}(\Delta t) \tag{22}$$

Equations (21) and (22) show that, starting from the knowledge of the eigenproperties of the equation (5), the corresponding eigenproperties of the matrix $\mathbf{D}_j$ can easily be evaluated without much effort. The numerical solution of equation (22) follows the step-by-step scheme in the form

$$\mathbf{m}_j(t_k+\Delta t) = \boldsymbol{\Theta}^{[j]}(\Delta t)\mathbf{m}_j(t_k) + \mathbf{L}_j(\Delta t)\mathbf{V}_j(t_k)\mathbf{q}_2(t_k) \tag{23}$$

where

$$\mathbf{L}_j(t) = \left[\boldsymbol{\Theta}^{[j]}(\Delta t) - \mathbf{I}_r^{[j]}\right]\mathbf{D}_j^{-1} \tag{24}$$

The inversion of $\mathbf{D}_j$ can be avoided by using the following relationship

$$\mathbf{D}_j^{-1} = \boldsymbol{\Psi}^{[j]}\boldsymbol{\Lambda}_j^{-1}\boldsymbol{\Psi}^{-[j]} \tag{25}$$

Notice that for classically damped systems $\boldsymbol{\Lambda}$ and $\boldsymbol{\Psi}$ are simply given as

$$\boldsymbol{\Lambda} = \begin{pmatrix} -\frac{1}{2}\boldsymbol{\Xi} + i\bar{\boldsymbol{\Omega}} & \mathbf{0} \\ \mathbf{0} & -\frac{1}{2}\boldsymbol{\Xi} - i\bar{\boldsymbol{\Omega}} \end{pmatrix} \tag{26}$$

$$\boldsymbol{\Psi} = \begin{pmatrix} \mathbf{I}_m & \mathbf{I}_m \\ -\frac{1}{2}\boldsymbol{\Xi} + i\bar{\boldsymbol{\Omega}} & -\frac{1}{2}\boldsymbol{\Xi} - i\bar{\boldsymbol{\Omega}} \end{pmatrix} \tag{27}$$

where i is the imaginary unit and $\bar{\boldsymbol{\Omega}}$ is a diagonal matrix listing the damping radian frequencies.

For a non-classically damped system a further complex eigenproblem should be solved; but it is more convenient to avoid the complex eigensolution and to evaluate the fundamental matrix and the load operator by means of the truncated Taylor expansion as follows

$$\bar{\boldsymbol{\Theta}}_j(\Delta t) = \sum_{k=0}^{N} \frac{1}{k!} (\Delta t \mathbf{D}_j)^k \tag{28}$$

and

$$\bar{\mathbf{L}}_j(\Delta t) = \Delta t \sum_{k=0}^{N-1} \frac{1}{(k+1)!} (\Delta t \mathbf{D}_j)^k \tag{29}$$

It is emphasized that for both classically and non-classically damped systems, using equation (23), the step-by-step procedure for evaluating moments of every order of the stochastic vector process $\mathbf{z}$ is unconditionally stable, if the matrix $\boldsymbol{\Theta}_j$ is evaluated in an exact form. Instead, using the approximate fundamental matrix $\bar{\boldsymbol{\Theta}}_j$ the step-by-step procedure becomes a conditionally stable one, and after simple algebra the stability criterion based on the spectral radii of $\bar{\boldsymbol{\Theta}}_j$ can be written as

$$\rho\left[\bar{\boldsymbol{\Theta}}_j(\Delta t)\right] = \left|\sum_{k=0}^{N} (\lambda_{max,j}\, \Delta t)^k\right| \leq 1 \tag{30}$$

where $\lambda_{max,j}$ is the maximum eigenvalue of $\mathbf{D}_j$ and, because of equation (21), is equal to j times the maximum eigenvalue of $\mathbf{D}$. It follows that, in order to ensure stability, the time step has to be selected as

$$\Delta t = \frac{2\,\pi}{4\,j\,\omega_m} \quad \text{for} \quad N = 7 \tag{31}$$

Hence, in the stochastic analysis of non-classically damped systems, if the fundamental and loading operators are evaluated in approximate form, the time step must be selected equal to the $j-$th part of the time step for the deterministic case. This is not a real limitation; indeed, the forcing vector in the stochastic analysis is $\mathbf{q}_2(t)$, that is a very smooth function which varies much more slowly than $\mathbf{F}(t)$, so that its frequency content is at very low frequency; on the other hand the natural frequencies of $\mathbf{D}_j$ are of order $j-$times the order of the natural frequencies of $\mathbf{D}$. It follows that, at the same order of accuracy, the number of modes selected for the stochastic analysis has to be less than the number of modes for the deterministic analysis.

NUMERICAL EXAMPLES

Two different numerical examples have been performed. In the first one the study of a classically damped structure highlights the fact that the stochastic analysis requires a lesser number of modes than that required in the deterministic case. In the second one a non-classically damped system is analyzed in order to show the applicability of the mode superposistion method, evaluating the fundamental matrix as a truncated series expansion.

The first system analyzed is the five-storey asymmetric classically damped framed building depicted in Fig. 2. The parameters selected are: mass centroid $X_G = Y_G = 300cm$, elastic modulus $E = 2.85 \times 10^5 kg/cm^2$, floor mass $m = 324 kg\, s^2/cm$ at each floor, moments of inertia with respect to the mass centroid $J_G = 77.76 \times 10^6 kg\, s^2/cm$ and damping ratio for all modes $\xi_i = 0.05$. The lowest five radian frequencies of the system are (in rad/s):

$$\omega_1 = 3.370 \quad \omega_2 = 4.147 \quad \omega_3 = 6.907 \quad \omega_4 = 10.541 \quad \omega_5 = 14.840$$

The structure is subjected to a ground motion in the x_0 direction and the deterministic analysis is performed for a single sample function of a non-stationary (band limited) white noise process obtained by means of the well-known generation formula

$$w(t) = \psi(t)\sqrt{2\, S_0 \Delta\omega} \sum_{k=1}^{M} \cos(\omega_k t + \theta_k) \tag{32}$$

where θ_k are random phases uniformly distributed between 0 and 2π, $\omega_k = k\,\Delta\omega$, $\Delta\omega = 0.067 rad/sec$, $S_0 = 1cm^2/sec^4\, rad$, $\psi(t) = \delta\, t\, exp(-t/\epsilon)$, $\delta = exp(1)/\epsilon$, $\epsilon = 4\, sec$, $M = 450$; the function is sampled for a time step $\Delta t = 0.02 sec$.

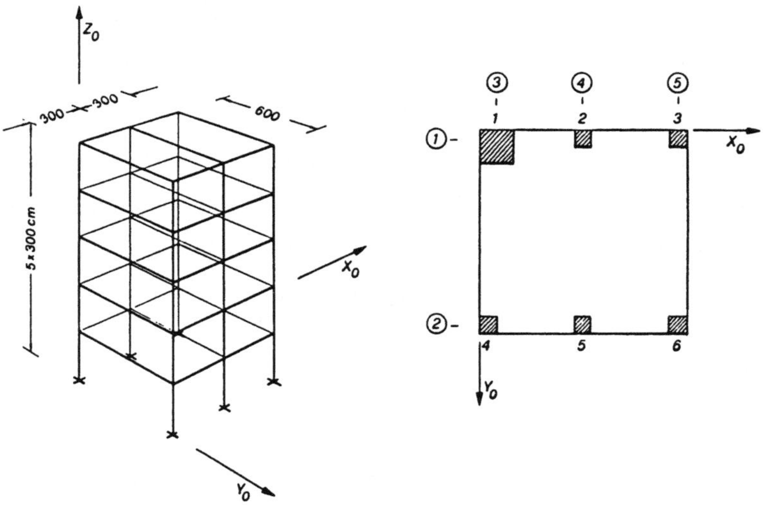

Fig. 2. *Five-storey building frame with 1-column* $90 \times 90cm$ *and all others* $30 \times 30cm$*, beams* $30 \times 50cm$

In Table III the percentage of error obtained for various number of modes is reported, in which

$$\varepsilon = \frac{\int_0^\infty |x_e - x_a|\, dt}{\int_0^\infty |x_e|\, dt} \times 100 \tag{33}$$

where x_e is the exact value of the displacement of the top floor of frame 1 in the x_0 direction and x_a is the displacement of the same frame evaluated by means of the step-by-step procedure described in equation (8), whose the fundamental matrix is given in equation (10).

TABLE III - Percentage of errors for deterministic analysis varying the number of modes

m	2	3	4	5
ε %	11.8 %	6.5 %	2.7 %	0.9 %

From Table III it is evident that using five modes very good accuracy is achieved, while decreasing the number of modes the error becomes considerable. If the fundamental matrix has been evaluated by means of the truncated Taylor expansion ($N = 7$), because the temporal step is smaller than the smallest fundamental period of the structure, the operator $\bar{\Theta}(\Delta t)$ is stable and the errors practically coincide with those evaluated by using the exact $\Theta(\Delta t)$.

The corresponding stochastic analysis is given assuming a strength of the white noise $q_2 = 2\pi S_0 \psi^2(t)$. In particular, for the second order moment analysis, the fundamental matrix has been evaluated as $\Theta^{[2]}(\Delta t)$ in which $\Theta(\Delta t)$ is the exact fundamental matrix of the deterministic analysis. The nodal response of the structure in terms of second order moments has been obtained for various values of the number of modes and for different values of the time step. The percentage of error between the exact standard deviation and the approximated one considering a reduced number of modes is plotted in Fig. 3. From this figure we can observe that the percentage of error in the stochastic analysis is smaller and smaller than the one in the deterministic analysis, even for a temporal step twenty-five times greater. Moreover, only three modes are sufficient to obtain very good accuracy and the percentage of error decreases as the time step decreases.

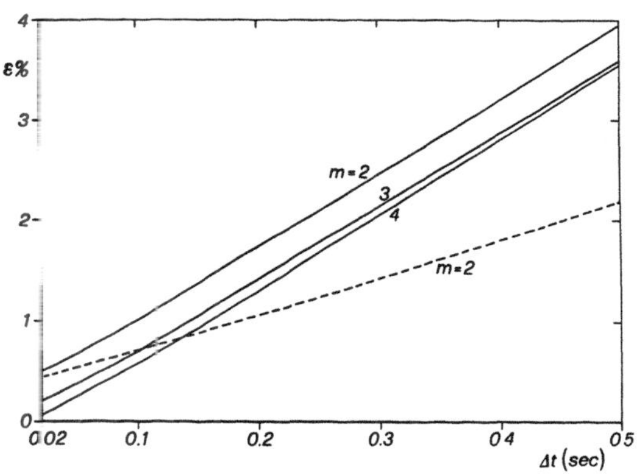

Fig. 3. Percentage of error for stochastic response varying the number of modes; solid line: standard deviation; dashed line : $\sqrt[4]{E[x^4]}$.

In spite of the fact that the system is linear and the input is Gaussian so that the fourth order moment could be evaluated by means of the knowledge of the second order moment, here the fourth order moment has also been evaluated by means of the step-by-step procedure in order to point out some numerical aspects. The percentage of error, evaluated between the exact $\sqrt[4]{E[x^4]}$ and the approximate one evaluated by using two modes, has also been plotted in the Fig. 3. We can observe that greater accuracy is achieved using two modes in comparison with the results obtained using four modes for the second order moment analysis.

In the second example a floodlight tower subject to a seismic input is analyzed. In Fig. 4*a* the schematic sketch of this structure is depicted. The main column is subdivided into 5 uniform beam elements of different length with different geometric characteristics in their cross-sections. The head structure can be considered as a frame with two lateral stiffening trusses. The latter are the only horizontal load-bearing structures of the head frame. The column and the head structure are connected by a horizontal frame. Since the seismic input is considered orthogonal to the front view of the head structure, the analysis can be performed considering the spatial structure as a plane one. In Fig. 4*b* the calculus schemes of the plane problem here considered are reported. The head structure is subdivided into 19 elements and the frame connecting it with the column is idealized by three springs. It follows that in the plane the whole structure has 32 degrees of freedom.

The seismic action is idealized as a non-stationary white noise filtered by a Tajimi-Kanai like filter. The parameters of the filter are $\xi_g = 0.6$ and $\omega_g = 5\pi \quad rad/sec$, while the strength of the white noise is assumed equal to $q_2 = 2\pi S_0 \psi^2(t)$. The first six natural radian frequencies of the whole structure are (in rad/sec)

$$\omega_1 = 2.302, \quad \omega_2 = 5.995 \quad \omega_3 = 19.306 \quad \omega_4 = 23.062 \quad \omega_5 = 50.488 \quad \omega_6 = 88.492$$

The truss and the column are considered as individually classically damped with a damping ratio $\xi = 0.01$. It follows that, according to substructure theory[21], the whole structure is a non-classically damped one.

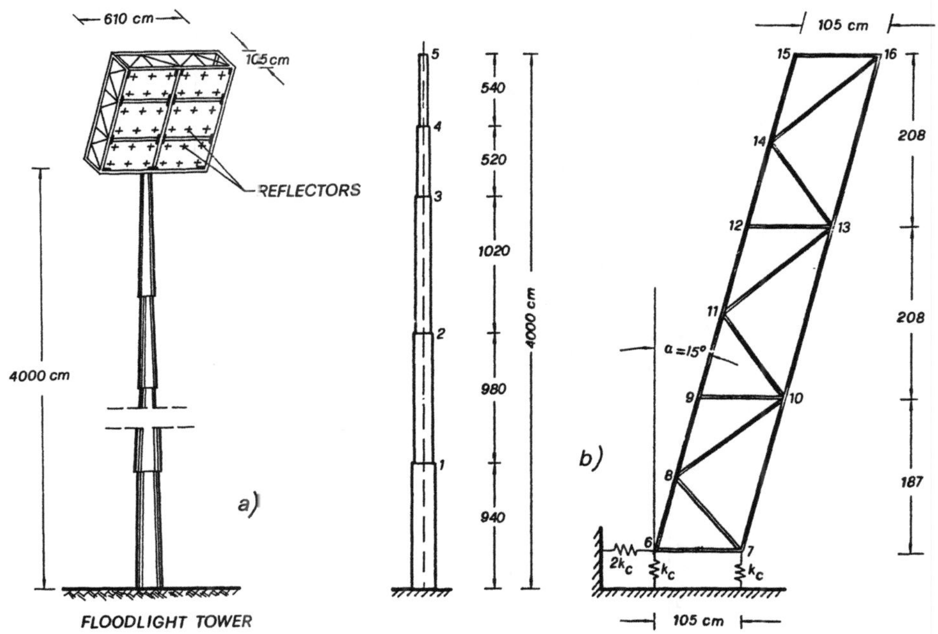

Figs. 4. a) Schematic sketch of the structure; b) calculus schemes of the main column and the head structure for the plane problem.

In Figs. 5 the mean square values of horizontal displacements of the top of the column and of the truss (node 5 and 16 of Fig. 4*b*) are depicted varying the number of modes included in the analysis. The numerical procedure is performed by evaluating the fundamental matrix $\bar{\Theta}^{[2]}(\Delta t)$ in an approximate manner considering $N = 8$ and $\Delta t = 0.01$. According to the accuracy studies one cannot appreciate the difference between the responses evaluated by using the exact fundamental matrix and the approximate one. These results show, once more, the great versatility of the mode superposition method in the stochastic analysis.

It is to be emphasized that we continually make reference to the number of modes in the deterministic case, so that in the stochastic analysis we take into account the m^2 combinations of the first m deterministic modes.

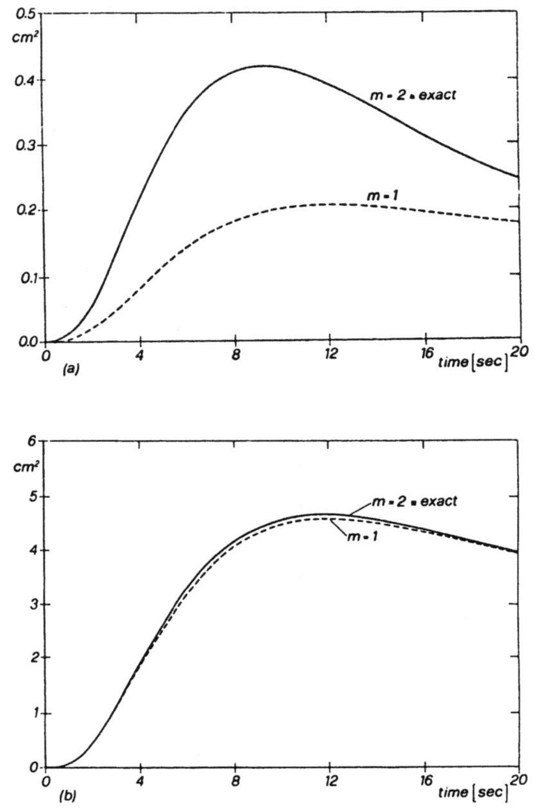

Fig. 5. Evolutionary mean square values of the horizontal displacement: a) node 5; b) node 16.

CONCLUSIONS

In this paper the validity of the modal analysis in both the deterministic and stochastic cases of structural systems subjected to non-stationary white noise input has been studied. In particular, for classically damped systems, the fundamental matrix for the deterministic analysis is given in a closed form, obtaining an unconditionally stable step-by-step procedure. The only source of numerical error is connected with the modelling of the forcing function (piece-wise constant in each temporal step). It follows that the size of the time step is only connected with the frequency content of the forcing function. The fundamental matrix for the stochastic analysis is strictly related to the fundamental matrix of the deterministic analysis (i.e. it is simply its Kronecker power) and the stochastic analysis can be performed in the same way as the deterministic one. However a remarkable difference lies on the fact that in the stochastic case the forcing function is smoother than in the deterministic case and that the eigenvalues in the stochastic analysis are greater than ones of deterministic case(the former are a Kronecker combination of the latter). It follows that, evaluating the fundamental matrix for the stochastic analysis as a Kronecker power of the deterministic one, a lesser number of modes needs to be evaluated with respect to the deterministic analysis. So we have two benefits: (i) increasing the order of moments, a lesser and lesser number of modes of the deterministic case can be included in the analysis; (ii) at parity of percentage of error, the temporal step can be $10 \div 20$ times greater than that of the corresponding temporal step utilized for the deterministic analysis.

In the case of a non-classically damped system, after the condensation of the equations of motion in the reduced space of the modal coordinates, the fundamental matrix can be evaluated using its truncated Taylor expansion. This gives a conditionally stable numerical procedure. However, this is not a limitation of the effectiveness of the method, especially for structural systems. In fact, in this case, the periods of vibration of the undamped system are only slightly different from the exact ones evaluated by means of complex eigenvalue analysis; furthermore, because the modal analysis has been performed, the various frequencies of the system are known, so that a fraction of the smallest period can be chosen in order to obtain a stable operator.

Extensive studies of accuracy have shown that if a time step ensuring the stability of the approximate fundamental matrix has been selected, then the results obtained practically coincide with those obtained with the exact fundamental matrix.

REFERENCES

1. Belytschko, T. Explicit time integration of structure-mechanical system, *Advanced Structural Dynamics*, by J. Donea ed., Applied Sc. Publs., London, 1980, 97-122

2. Houbolt, J.C. A recurrence matrix solution for the dynamic response of elastic aircraft, *Journal of the Aeronautical Science*, 1950, **17**, 540-550
3. Newmark, N.M. A method of computation for structural dynamics, *Proc. of American Society Civil Engineers*,1959, **85**(EM3),67-94
4. Bathe, K.J. and Wilson, E.L. Stability and accuracy analysis of direct integration methods, *Earthquake Engineering and Structural Dynamics*, 1973, **1**, 283-291
5. Bathe, K.J. *Finite element procedures in engineering analysis*, Prentice-Hall, Englewood Cliffs, New Jersey, 1982
6. Di Paola, M., Joppolo, M. and Muscolino, G. Stochastic seismic analysis of multidegree-of-freedom systems, *Engineering Structures*, 1984, **6**, 113-118
7. Borino, G. and Muscolino, G. Mode-superposition methods in dynamic analysis of classically and non-classically damped linear system, *Earthquake Engineering and Structural Dynamics*, 1986, **14**, 705-717
8. Muscolino, G. Mode-superposition methods for elastoplastic systems, *Journal of Engineering Mechanics, ASCE*, 1989, **115**, 2199-2215
9. To, C.W.S. The stochastic central difference method in structural dynamics, *Computers and Structures*, 1986,**23**(6), 813-818
10. To, C.W.S. Direct integration operators and their stability for random response of multidegree-of-freedom systems, *Computers and Structures*, 1988, **30**, 865-874
11. To, C.W.S. An implicit direct integrator for random response of multidegree-of-freedom systems, *Computers and Structures*, 1989, **33**, 73-77
12. Di Paola, M. Moments of non-linear systems, *Probabilistic Methods in Civil Engineering, Proceedings of the fiveth ASCE Special Conference*, Blacksburg (VA), 1988, 285-288
13. Di Paola, M. and Muscolino, G. Differential moment equations of FE modelled structures with geometrical non-linearities, *International Journal of Non-Linear Mechanics*, 1990, **25**, 363-373.
14. Di Paola, M., Falsone, G. and Muscolino, G. Random analysis of geometrically non-linear FE modelled structures under seismic actions, *Structural Safety*, 1990, **8**, 209-222.

15. Itô, K. On a formula concerning stochastic differential, *Nagoya Math. Journal*, 1951, **3**, 55-65

16. Jazwinski, A.H. *Stochastic processes and filtering theory*, Academic Press, New York, N.Y., 1970

17. Itô, K. Lectures on stochastic processes, *Tata Institute Fundamental Research*, Bombay, India, 1961

18. Ma, F. Extension of second moment analysis to vector valued and matrix valued functions, *Internationa Journal of Non-Linear Mechanics*, 1987, **22**(3), 251-260

19. Meirovitch, L. *Computational methods in structural dynamics*, The Netherlands, Sijthoff-Noordhoff Int. Publ., 1980

20. Muller, P.C. and Schiehlen, W.O. *Linear vibrations* , The Netherlands, Martinus Nijhoff Publishers, 1985

21. Muscolino, G. Dynamic response of multiply connected primary-secondary systems, *Earthquake Engineering and Structural Dyamics*, 1990, **19**, 205-216.

Linear Viscoelastic Analysis with Random Material Properties

HARRY H. HILTON
Department of Aeronautical and Astronautical Engineering
University of Illinois, Urbana-Champaign, 104 S. Mathews
Urbana, IL 61801-2997

JOHN HSU
Asia Cement Corp., Singapore

JOHN S. KIRBY
Advanced Materials and Survivability, McDonnell Douglas Corp.
Huntington Beach, CA

ABSTRACT

Analytical studies are presented which extend the elastic-viscoelastic analogies to stochastic processes caused by random linear viscoelastic material properties. Separation of variable as well as integral transform correspondence principles are formulated and discussed in detail. The statistical differential equation of the moment characteristic functional is derived, but rather than solving the highly complex functional equation, solutions are formulated in terms of first and second order statistical properties. Both Gaussian and beta distributions are considered for the probability density distributions of creep and relaxation functions and their effectiveness is evaluated.

In order to illustrate the developed general theory, specific examples of beam bending and pressurized hollow cylinders are solved. The influence of various parameters contributing to the creep and relaxation correlation functions is evaluated and the relationship between deterministic and stochastic bounds is also investigated. It is shown that deterministic bounds based on material data spread are unrealistic in the presence of random viscoelastic properties, since the do not correctly predict the limits of this stochastic process.

INTRODUCTION

The inherent nature of biological materials and the inability to manufacture high polymers, rubberlike materials, concrete and high temperature metals to close mechanical property specifications results in the general usage of materials whose relaxation and creep properties show large degrees of scatter. Since such scatter of viscoelastic property data falls within a band of relaxation and/or creep functions, distributions can be applied to these functions at all times. Additional uncertainties are introduced into such

Y. K. Lin · I. Elishakoff (Eds.)
Stochastic Structural Dynamics 1
New Theoretical Developments
2nd International Conference on Stochastic Structural Dynamics
May 9–11, 1990, Boca Raton, Florida

problems by the presence of random loads, aging, temperatures, moisture content and geometry. The stress and strain analysis of real materials under actual conditions must, therefore, lead to statistical solutions based on random input states.

Compared to the extensive deterministic viscoelastic literature, considerably less work has been published on random viscoelasticity. Eringen[1] has studied the response of simple Maxwell or Kelvin models to statistical inertia effects, while Bienek[2] has investigated the influence of random loads on first order nonlinear and general linear deterministic viscoelastic materials under creep conditions. Huang and Cozarelli[3-5] have analyzed a number of problems (beams, trusses) with random material effects using the creep power law. Parkus[6,7] and Ziegler[8,9] have extensively investigated the effects of random surface temperatures on deterministic viscoelasticity. Parkus *et al* [10,11] have also studied thermoheologically simple materials under random temperatures, resulting in stochastic reduced time processes. The influence of random temperatures on one dimensional nonlinear viscoelastic materials under dynamic conditions has been analyzed by Cozarelli and Chang[12]. Finally, Molyneaux and Beran[13] have considered the statistical properties of stress and strain fields in media with small random variations in elastic coefficients. Bazant *et al* [14,15] have analyzed the random effects of creep, shrinkage, humidity and temperature on stresses and deformations in concrete.

The statistical mathematical approach used in Reference 13 is extended in the present paper in conjunction with the formulation of elastic-viscoelastic analogies applicable to random viscoelastic material property time functions. The basic statistical philosophy used here is similar to that used in formulating the statistical theory of turbulence and even though the main objective of this paper is to treat random material viscoelastic property boundary value problems, statistical behavior introduced through boundary conditions is included whenever economically possible.

Since the statistical formulation hinges on deterministic viscoelasticity, it is pertinent to first briefly review the latter. Deterministic linear viscoelastic solutions are obtained by either direct methods, separation of variables or integral transform techniques. Intrinsic to the ease of solution is the question whether or not viscoelastic properties are time dependent because of aging, transient temperatures, moisture content, etc.[16] If they are, then the stress-strain relations are representable as either differential equations with variable coefficients or as nonconvolutional type integral equations. Direct solutions of classical problems may be found, for instance, in References 17-21, while the other two techniques yield elastic-viscoelastic analogies of an exact nature for time independent material properties[22-24] and of approximate ones for time dependent viscoelastic materials.[25,26] The Laplace and Fourier transforms arising in the transform solutions cannot in many instances be readily inverted analytically, however, approximate numerical schemes are available[27,28]. The beauty of the analogies is that viscoelastic solutions are obtainable directly from equivalent elastic ones.

The purpose of this paper is to formulate elastic-viscoelastic analogies for random material properties. To accomplish this end one must assume that all statistical quantities are measurable, at least in principle, and then interpret viscoelastic boundary value problems in terms of these quantities. Ideally, one would like to solve the characteristic moment functional equations,[29] since random processes are definable by their characteristic functionals. For linear viscoelasticity, these equations are derived in the form of displacement field equations, but unfortunately mathematical techniques are currently unavailable for their complete solutions. Consequently, the present analysis considers only the lowest order statistical equations corresponding to random time variations in material properties.

Thus, the first order statistical equations allow the determination of the mean functions of the stress and strain fields in terms of the material property mean functions, while the second order statistics fix the two-time stress and strain correlation functions in terms of material property correlations. From these statistical equations the general viscoelastic solutions of the stress, displacement and mixed boundary value problems are obtained by correspondence with deterministic elastic equivalent problems.

ANALYSIS

Fundamental Concepts

The stress-strain relations of a generalized, constant temperature, homogeneous, isotropic, linear viscoelastic medium can be expressed in Cartesian tensors by

$$\sigma_{kl}(x,t) = \delta_{kl} \int_0^t \phi_v(t-s)\, \varepsilon_{,s}(x,s)\, ds + 2 \int_0^t \phi\,(t-s)\, [\varepsilon_{kl,s}(x,s) - \delta_{kl}\varepsilon_{,s}(x,s)]ds \tag{1}$$

where the mean strain* $\varepsilon = \varepsilon_{kk}/3$ and ϕ and ϕ_v are the relaxation functions contributing to changes in shape and volume respectively.[16] Under small deformations one may readily substitute for strains in terms of the displacements u_k and eliminate the stresses through the equilibrium equations, yielding the governing displacement relations

$$L\{\phi, \phi_v, u\} = \int_o^t \{[\phi_v(t-s) + \phi(t-s)]\, u_{l l,ks}(x,s)/3 + \phi(t-s)\, u_{k,lls}(x,s)\}\, ds + f_k - \rho\, u_{k,tt}$$
$$= 0 \qquad (2)$$

where f_k are the body forces other than inertia and with boundary conditions consisting of specified forces and/or displacements over appropriate portions of the boundary surfaces. The density ρ will be considered a deterministic quantity only to focus full attention on the influence of the relaxation (or creep) functions, which are random functions of time.

A random function or process is a family of random variables denoted by $[X(.,t) : t\varepsilon T]$ or simply by $X(t)$ where the index set T represents the observation times.[30] Such functions $X(t)$ can be specified either by their probability density functional $P[X(t)]$ or by their characteristic functional $M[Y(t)]$. The relation between P and M is

$$M[Y(t)] = \int_R \exp\left[\, i \int_T Y(t)\, X(t)\, dt\right] P\,[X(t)]\, dX(t) \qquad (3)$$

where R is an integral over the infinite-dimensional function space and $\int_T Y(t)X(t)dt$ is an integral over all t values for which $X(t)$ is defined.

A random process can also be described by all its moment functions which are defined by

$$\eta_X(t_1) = \overline{X(t_1)} = \int_R X(t_1)\, P_1[X(t_1)]\, dX(t_1) \qquad (4)$$

$$R_{XX}(t_1,t_2) = \overline{X(t_1)X(t_2)} = \int_R X(t_1)\, X(t_2)\, P_2[X(t_1),X(t_2)]\, dX(t_1)dX(t_2) \qquad \text{etc.}$$

* The Einstein summation convention is used, i.e., $\varepsilon_{kk} = \varepsilon_{11} + \varepsilon_{22} + \varepsilon_{33}$.

where $P_1[X(t_1)]$, $P_2[X(t_1, X(t_2)]$, ... are the probability densities of the random process X(t) of the appropriate order and $\eta_x(t_1)$ and $R_{xx}(t_1, t_2)$ are respectively the mean and autocorrelation functions.

If one is able to obtain the statistical descriptions of $\phi(t)$ and $\phi_v(t)$ of Eq. (1) economically, then averaging Eq. (2) results in $\overline{L\{\phi, \phi_v, u\}} = 0$ where in general $\overline{\phi u} \neq \overline{\phi}\,\overline{u}$. Clearly, this does not allow one to determine $\overline{u_k(x,t)}$ in terms of $\overline{\phi(t)}$ and $\overline{\phi_v(t)}$ alone, since $\overline{u}_k$ depends on the entire probability density functional $P[\phi(t), \phi_v(t)]$. Similarly, the same applies to any of the moments of Eq. (4). It is not generally possible to solve either the set of functional equations or the infinite set of moment equations and the random process can only be characterized approximately.

The common assumption made in the literature[13,29] in order to simplify the statistical moment equations is that the variations from the mean are relatively small. Consequently, assuming that the material property variables $\phi(t)$ and $\phi_v(t)$ are uncorrelated with the stresses and displacements, one obtains

$$\overline{[\phi(t) + \phi_v(t)]\, A(x,t)} = [\overline{\phi(t)} + \overline{\phi_v(t)}]\, \overline{A(x, t)} \tag{5}$$

where for A read σ_{kl}, ε_{kl} or u_k and where the bar indicates ensemble averages in the sense of Eq. (4).

Extension of Alfrey's Analogy

The deterministic separation of variable solution first applied by Alfrey has been extended to thermal stress conditions[16,24-26]. In its deterministic form, it applies to incompressible, linear homogeneous viscoelasticity with separable stress or displacement boundary conditions and in the absence of all body forces. The corresponding statistical problem can be briefly stated by:

(a) stress strain relations

$$\overline{P}\,\{\overline{S_{kl}}\} = 2\,\overline{Q}\,\{\overline{\varepsilon_{kl}}\} \tag{6}$$

where $\overline{S}_{kl} = \overline{\sigma_{kl}} - \delta_{kl}\,\overline{\sigma_{jj}}/3$ and the differential operators are

$$\overline{P} = \sum_{n=o}^{r} \overline{a_n(t)}\frac{\partial^n}{\partial t^n} \tag{7}$$

$$\overline{Q} = \sum_{n=o}^{s} \overline{b_n(t)} \frac{\partial^n}{\partial t^n}$$

(b) the boundary conditions on the surface Γ are either

$$\overline{X_k}(x, t) = \overline{g_1(t)}\ X_k^*(x) \qquad \text{on } \Gamma(x) \tag{8}$$

or

$$\overline{U_k(x,t)} = \overline{h_2(t)}\ U_k^*(x) \qquad \text{on } \Gamma(x) \tag{9}$$

(c) the solutions are of the form

$$\overline{\sigma_{kl}(x,t)} = \overline{g_j(t)}\ \sigma_{kl}^*(x) \tag{10}$$

$$\overline{u_k\ (x,t)} = \overline{h_j(t)}\ u^*_k\ (x)$$

with the time functions determined by

$$\overline{P}\ \{\overline{g_j(t)}\} = \overline{Q}\ \{\overline{h_j(t)}\} \tag{11}$$

with $j = 1$ or 2.

The starred quantities in Eq. (10) represent the equivalent deterministic static elastic solutions with unit shear modulus corresponding to the starred boundary conditions of either Eq. (8) or (9). Depending on whether $\overline{g_1}$ or $\overline{h_2}$ are prescribed in (8) or (9), either $\overline{h_1}$ or $\overline{g_2}$ are obtained from (11). Obtaining solutions to the differential equation (11) can be clumsy, but such difficulties are readily overcome by using instead the integral stress-strain relations of Eq. (1). For an incompressible material, the deterministic stress-strain relations (1) can be written as [16]

$$2\varepsilon_{kl}\ (x,t) = \psi(o)S_{kl}\ (x,t) - \int_o^t \psi'(t-s)\ S_{kl}(x,s)\ ds \tag{12}$$

where the prime indicates differentiation with respect to s and ψ is the creep function. After separating variables and averaging, Eq. (11) is replaced by

$$\overline{h_1(t)} = \overline{\psi(o)}\ \overline{g_1(t)} - \int_o^t \overline{\psi'(t-s)}\ \overline{g_1(s)}\ ds \tag{13}$$

$$\overline{g_2(t)} = \overline{\phi(o)}\ \overline{h_2(t)} - \int_o^t \overline{\phi'(t-s)}\ \overline{h_2(t)}\ ds \tag{14}$$

One can now readily determine the two-point correlation functions with respect to time at the same spatial point. For the separable solutions these are

$$\overline{\sigma_{kl}(x,t_1)\sigma_{mn}(x,t_2)} = \overline{g_j(t_1)g_j(t_2)}\ \sigma^*_{kl}(x)\sigma^*_{mn}(x)$$

$$\overline{u_k(x,t_1)u_m(x_1,t_2)} = \overline{h_j(t_1)h_j(t_2)}\ u^*_k(x)u^*_m(x) \tag{15}$$

where the first boundary value problem $\overline{g_1(t_1)g_1(t_2)}$ is prescribed by the boundary conditions and the $\overline{h}$ functions in (15) are given with the help of Eq. (13)

$$\overline{h_1(t_1)h_1(t_2)} = \overline{\psi^2(o)}\ \overline{g_1(t_1)g_1(t_2)} - \int_o^{t_1} \overline{\psi(o)\psi'(t_1 - s_1)g_1(s_1)}\ \overline{g_1(t_2)}\ ds_1 \tag{16}$$

$$- \int_o^{t_2} \overline{\psi(o)\psi'(t_2 - s_2)g_1(t_1)g_1(s_2)}\ ds_2 + \int_o^{t_1}\int_o^{t_2} \overline{\psi'(t_1 - s_1)\ \psi'(t_2 - s_2)}\ \overline{g_1(s_1)g_1(s_2)}ds_2 ds_1$$

Thus, the response correlation function $\overline{h_1(t_1)h_1(t_2)}$ can be determined provided $\overline{\psi(t_1)\psi(t_2)}$ is known. Unfortunately, material correlation functions are not available in the literature, but in a subsequent section an approach for their estimation is presented.

Similarly, using Eqs. (14) and (15) one can obtain the stress correlation functions $\overline{g_2(t_1)g_2(t_2)}$ for the separable displacement boundary value problem. Thus, the first and second order statistics for random homogeneous incompressible viscoelastic material properties without inertia and for separable stress or displacement boundary conditions have been derived.

Extension of the Correspondence Principle

The deterministic elastic-viscoelastic correspondence principle was first derived by Read[23] for compressible, homogeneous, isotropic, time independent linear material properties including inertia effects and mixed boundary value problems. It has subsequently been extended to anisotropic, non-homogeneous, temperature dependent media[24-26]. In all cases it relates the Laplace or Fourier transforms of the viscoelastic variables to corresponding transforms of elastic variables but with the transforms of the viscoelastic moduli, boundary conditions and thermal expansions substituted directly into the elastic solution.

Let $\hat{\sigma}_{kl}(x,\omega)$ be the Fourier transform of $\sigma_{kl}(x,t)$. For time independent deterministic viscoelasticity, in the absence of thermal expansions, the deterministic correspondence principle may then be stated as

$$\sigma_{kl}(x,t) = \int_{-\infty}^{\infty} \hat{\sigma}^{e}_{kl}(x, \omega, \hat{X}, \hat{U}, \hat{G}, \hat{K})\, e^{i\omega t}\, d\omega \tag{17}$$

where the $\hat{G}$ and $\hat{K}$ are the complex viscoelastic shear and bulk moduli[10], and the boundary conditions are

$$\begin{aligned} \sigma_{kl}(x,t)\, n_l(x) &= X_k(x,t) && \text{on } \Gamma_1(x) \\ u_k(x,t) &= U_k(x,t) && \text{on } \Gamma_2(x) \end{aligned} \tag{18}$$

with n_l the components of the unit outward normal on $\Gamma = \Gamma_1 + \Gamma_2$, the boundary surface.

Not all random functions satisfy the existence conditions of the Fourier transforms. For instance, a weakly homogeneous function[30] would fail the test. However, one can safely assume that the viscoelastic material properties are strongly homogeneous and that the transforms and their inverse exist in the mean square. Consequently, one can define the mean and correlation functions of random stress, strain or displacement functions in terms of the mathematical expectations, i.e.

$$\overline{\sigma_{kl}(x,t)} = E\{\sigma_{kl}(x,t)\} \tag{19}$$

$$\overline{\sigma_{kl}(x,t_1)\sigma_{kl}(x,t_2)} = E\{\sigma_{kl}(x,t_1)\sigma_{kl}(x,t_2)\} \tag{20}$$

with σ_{kl}, ε_{kl} or u_k defined by Eq. (17).

STATISTICAL CHARACTERIZATION OF VISCOELASTIC PROPERTIES

It can be readily shown that generalized linear deterministic characterization of viscoelastic properties leads to relaxation functions given in terms of Prony series[16]

$$\phi(t) = \sum_{m=0}^{M} a_m\, e^{-\alpha_m t} \qquad \alpha_m > 0 \tag{21}$$

with a similar expression for the creep function $\psi(t)$. The parameters a_m and the inverse relaxation times α_m must, of course, be determined experimentally. Only limited deterministic data is available, such as for instance found in References 31 and 32 which may be considered representative of mean values without giving any probability distribution information as a function of time. However, in the absence of statistical experimental data and for the purposes of the present analysis, one can generate typical distributions based on such mean curves.

For instance, one could use a Gaussian distribution at any time and calculate the upper and lower values of $\phi(t)$ based on say 3 standard deviations of the mean which would then include 99.73% of all possible data. Such Gaussian distributions for a random variable w are expressed by

$$p_w(w) = \frac{1}{2\pi\sqrt{\sigma_{wn}}} \exp\left[-\frac{1}{2}\left(\frac{w-\eta_{wn}}{\sigma_{wn}}\right)^2\right] \qquad -\infty \leq w \leq \infty \tag{22}$$

where η_{w_n} and σ_{wn} are Gaussian mean values and standard deviations of w and

$$w = w(t) = [\phi - \phi_L(t)] / [\phi_U(t) - \phi_L(t)] \tag{23}$$

Using the results of Reference 32 for solid propellants, the upper and lower bounds for M = 13 are given by $a_m = A_m \pm 3\, D_m$ respectively for Eq. (21) and the values of the coefficient are tabulated in Table I with the curves shown in Fig. 1. Unfortunately, Gaussian distributions suffer from a serious deficiency in that they extend over the infinite domain of ϕ at all times and tend to lead to negative ϕ values at small but still significant probabilities. This has the effect of predicting for relatively large dispersions, say, uniaxial tension for compressive strains (or vice versa) which, of course, is physically unrealistic. Consequently, a probability density distribution extending over a finite domain, such as $\phi_L(t) \leq \phi(t) \leq \phi_U(t)$ is the preferred choice. A random variable w is said to have a Beta distribution[35] if its probability density distribution is

$$p_w(w) = \begin{cases} w^{a-1}(1-w)^{b-1} / B(a,b)\,; & 0 \leq w \leq 1 \\ 0 \quad ; & \text{elsewhere} \end{cases} \tag{24}$$

and B(a,b) is the Beta function. The corresponding mean value and standard deviation are

$$\eta_{WB} = a/(a+b)$$
$$\sigma^2_{WB} = ab/(a+b)^2(a+b+1) \tag{25}$$

The exponents of a and b of (24) may be determined either from Eqs. (25) with η_{WB} and σ^2_{WB} set equal to the corresponding experimental data values or by curve fitting Eq. (24) to the actual data points by the method of least squares.

Another significant property that must be defined in order to characterize statistical viscoelastic material properties is the autocorrelation function, $R_{XX}(t_1, t_2)$, for the r.v. $X(.,t_1)$ and $X(., t_2)$. Its general properties are well known[30] and an analytical expression for a random process such as

$$R_{XX}(t_1, t_2) = \sigma_X(t_1)\, \sigma_X(t_2)\, e^{-\alpha|t_1-t_2|} \cos \omega(t_1 - t_2) \tag{26}$$

exhibits the two basic physical ingredients, namely exponential decay and oscillation. Since the experimental data in Fig. 1 shows nonparallel slopes for the upper and lower bounds of the relaxation function, time function $\sigma_X(t)$ must be included in Eq. (26).

APPLICATIONS TO SPECIFIC VISCOELASTIC PROBLEMS

Bending of Beams (White Noise)

Consider a prismatic cantilever beam of length l in the x_2 direction and loaded at the free end with a random load $g_1(t)$ parallel to the x_3 axis. By applying the extended Alfrey's analogy to this random load problem one obtains

$$\overline{\sigma_{22}(x_2, x_3, t)} = [(x_2 - l)\, x_3 / I]\, \overline{g_1(t)} = \sigma^*_{22}(x_2, x_3)\, \overline{g_1(t)}$$
$$\overline{u_{33}(x_2, t)} = [- (x_2^3 - 3l^2 + 2l^3)\, x_3 / 6I]\, \overline{h_1(t)} = u^*_{33}(x_2)\, \overline{h_1(t)} \tag{27}$$

with $\overline{h_1(t)}$ defined by Eq. (13). If, for instance, the end load $g_1(t)$ is a white noise, then

$$\overline{g_1(t)} = 0\,, \qquad \overline{g_1(t_1) g_1(t_2)} = 2\pi K \delta\, (t_2 - t_1) \tag{28}$$

where K is the spectral density of the white noise and $\delta(t)$ is the Dirac delta function. The displacement correlation function $\overline{h_1(t_1)h_1(t_2)}$ can be determined from Eq. (16).

For this example let the creep function be a random process such that its correlation function is

$$\overline{\psi(t_1 - \xi_1)\psi(t_2 - \xi_2)} = \sigma_\psi^2 \exp[-c\,|\,t_2 - \xi_2 + t_1 - \xi_1\,|\,] \tag{29}$$

where σ_ψ is the standard deviation of the process $\psi(t)$. (See Eq. (26) for $\omega = 0$.) Then the displacement correlation function becomes

$$\overline{h_1(t_1)h_1(t_2)} = c\sigma_\psi^2\, e^{-ct_2} \sinh ct_1 \qquad t_2 \geq t_1 \tag{30}$$

and the variance function is given by

$$\overline{h_1^2(t)} = c\sigma_\psi^2\,(1 - e^{-2ct})/2 \tag{31}$$

For this example, one can conclude that for small times the displacement response is nonstationary since Eq. (30) does not indicate a function of $t_2 - t_1$ and for large t_1 and t_2, the correlation function tends to $e^{-c(t_2 - t_1)}$. Furthermore, as seen from Eq. (31), the mean square response $\overline{u_{33}^2(x_2, t)}$, becomes a constant for large times.

Hollow Pressurized Cylinder

Consider next an infinitely long, hollow, circular, homogeneous viscoelastic cylinder with a rigid outer case at $r = r_o$ and under uniform internal pressure $\Pi(t)$ at $r = r_i$. This classic plane strain problem has a deterministic solution as shown in References 16 and 18. The Laplace transforms* of the viscoelastic radial and hoop stresses, obtained from corresponding elastic ones, are

$$\tilde{\sigma}_{rr}(r,p) = [-\tilde{Z}(p)R_3 + R_4]\tilde{\Pi}(p)/\tilde{Z}_1(p) \tag{32}$$

$$\tilde{\sigma}_{\theta\theta}(r,p) = [\tilde{Z}(p)R_4 - R_3]\tilde{\Pi}(p)/\tilde{Z}_1(p) \tag{33}$$

---

* The Laplace transform is defined as $\tilde{f}(p) = \int_0^\infty f(t)e^{-pt}dt$.

where

$$\widetilde{Z_1}(p) = \widetilde{Z}(p)R_1 - R_2$$

$$R_3 = (r_o/r)^2 + 1 \quad R_4 = (r_o/r)^2 - 1 \tag{34}$$

$$R_1 = R_3(r_i) \quad R_2 = R_4(r_i)$$

and the function $\widetilde{Z}(p)$ depends solely on the transform of the cylinder relaxation functions

$$\widetilde{Z}(p) = [4\tilde{\phi}(p) + \tilde{\phi}_v(p)]/[\tilde{\phi}_v(p) - 2\tilde{\phi}(p)] \tag{35}$$

The experimental viscoelastic material property data considered in the previous section indicates that the volumetric properties are essentially elastic, i.e. $\tilde{\phi}_v = K_o$ the bulk modulus**. For a constant internal pressure $\Pi = \Pi_o$, Eqs. (32) and (33) can be readily inverted and the resulting average function can be found to be

$$\overline{\sigma_{ij}(r,t)} = [R_6 \int_o^t \frac{\partial}{\partial \xi} \overline{\phi(t-\xi)}\, \overline{\sigma_{ij}(r,\xi)}\, d\xi + \overline{f_{ij}(r,t)}] / \overline{D}_1 \tag{36}$$

where

$$R_5 = 4R_1 + 2R_2 \quad R_6 = R_5(r_i) \quad R_7 = 4R_4 + 2R_3$$

$$\overline{D}_1 = R_6\, \overline{\phi(o)} + 2K_o \tag{37}$$

$$f_{ij}(r,t) = \Pi_o \left[\begin{matrix} -R_6 \\ +R_7 \end{matrix} \; \overline{\phi(t)} \; -2K_o \right] / \overline{D}_1$$

and where the multiplier $(-R_6)$ refers to $i=j=r$ and R_7 to $i=j=\theta$. From Eq. (20), the radial stress correlation function becomes:

** The deterministic value of the elastic bulk modulus is used. If the dilation modulus ϕ_v is found to be stochastic then the correlation between ϕ and ϕ_v must be considered uncorrelated.

$$\overline{\sigma_{rr}(r,t_1)\sigma_{rr}(r,t_2)} = \bar{\mu}_o \int_o^{t_1}\int_o^{t_1} \frac{\partial}{\partial \xi_1}\frac{\partial}{\partial \xi_2} \overline{\phi(t_1-\xi_1)\phi(t_2-\xi_2)}\ \overline{\sigma_{rr}(r,\xi_1)\sigma_{rr}(r,\xi_2)}\ d\xi_2 d\xi_1$$

$$+ [\overline{\sigma_{rr}(r,t_1)}\ \overline{F(r_1 t_2)} + \overline{F(r,t_1)}\ \overline{\sigma_{rr}(r,t_2)} + \overline{f(r,t_1)f(r,t_2)}]/\overline{D}_2 \qquad (38)$$

where $\bar{f} \equiv \bar{f}_{rr}$ and

$$\bar{\mu}_o = R_5^2/\overline{D}_2 \qquad \overline{D}_2 = R_5\,\overline{\phi^2(o)} + 4\,R_5K_o\,\overline{\phi(o)} + 4K_o^2$$

$$\overline{F(r,t)} = -\{R_1R_3\,\overline{\phi(o)\phi(t)} + R_2R_3K_o\,\overline{\phi(t)} + R_1R_4K_o\,\overline{\phi(o)} + R_2R_4K_o^2\} \qquad (39)$$

$$\overline{f(r,t_1)f(r,t_2)} = \Pi_o^2\,\{R_6^2\,\overline{\phi(t_1)\phi(t_2)} + 2R_6K_o\,[\overline{\phi(t_1)} + \overline{\phi(t_2)}] + 4\,K_o^2\}/\overline{D}_1^2$$

Furthermore, by setting $t_1 = t_2 = t$ in Eq. (38), the mean square radial stress response is found to be

$$\overline{\sigma_{rr}^2(r,t)} = \bar{\mu}_o \int_o^t\int_o^t \frac{\partial}{\partial \xi_1}\frac{\partial}{\partial \xi_2} \overline{\phi(t-\xi_1)\phi(t-\xi_2)}\ \overline{\sigma_{rr}(r,\xi_1)\,\sigma_{rr}(r,\xi_2)}\ d\xi_2 d\xi_1 \qquad (40)$$

$$+ [\overline{f^2(r,t)} + 2\overline{\sigma_{rr}(r,t)}\ \overline{F(r,t)}]/\overline{D}_2$$

Similar expressions can be derived for the corresponding statistical hoop stress functions. Eq. (36) is a Volterra integral equation of the second kind in a single variable, whereas Eq. (38) is one in two variables[33]. Solutions to these integral equations can be obtained numerically for known relaxation functions $\phi(t)$ and bulk moduli K_o.

Eq. (36) can be recast in the general form

$$q(t) + \lambda_o \int_o^t H(t-\xi)\,q(\xi)\,d\xi = N(t) \qquad (0 \le t \le t_3) \qquad (41)$$

where the kernel $H(t-\xi)$ and the function $N(t)$ are known, continuous and bounded in [0, t3] and $q(t)$ is the unknown function. This integral equation can be solved by Picard's successive approximation method and results in

$$q(t) = N(t) - \lambda_o \int_o^t M(t-\xi)\, N(\xi)\, d\xi \tag{42}$$

where the resolvent kernel M is defined by a series of iterated kernels H_k

$$M(t-\xi) = \sum_{k=o}^{\infty} \lambda_o^k\, H_{k+1}(t-\xi) \tag{43}$$

with the recurrence relation

$$H_1(t-\xi) = H(t-\xi)$$

$$H_{k+1}(t-\xi) = \int_\xi^t H(t-s)\, H_k(s-\xi)\, ds \qquad (k \geq 1) \tag{44}$$

Except for the most simple functions H, such calculations can be extremely tedious before proper convergence is achieved. However, the use of finite differences in Eq. (41), results in a set of algebraic equations, which can be easily evaluated on a high speed computer and are of the form

$$q(t_{n+1}) = \frac{N(t_{n+1}) - \frac{\lambda_o}{2} q(t_n) \int_{t_n}^{t_{n+1}} H(t_{n+1}-\xi)d\xi - \frac{\lambda_o}{2} \sum_{i=1}^{n-1} [q(t_{i+1}) + q(t_i)] \int_{t_i}^{t_{i+1}} H(t_{n+1}-\xi)d\xi}{1 + \frac{\lambda_o}{2} \int_{t_n}^{t_{n+1}} H(t_{n+1}-\xi)d\xi} \tag{45}$$

with the time scale divided into n intervals such that $t_1=0$ and $t_{n+1}=t$.

The initial condition is given by

$$q(o) = q(t_1) = N(o) \tag{46}$$

and $q(t_n)$ values can be achieved from Eq. (45) by successive computations for n=1,2,3---, where each $q(t_n)$ is determined by previously computed values.

Similarly, the radial stress correlation function can be computed from

$$\overline{D_4}\, \overline{\sigma_{rr}(r,t_{m+1})\sigma_{rr}(r,t_{n+1})} = \overline{-f(r,t_{m+1})f(r,t_{n+1})} + \overline{\sigma_{rr}(r,t_{m+1})}\, \overline{F(r,t_{n+1})} \tag{47}$$

$$+ \overline{\sigma_{rr}(r,t_{n+1})}\, \overline{F(r,t_{m+1})} + \frac{\mu_o}{2} \overline{\sigma_{rr}(r,t_m)\sigma_{rr}(r,t_n)} \int_{t_m}^{t_{m+1}} \int_{t_n}^{t_{n+1}} \hat{\phi}(t_{m+1}-\xi_1,\ t_{n+1}-\xi_2)\, d\xi_2 d\xi_1$$

$$+\frac{\overline{\mu}_o}{2}\sum_{i=1}^{m-1}\left[\overline{\sigma_{rr}(r,t_{i+1})\sigma_{rr}(r,t_{n+1})}+\overline{\sigma_{rr}(r,t_i)\sigma_{rr}(r,t_{n+1})}\right]\int_{t_i}^{t_{i+1}}\int_{t_n}^{t_{n+1}}\hat{\phi}\,(t_{m+1}-\xi_1,t_{n+1}-\xi_2)\,d\xi_2 d\xi_1$$

$$+\frac{\overline{\mu}_o}{2}\sum_{j=1}^{n-1}\left[\overline{\sigma_{rr}(r,t_{m+1})\sigma_{rr}(r,t_{j+1})}+\overline{\sigma_{rr}(r,t_m)\sigma_{rr}(r,t_j)}\right]\int_{t_m}^{t_{m+1}}\int_{t_j}^{t_{j+1}}\hat{\phi}\,(t_{m+1}-\xi_1,t_{n+1}-\xi_2)\,d\xi_2 d\xi_1$$

$$+\frac{\overline{\mu}_o}{2}\sum_{i=1}^{m-1}\sum_{j=1}^{n-1}\left[\overline{\sigma_{rr}(r,t_{i+1})\sigma_{rr}(r,t_{j+1})}+\overline{\sigma_{rr}(r,t_i)\sigma_{rr}(r,t_j)}\right]\int_{t_i}^{t_{i+1}}\int_{t_j}^{t_{j+1}}\hat{\phi}\,(t_{m+1}-\xi_1,t_{n+1}-\xi_2)\,d\xi_2 d\xi_1$$

Finally, form m=n one obtains from Eq. (47) the mean square response of the radial stress

$$\overline{D}_3\,\overline{\sigma_{rr}^2(r,t_{m+1})}=-\overline{f^2(r,t_{m+1})}+2\,\overline{\sigma_{rr}(r,t_{m+1})\,F(r,t_{m+1})} \tag{48}$$

$$+\frac{\overline{\mu}_o}{2}\overline{\sigma_{rr}^2(r,t_m)}\int_{t_m}^{t_{m+1}}\int_{t_m}^{t_{m+1}}\frac{\partial}{\partial\xi_1}\frac{\partial}{\partial\xi_2}\overline{\phi(t_{m+1}-\xi_1)\phi(t_{m+1}-\xi_2)}\,d\xi_2 d\xi_1$$

$$+\mu_o\sum_{i=1}^{m-1}\left[\overline{\sigma_{rr}(r,t_{i+1})\sigma_{rr}(r,t_{m+1})}+\overline{\sigma_{rr}(r,t_i)\sigma(r,t_m)}\right]\int_{t_m}^{t_{m+1}}\int_{t_i}^{t_{i+1}}\hat{\phi}\,(t_{m+1}-\xi_1,t_{m+1}-\xi_2)d\xi_2 d\xi_1$$

$$+\frac{\overline{\mu}_o}{2}\sum_{i=1}^{m-1}\sum_{j=1}^{m-1}\left[\overline{\sigma_{rr}(r,t_{i+1})\sigma_{rr}(r,t_{j+1})}+\overline{\sigma_{rr}(r,t_i)\sigma_{rr}(r,t_j)}\right]\int_{t_i}^{t_{i+1}}\int_{t_j}^{t_{j+1}}\hat{\phi}\,(t_{m+1}-\xi_1,t_{m+1}-\xi_2)d\xi_2 d\xi_1$$

where

$$\overline{D}_3=\overline{D}_2-\overline{\mu}_o/2\int_{t_m}^{t_{m+1}}\int_{t_m}^{t_{m+1}}\phi(t_{m+1}-\xi_1,\,t_{m+1}-\xi_2)\,d\xi_2\,d\xi_1$$

$$\overline{D}_4=\overline{D}_2-\overline{\mu}_o/2\int_{t_m}^{t_{m+1}}\int_{t_n}^{t_{n+1}}\phi(t_{m+1}-\xi_1,\,t_{n+1}-\xi_2)\,d\xi_2\,d\xi_1 \tag{49}$$

$$\hat{\phi}(t_m-\xi_1, t_n-\xi_2) = \frac{\partial}{\partial \xi_1}\frac{\partial}{\partial \xi_1}\overline{\phi(t_m-\xi_1)\,\phi(t_n-\xi_2)}$$

with initial conditions

$$\overline{\sigma_{rr}(r,o)\sigma_{rr}(r,t_{m+1})} = [\overline{\sigma_{rr}(r,o)}\;\overline{F(r,t_{m+1})} + \overline{F(r,o)}\;\overline{\sigma_{rr}(r,t_{m+1})} + \overline{f(r,o)}\;\overline{f(r,t_{m+1})}]/\overline{D}_2$$

$$\overline{\sigma_{rr}{}^2(r,o)} = [\overline{f^2(r,o)} + 2\overline{\sigma_{rr}(r,o)}\;\overline{F(r,o)}]/\overline{D}_2 \qquad (50)$$

and with two dimensional domain divided into t_i and t_j, where $i = 1,2,\text{---}m+1$ and $j = 1,2,\text{---}n+1$ and such that $t_1=0$ and t_{m+1} and t_{n+1} are the final times t. The results of these computations are discussed in the next section.

DISCUSSIONS AND CONCLUSIONS

Numerical solutions of the foregoing results were obtained by writing double precision FORTRAN programs and utilizing an University of Illinois IBM mainframe. The deviatoric relaxation function $\phi(t)$ was characterized by a Prony series and its coefficients are presented in Table I. Its distribution was represented as either by the Gaussian or beta forms each in identical non-dimensionalized forms over the entire time domain. The beta and normal distributions were matched over the 6σ interval by enforcing equal means and standard deviations or by curve fitting the beta on the normal distributions. Several equal values of the exponents a and b of Eq. (24) were used to give symmetric distributions with respect to the mean relaxation function $\overline{\phi}(t)$. The time increments in Eq. (45) were selected by dividing each decade of time into ten steps in the chosen total interval $10^{-7} \leq t \leq 10^8$ seconds. However no notable differences resulted for calculations with doubled step size, indicating "convergence" for the process. Results for $t < 10^{-7}$ secs. essentially correspond to the initial elastic conditions. Since References 31 and 33 indicate the typical uncertainty generally found in the literature regarding proper values of high polymer bulk moduli, two representative values were chosen for illustrative purposes, i.e. K_0 equal to 30,000 and 100,000 psi.

Figure 2 shows the mean radial stress $\overline{\sigma_{rr}(r_o,t)}$ at the outer radius for two values each of the radius ratio and the bulk modulus. It must be remembered that, as is the case in the deterministic solutions,[16,18] so also are the statistical stress values highly influenced by the rigid case and by the fact that $K_o >> \phi(t)$. This, of course, explains the stabilizing trend of the solution with time, i.e. radial and hoop stresses approach the inner circle pressure value Π_o with time. It is seen that for the smaller bulk modulus, i.e. more elastically compressible material, the mean radial stress at the outer radius has a smaller initial value and, consequently, takes longer to reach its steady state value. A similar effect can be observed as the radius ratio r_o/r_i is increased. The mean hoop stress-function for the outer radius (Fig. 3) behaves similarly. Note that the deterministic pattern for these stresses is preserved - namely, radial stresses are compressive for all times and hoop stresses start in tension and become compressive as time increases.

The relaxation correlation function, which appears in Eqs. (47) to (49), was assumed in keeping with Eq. (26) to have the form

$$\overline{\phi(t_{m+1} - \xi_1)\phi(t_{n+1}-\xi_2)} = \sigma_\phi(t_{m+1} - \xi_1)\sigma_\phi(t_{n+1}-\xi_2) \exp\left[-c\,|\,t_{n+1}-\xi_2-t_{m+1}+\xi_1\,|\right] \quad (51)$$

where $\sigma_\phi(t-\xi)$ is the standard deviation function of the relaxation function $\phi(t)$. The mean square function $\overline{\sigma^2_{rr}(r_o,t)}/\Pi_o^2$ is shown in Fig. 4 for two sets of equal (a=b) beta function exponents, indicating practically no differences for this cylinder problem.

The standard deviation of the random stress function $\sigma_{rr}(r_o,t)$, which is a measure of the spread of the mean function $\overline{\sigma_{rr}(r_o, t)}$ can be calculated from

$$\sigma_{\sigma_{rr}} = \left|\overline{\sigma^2_{rr}(r_o,t)} - [\,\overline{\sigma_{rr}(r_o,t)}\,]^2\right|^{1/2} \quad (52)$$

and is shown in Fig. 5 normalized with respect to the constant pressure Π_o. It can be seen that this standard deviation is strongly influenced by the constant c of Eq. (51) and by the bulk modulus K_o. From Eq. (25) it can be seen that the beta function exponents a and b are directly related to the means and standard deviations of $\phi(t)$, giving $\sigma_\phi{}^2$ equal to .0714 and .0227 for a=b of 3 and 5 respectively and η_ϕ = .5 for both pairs of exponent values. It is evident from Fig. 5 that if the material property data has a wider dispersion (a=b=3), then so does the stress response. Furthermore, the graphs indicate that the standard deviation function of $\sigma_{rr}(r_o, t)$ increases from its initial value to a maximum in approximately 4 time decades and then gradually decreases with time because of the asymptotic nature of this solution. The degree of material property scatter of $\phi(t)$ initially has the greatest influence on the response, implying that response fluctuations are initially more pronounced but they decay with time for this problem to become insensitive to a and b values as the steady state solution is approached. Finally, the effect of smaller c values in Eq. (51) is to produce smaller standard deviations in the stress response.

Of course, one of the inherent difficulties is the unavailability of published statistical material property data. The present analysis indicates that meaningful solutions can be obtained on a statistical basis and, therefore, the need to gather such data becomes imperative. In the meantime, although no information exists on such critical parameters as η_ϕ, σ_ϕ, c and ω - See Eqs. (25) and (26) - one can, nevertheless, as has been shown, draw conclusions on their effects and influence on the selection and manufacture of viscoelastic materials based on such statistical considerations. Additionally, one also needs to address oneself to the estimation of upper and lower response bounds. The simplest procedure would be to calculate deterministic responses based on the upper and lower functions of ϕ_V and ϕ_L of Fig. 1 and ignore the statistical analysis. However, this leads to false conclusions as can be seen from the following example.

Consider the same cylinder that was analyzed in the previous section, but without a case, subjected to a time dependent slowly varying pressure $\Pi(t)$ and without resulting inertia forces. This problem can be solved by the separation of variable technique as outlined in the previous section,[16] but the details of the solution will not be reproduced here because of lack of space. A typical strain response is shown in Fig. 6 for a short time interval to illustrate the points in question.

The five separate curves shown are the deterministic responses of the system for the upper, lower and mean viscoelastic material property curves and the probabilistic response drawn at a distance of three standard deviations from the average viscoelastic response. Thus, these last two curves should encompass over ninety-nine percent of all solutions. However, the probabilistic curves cross the deterministic bounds as shown in the plot and, therefore, the solutions based on bounding curves for the data do not bound all of the possible solutions.

A clue to this behavior pattern can be found in the stress-strain relations (12) and (13) which can be written as

$$2\varepsilon_{kl}(x,t) = \int_o^t \psi(t-s)\, \frac{\partial S_{kl}(x,s)}{\partial s}\, ds \tag{53}$$

for an incompressible material and with a similar relation for the mean functions of Eq. (13). Since these are integral relations, particularly when put in the form of Eq. (53), there is no reason to expect the deterministic relation (12) to yield the same results as the statistical relations (13) to (16) even for only the mean functions. Indeed, when looking to the upper and lower response bounds one needs to specify data distributions such as Eq. (21) and correlation functions such as Eqs. (26) and (29) which do not solely depend on the bounds of the data curves $\phi(t)$ or $\psi(t)$.

In order to demonstrate this effect, the above problem was analyzed with a deterministic forcing function $g(t) = 1 - \cos \omega^* t$. Different values of the coefficients c and ω in Eq. (26) were considered and the results are presented in Table II. The column designated R is the ratio of the response standard deviation divided by one sixth of the spread of the upper and lower bound deterministic solutions. The α_n are inverses of relaxation times of Eq. (21). It is noted that for small c values the ratio R is larger than unity and this phenomena occurs for combinations of Gaussian dispersions where compressive response due to tension inputs (and vice versa) starts being predicted with reasonable probabilities. Clearly, in these regions the parametric variations have produced dispersions which have become larger than can be physically accepted. The last column in Table II indicates the time interval over which the correlation coefficient remains equal to unity, i.e. the length of time over which the random process remains completely correlated. It is seen that the material property functions $\phi(t)$ or $\psi(t)$ would remain completely correlated over longer time periods as c decreases. From Fig. 1, it can be inferred that the longer periods are unreasonable and, therefore, the parameter c in Eq. (26) must be restricted in size depending on the value of the frequency ω.

Two other illustrative examples are of interest for this problem and the results are shown in Table III for forcing functions $g(t) = 10^{-3}t - 10^{-6}t^2$ and $g(t) = \delta(t)$. It should be noted that when ω is zero in the correlation function (26), the probabilistic results are considerably more insensitive to c values. Of course, all these examples were for the sake of simplicity, considered to be free of inertia forces which influences the numerical values, but not the behavioral principles of probabilistic and deterministic patterns. From the previous examples, one can conclude that only stochastic analysis can provide answers to proper bounds and deterministic upper and lower bound predictions are misleading and incorrect.

In general, it may be concluded that the system of statistical equations which needs to be solved is so complex, that the use of beta distributions in preference to Gaussian ones does not result in sufficient additional complexity to rule out their use. Since beta distributions more accurately reflect the scatter of viscoelastic data between finite bounds and do not lead to unrealistic predictions associated with Gaussian distributions for larger data spread, the use of the former is recommended.

The illustrative examples treat essentially quasi-static viscoelastic problems for illustrative simplicity purposes, although the general analysis covers dynamic conditions. In dynamic problems, it is not common to find material property functions ϕ and ϕ_V as arguments of elementary exponential, trigonometric, Bessel, etc. functions.[16,21] This complicates not only the evaluation of deterministic responses through transform inversions, but also greatly increases the computational complexity of first and second order statistics necessitating the use of considerable more laborious numerical techniques than those used to solve the quasi-static problems presented here. Furthermore, only linear viscoelastic stress strain relations were considered and although nonlinear relations can be represented piecewise by linear ones[36], no analytical information is available describing any relations between linear and nonlinear statistical responses.

REFERENCES

1 Eringen, A. C. Stochastic loads. *Handbook of Engineering Mechanics* McGraw Hill Book Co., 1962, Ch. 18

2 Bieniek, M. P. Creep under random loading. *AIAA J.* 1965, **3**, 1559-1561

3 Cozarelli, F. A. and Huang, W. N. Effect of random material parameters on nonlinear steady creep solutions. *Int. J. Solids and Structures* **2**, 1971, 1477-1494

4 Huang, W. N. and Cozarelli, F. N. Steady creep bending in a beam with random material parameters. *J. Franklin Institute* 1972, **294,** 323-339

5 Huang, W. N. and Cozarelli, F. N. Damped lateral vibration in an axially creeping beam with random material parameters. *Int. J. Solids and Structures* 1973, **9**, 765-788

6 Parkus, H. Warmespannungen bei zufellsabhanginger Oberflachen-temperatur. *ZAMM* 1962, **42**, 499-507

7 Parkus, H. On the lifetime of viscoelastic structures in a random temperature field. *Recent Progress in Applied Mechanics* 1967, Wiley, N.Y., 391-397

8 Ziegler, F. Zufallige Temperaturschwankungen und ihr Einfluss auf die Lebensdauer eines Druckstabes aus nichtlinear-viskoelastischem Material. *ZAMM* 1972, **52**, 176-178

9 Ziegler, F. Snap-through buckling of a viscoelastic von Mises truss in a random temperature field. *J. Appl. Mech.* **36**, 1969, 338-340

10 Parkus, H. and Zeman, J. L. Note on the behavior of thermorheologically simple materials in random temperature fields. *Acta Mechanica* 1970, **9**, 152-157

11 Parkus, H. and Zeman, J. L. Some stochastic problems of thermoviscoelasticity. *IUTAM Symposium on Thermoinelasticity* Springer, N.Y., 1970, 226-240

12 Cozarelli, F. A. and Chang, W. P. Wave front stress relaxation in viscoelastic materials with random temperature distributions. *Acta Mechanica* 1975, **22**, 11-30

13 Molyneux, J. and Beran, M. J. Statistical properties of the stress and strain fields, in a medium with small random variations in elastic coefficients. *J. Math. Mech.* 1965, **14**, 337-351

14 Bazant, Z. P. and Xi, Y. Probabilistic prediction of creep and shrinkage in concrete structures: combined sampling and spectral approach. *5th Int. Conf. on Structural Safety and Reliability (ICOSSAR)* A. H. S. Ang, and Shinozuka, M. and Schueller, G. I. eds., 1989, **1**, 803-808

15 Bazant, Z. P. Response of aging linear systems to ergodic random input. *J. Eng. Mech. ASCE* 1986, **112**, 322-350

16 Hilton, H. H. Viscoelastic analysis. *Engineering Design for Plastics* Reinhold Publ. Corp, New York, 1964, 199-276

17 Hilton, H. H. Thermal stresses in thick walled cylinders exhibiting temperature dependent viscoelastic properties of the Kelvin type. *Proc. Second U.S. Nat. Congress on Appl. Mech.* 1954, 547-553

18 Lee, E. H. and Rogers, T. G. Solution of viscoelastic stress analysis problems using measured creep and relaxation functions. *J. Appl. Mech.* 1963, **30**, 127-133

19 Morland, L. W. and Lee, E. H. Stress analysis for linear viscoelastic materials with temperature variation. *Trans. Society of Rheology* 1960, **4**, 233-263

20. Muki, R. and Sternberg, E. On transient thermal stresses in viscoelastic materials with temperature dependent properties. *J. Appl. Mech.* 1961, **28**, 193-207

21 Hunter, S. C. Tentative equations for the propagation of stress, strain and temperature fields in viscoelastic solids. *J. Mechanics and Physics of Solids* 1961, **9**, 39-51

22 Alfrey, T. Nonhomogeneous stress in viscoelastic media. *Q. Appl. Math.* 1944, **2**, 113-119

23 Read, W. T. Stress analysis for compressible viscoelastic materials. *J. Appl. Physics* 1950, **21**, 671-674

24 Hilton, H. H. and Dong, S. B. An analogy for anisotropic, nonhomogeneous linear viscoelasticity including thermal stresses. *Proc. Eighth Midwestern Mechanics Conf.* 1964, 58-73

25 Hilton, H. H. and Russell, H. G. An extension of Alfrey's analogy to thermal stress problems in temperature dependent linear viscoelastic media. *J. Mechanics and Physics of Solids* 1961, **9**, 152-164

26 Hilton, H. H. and Clements, J. R. Formulation and evaluation of approximate analogies for temperature dependent linear viscoelastic media. *Proc. Conference on Thermal Loading and Creep* Inst. Mech. Eng. London, 1964, 6:17-6:24

27 Schapery, R. A. Approximate methods of transform inversion for viscoelastic stress analysis. *Proc. Fourth U.S. Nat. Congress of Appl. Mech.* 1962, **2**, 1075-1085

28. Cost, T. L. Approximate Laplace inversion in viscoelastic stress analysis. *AIAA J.*, 1964, **2**, 2157-2166

29 Beran, M. J. *Statistical Continuum Theories.* Interscience Publ., 1968

30 Lin, Y. K. *Probabilistic Theory of Structural Dynamics,* McGraw Hill Book Co., 1967

31 Hilton, H. H., Majerus, J. N. and Tamekuni, M. Analytical formulation of generalized characterization for linear viscoelastic materials from uni- and multi-axial creep and relaxation data. *ICRPG Proceedings* 1964, **2**, 114-128

32 Zak, A. R. Structural analysis of realistic solid-propellant materials. *J. Spacecraft* 1968, **5**, 270-275

33 Tricomi, F. G. *Integral Equations*. Interscience Publishers, 1957

34 Hilton, H. H. On the representation of nonlinear creep by a linear viscoelastic model. *J. Aerospace Sciences* 1959, **26**, 311-312

35 Wen, Y. K. *Structural Load Modeling and Combination for Performance and Safety Evaluation*. Elsevier, 1990, 19-20

TABLE I

Viscoelastic Coefficients of the Upper and Lower Relaxation Functions - Eq. (21).

m	A_m	D_m	$\alpha_m \sec^{-1}$
0	$.925^{2}$†	$.10833^{2}$	.0
1	$.154^{5}$	1.88833^{3}	$.5^{5}$
2	$.570^{4}$	0.0	$.5^{4}$
3	$.417^{4}$	3.70^{2}	$.5^{3}$
4	$.193^{4}$	$.16^{3}$	$.5^{2}$
5	$.932^{3}$	$.100167^{3}$	$.5^{1}$
6	$.262^{3}$	$.265^{2}$	.5
7	$.266^{3}$	$.113^{2}$	$.5^{-1}$
8	$-.871^{1}$	$.40167^{1}$	$.5^{-2}$
9	$.272^{2}$	$.2533^{1}$	$.5^{-3}$
10	$.230^{2}$	.367	$.5^{-4}$
11	$.124^{2}$	.1833	$.5^{-6}$
12	$.168^{2}$	$.1^{1}$	$.5^{-6}$
13	$.490^{1}$	$.10967^{1}$	$.5^{-7}$

†$.925^{2} = .925 \times 10^{2}$

TABLE II

Comparison of Stochastic & Deterministic Strains for Pressurized Cylinder with $\omega^* = 4\pi \times 10^{-8}$ & $g(t) = 1 - \cos \omega t$

$c \sec^{-1}$		ω	R	Time (secs) for R = 1
5^{4}†	(α_2)	ω^*	2.66^{-6}	10^{-5}
5^{2}	(α_3)	ω^*	2.66^{-5}	10^{-5}
5	(α_4)	ω^*	2.66^{-4}	.005
5^{-2}	(α_5)	ω^*	2.66^{-3}	.8
5^{-4}	(α_6)	ω^*	.0266	100
5^{-6}	(α_7)	ω^*	.266	160
5^{-8}	(α_8)	ω^*	1.65	1.7^{4}
4.5^{-8}	$(.9\alpha_8)$	ω^*	1.68	
5.5^{-8}	$(1.1\alpha_8)$	ω^*	1.58	
5^{-8}	(α_8)	$.9\omega^*$	1.635	
5^{-8}	(α_8)	$1.1\omega^*$	1.612	

† See Table I

TABLE III

Comparison of Stochastic and Deterministic Strains for Pressurized Cylinder with ω = o

$c\ sec^{-1}$	R for $g(t) = 10^{-3}t - 10^{-6}t^2$	R for $g(t) = \delta(t)$
0 (α_1)	1.0	.635
$5^{4\dagger}$ (α_1)	.515	.835
5^2 (α_2)	.515	.835
5 (α_4)	.515	.820
5^{-2} (α_5)	.515	.653
5^{-4} (α_6)	.515	.637
5^{-6} (α_7)	.515	.635
5^{-8} (α_8)	.778	.635
4^{-8}	.812	.636
6^{-8}	.745	.635

fig.1. Relaxation Curve with ±3σ Dispersion

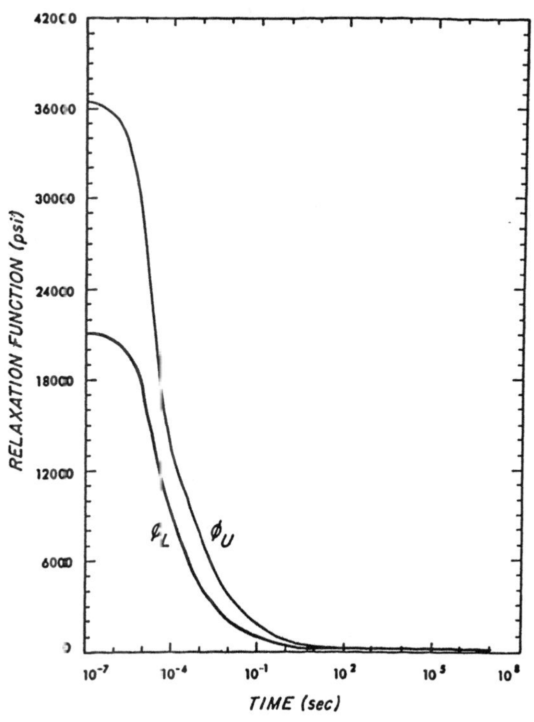

fig. 2. Mean Radial Stresses at Outer Radius with a = b

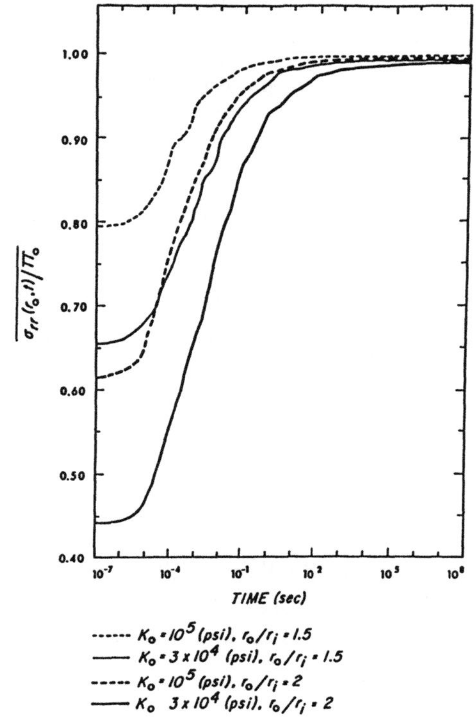

fig. 3. Mean Hoop Stresses at Inner Radius with a = b

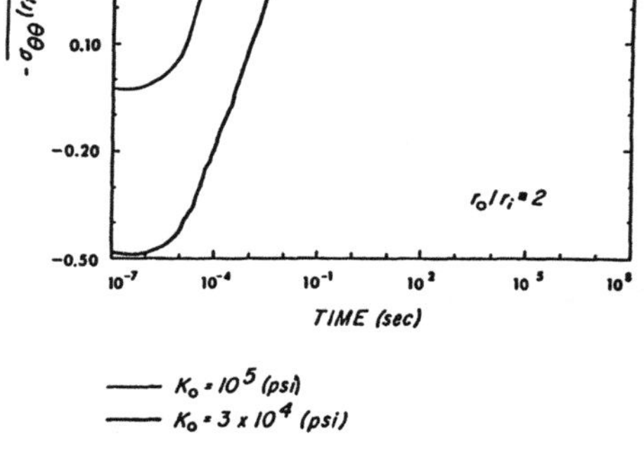

fig. 4. Mean-Square Response of $\sigma_{rr}(r_o,t)/\Pi_o$ with $K_o = 10^5$ (psi) and $C = .25 \times 10^5$

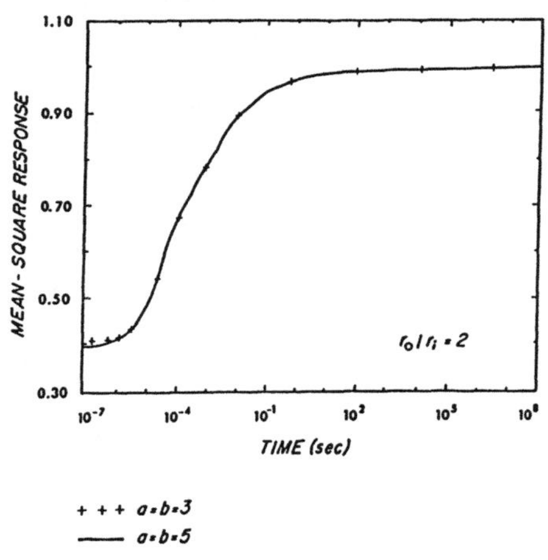

fig. 5. Standard Deviations of Radial Stress at Outer Radius with $r_o/r_i = 2$

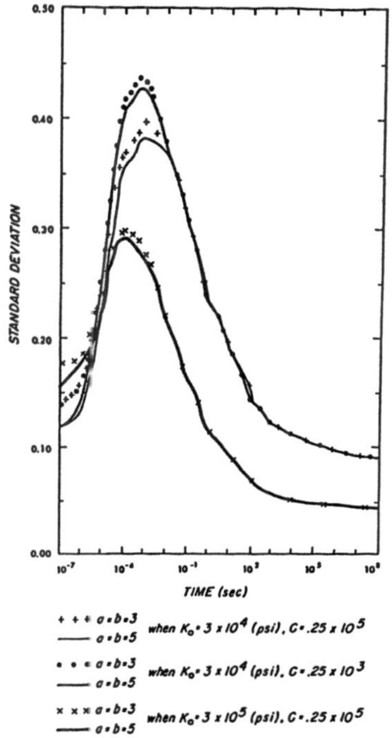

fig. 6. Typical Solution as Strain Passes Through Zero

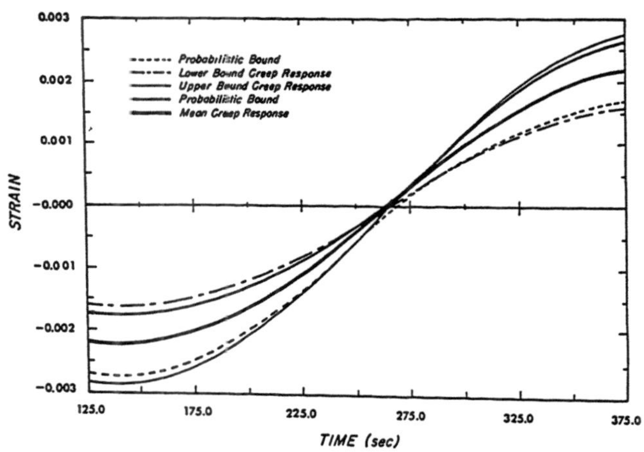

Nonstationary Response of Linear Systems under Uncorrelated Parametric and External Excitations

Z. K. Hou and W. D. Iwan

California Institute of Technology
Pasadena, CA 91125
U. S. A.

Abstract

An explicit solution for the nonstationary second moment response and the evolutionary spectral density of the response of linear systems subjected to uncorrelated parametric and external excitations is obtained by a simplified state-variable approach. The excitations are modeled as suddenly applied white noise processes. The solution is expressed as a sum of the solution for the external excitation alone and a correction term which accounts for the effect of the parametric excitation. In the limiting stationary case, the total solution is proportional to the solution for the external solution alone. However, the proportionality does not hold during the nonstationary stage. The solution presented is conditionally stable. The effect of the parametric excitation on the second moment response is significant in certain cases, especially near the region of instablity where the intensity of the parametric excitation could be very small relative to that of the external excitation. It is also shown that the solution under the uncorrelated parametric and external excitations is exponentially stable for a general form of envelopes employed in engineering practice. Some numerical results are presented.

Introduction

Considerable attention has recently been drawn to the dynamic analysis of structures subjected to random parametric excitation. The studies may be classified into two catogories: quantitative approaches and qualitative approaches. The probability density function as well as the moments of the response have been predicted by various quatitative methods such as the Fokker-Plank equation approach, the method of stochastic averaging, the truncated hierachy technique, and the Voltera integral equation of the second kind. While some explicit solutions for moments of the response are available in the stationary case, numerical techniques are generally needed to obtain a nonstationary solution. The qualitative approach was developed to investigate the stability of the response, partly due to the importance of the stability problem and partly due to the difficulty in obtaining exact solutions for the response. Extensive studies have been conducted on the stability analysis of various systems. Different definitions of stability have been employed, such as the almost sure stability and the mean square stability.

Y. K. Lin · I. Elishakoff (Eds.)
Stochastic Structural Dynamics 1
New Theoretical Developments
2nd International Conference on Stochastic Structural Dynamics
May 9–11, 1990, Boca Raton, Florida

Detailed reviews of parametric random vibration may be found in Ibrahim [1], Kozin [2] and To [3].

In the present study, consideration is directed toward the following random differential equation

$$\frac{d^2}{dt^2}x(t) + 2\varsigma_0\omega_0\frac{d}{dt}x(t) + \omega_0^2(1 - v(t))x(t) = h(t) \tag{1}$$

in which $x(t)$ represents the displacement, ς_0 and ω_0 denote the fraction of critical damping and natural frequency of the system, and $v(t)$ and $h(t)$ are respectively the parametric and external excitations. Eq. (1) has been used by Lin & Shih [4] to study seismic behavior of slender column structures subjected to combined horizontal and vertical components of earthquake ground motion. An exact solution for stationary mean square response of Eq. (1) has been found by Benaroya & Rehak [5], and numerical solutions for nonstationary second moment response are presented in [4] for uncorrelated, nonstationary, combined excitations. However, explicit solutions for the nonstationary second moment response and the evolutionary spectral density of the response are not available to the authors' knowledge.

This paper presents explicit solutions for the second moment response and the evolutionary spectral density of the response of Eq. (1) under combined, uncorrelated parametric and external excitations by using the simplified state-variable approach [6,7]. The final solution can be expressed as a sum of the solution for the external solution alone and a correction term resulted from the parametric excitation. The mean square stability of the solution is discussed and some numerical results are presented.

Formulation

Consider a simple linear system subjected to combined parametric and external excitations, as desribed by Eq. (1). The excitations are modeled as modulated white noises having the same envelope. That is

$$\begin{aligned} h(t) &= \eta(t)w_1(t) \\ v(t) &= \eta(t)w_2(t) \end{aligned} \tag{2}$$

where $w_1(t)$ and $w_2(t)$ are assumed to be zero-mean white noise processes with the properties:

$$\begin{aligned} E[w_1(t)w_1(t+\tau)] &= S_{11}\delta(\tau) \\ E[w_2(t)w_2(t+\tau)] &= S_{22}\delta(\tau) \\ E[w_1(t)w_2(t+\tau)] &= S_{12}\delta(\tau) \end{aligned} \tag{3}$$

in which S_{11} and S_{22} are constants representing intensities of the two white noise prosesses and S_{12} specifies the correlation of the parametric and extermal excitations. In the present study, the parametric and external excitations are assumed to be uncorrelated, namely,

$$S_{12} = 0 \tag{4}$$

Discussion on correlated excitations is contained in a companion paper [8]. $\eta(t)$ is assumed to be the unit step envelope function defined by

$$\eta(t) = \begin{cases} 1, & \text{if } t \geq 0; \\ 0, & \text{otherwise.} \end{cases} \tag{5}$$

Eq. (1) may be written in terms of state variables as

$$\frac{d}{dt}\mathbf{Y}(t) = \mathbf{A}\mathbf{Y}(t) + \mathbf{F}(t) + \mathbf{B}(t)\mathbf{Y}(t) \tag{6}$$

where

$$\begin{aligned} \mathbf{Y}(t) &= \begin{pmatrix} x(t) \\ \dot{x}(t) \end{pmatrix} & \mathbf{A} &= \begin{pmatrix} 0 & 1 \\ -\omega_0^2 & -2\varsigma_0\omega_0 \end{pmatrix} \\ \mathbf{F}(t) &= \begin{pmatrix} 0 \\ h(t) \end{pmatrix} & \mathbf{B}(t) &= \begin{pmatrix} 0 & 0 \\ v(t) & 0 \end{pmatrix} \end{aligned} \tag{7}$$

For simplicity, the initial conditions are assumed to be zero with probability 1.

The general solution of Eq. (6) can be expressed as

$$\mathbf{Y}(t) = \mathbf{Y}^{(0)}(t) + \int_0^t \mathbf{\Phi}(t-s)\mathbf{B}(s)\mathbf{Y}(s)ds \tag{8}$$

where $\mathbf{\Phi}(t)$ is the fundamental solution of the system and $\mathbf{Y}^{(0)}(t)$ is defined as

$$\mathbf{Y}^{(0)}(t) = \int_0^t \mathbf{\Phi}(t-s)\mathbf{F}(s)ds \tag{9}$$

which is the solution when the parametric excitation is absent.

It follows that the mean value of the response is given as

$$E[\mathbf{Y}(t)] = \int_0^t \mathbf{\Phi}(t-s)E[\mathbf{B}(s)\mathbf{Y}(s)]ds \tag{10}$$

and the correlation matrix of the response is expressed as

$$\begin{aligned} \mathbf{Q}(t_1,t_2) &= E[\mathbf{Y}(t_1)\mathbf{Y}^T(t_2)] \\ &= E[\mathbf{Y}^{(0)}(t_1)\mathbf{Y}^{(0)T}(t_2)] + \int_0^{t_1} \mathbf{\Phi}(t_1-s_1)E[\mathbf{B}(s_1)\mathbf{Y}(s_1)\mathbf{Y}^{(0)T}(t_2)]ds_1 \\ &+ \int_0^{t_2} E[\mathbf{Y}^{(0)}(t_1)\mathbf{Y}^T(s_2)\mathbf{B}^T(s_2)]\mathbf{\Phi}^T(t_2-s_2)ds_2 \\ &+ \int_0^{t_1}\int_0^{t_2} \mathbf{\Phi}(t_1-s_1)E[\mathbf{B}(s_1)\mathbf{Y}(s_1)\mathbf{Y}^T(s_2)\mathbf{B}^T(s_2)]\mathbf{\Phi}^T(t_2-s_2)ds_1ds_2 \end{aligned} \tag{11}$$

It has been shown by Samuels [9] and later confirmed by Benaroya and Rehak [5][10] that $\mathbf{Y}(t)$ and $\mathbf{B}(t)$ are uncorrelated in the case that the excitations are stationary white noise processes and this conclusion can be extended to the case of nonstationary modulated white noise. Therefore, the following holds.

$$\begin{aligned} &E[y_i(t)b_{jk}(t)] = E[y_i(t)]E[b_{jk}(t)] \\ &E[b_{ij}(t_1)b_{mn}(t_2)y_k(t_1)y_l(t_2)] = E[b_{ij}(t_1)b_{mn}(t_2)]E[y_k(t_1)y_l(t_2)] \end{aligned} \tag{12}$$

which leads to

$$E[\mathbf{Y}(t)] = \mathbf{O} \tag{13}$$

and

$$\mathbf{Q}(t_1,t_2) = \mathbf{Q}^{(0)}(t_1,t_2) + \mathbf{Q}^{(1)}(t_1,t_2) \tag{14}$$

where $\mathbf{Q}^{(0)}(t_1,t_2)$ is the solution when the parametric excitation is absent and $\mathbf{Q}^{(1)}$ is the correction term resulting from the parametric excitation.

Using the simplified state-variable approach [7], $\mathbf{Q}^{(i)}(t)$, $i = 0, 1$ may be expressed as

$$\begin{aligned} \mathbf{Q}^{(0)}(t_1,t_2) &= S_{11}\mathbf{L}\int_0^{min(t_1,t_2)} \mathbf{P}(t_1-\tau)\mathbf{P}^T(t_2-\tau)d\tau\mathbf{L}^T \\ \mathbf{Q}^{(1)}(t_1,t_2) &= S_{22}\mathbf{L}\int_0^{min(t_1,t_2)} Q_{11}(\tau,\tau)\mathbf{P}(t_1-\tau)\mathbf{P}^T(t_2-\tau)d\tau\mathbf{L}^T \end{aligned} \tag{15}$$

where

$$\mathbf{P}(\tau) = e^{-\varsigma_0\omega_0\tau}\begin{pmatrix}\cos\omega_{0d}\tau \\ \sin\omega_{0d}\tau\end{pmatrix} \qquad \mathbf{L} = \begin{pmatrix} 0 & \dfrac{1}{\omega_{0d}} \\ 1 & -\dfrac{\varsigma_0\omega_0}{\omega_{0d}} \end{pmatrix} \tag{16}$$

in which ω_{0d} is the damped natural frequency.

Stationary Correlation Matrix

As a special case, the stationary correlation matrix of the response may be obtained by taking the limit in Eq. (14) as time approaches infinity. Assume that

$$t_2 = t_1 + \tau \tag{17}$$

and denote the stationary correlation matrix as

$$\mathbf{R}(\tau) = \mathbf{R}(t_2 - t_1) = \mathbf{Q}(t_1,t_2) \tag{18}$$

Letting t_1 approach infinity in the solution (14) yields

$$\mathbf{R}(\tau) = \mathbf{R}^{(0)}(\tau) + \mathbf{R}^{(1)}(\tau) \tag{19}$$

where

$$\mathbf{R}^{(0)}(\tau) = S_{11}\mathbf{L}\int_0^{+\infty}\mathbf{P}(s)\mathbf{P}^T(s+\tau)ds\mathbf{L}^T \tag{20}$$

is the correlation matrix for the external load alone and

$$\mathbf{R}^{(1)}(\tau) = S_{22}R_{11}(0)\mathbf{L}\int_0^{+\infty}\mathbf{P}(s)\mathbf{P}^T(s+\tau)ds\mathbf{L}^T \tag{21}$$

is the correction term. Note that $R_{11}(\tau)$ denotes the first row, first column component of $\mathbf{R}(\tau)$. Substituting Eqs. (20) and (21) into Eq. (19) yields

$$\mathbf{R}(\tau) = (1+\frac{S_{22}}{S_{11}}R_{11}(0))\mathbf{R}^{(0)}(\tau) \tag{22}$$

The explicit solution for $\mathbf{R}^{(0)}(\tau)$ may be written as

$$\mathbf{R}^{(0)}(\tau) = S_{11}\mathbf{L}\mathbf{I}(\tau)\mathbf{L}^T \tag{23}$$

where the components of $\mathbf{I}(\tau)$ are

$$\begin{aligned}
I_{11}(\tau) &= \frac{e^{-\varsigma_0\omega_0\tau}}{4}(\frac{1+\varsigma_0^2}{\varsigma_0\omega_0}\cos\omega_{0d}\tau - \frac{\omega_{0d}}{\omega_0^2}\sin\omega_{0d}\tau)\\
I_{12}(\tau) &= \frac{e^{-\varsigma_0\omega_0\tau}}{4}(\frac{\omega_{0d}}{\omega_0^2}\cos\omega_{0d}\tau + \frac{1+\varsigma_0^2}{\varsigma_0\omega_0}\sin\omega_{0d}\tau)\\
I_{21}(\tau) &= \frac{e^{-\varsigma_0\omega_0\tau}}{4}(\frac{\omega_{0d}}{\omega_0^2}\cos\omega_{0d}\tau - \frac{1-\varsigma_0^2}{\varsigma_0\omega_0}\sin\omega_{0d}\tau)\\
I_{22}(\tau) &= \frac{e^{-\varsigma_0\omega_0\tau}}{4}(\frac{1-\varsigma_0^2}{\varsigma_0\omega_0}\cos\omega_{0d}\tau + \frac{\omega_{0d}}{\omega_0^2}\sin\omega_{0d}\tau)
\end{aligned} \tag{24}$$

The solution for $\mathbf{R}(\tau)$ can be obtained by first solving for $R_{11}(0)$. Letting $\tau = 0$ in Eq. (22) gives

$$(1-\frac{S_{22}}{S_{11}}R_{11}^{(0)}(0))R_{11}(0) = R_{11}^{(0)}(0) \tag{25}$$

where $R_{11}^{(0)}(0)$ is evaluated from Eq. (20). In order to have a finite positive solution for $R_{11}(0)$, it is required that

$$1-\frac{S_{22}}{S_{11}}R_{11}^{(0)}(0) > 0 \tag{26}$$

If the above inequality holds,

$$R_{11}(0) = \frac{S_{11}}{4\varsigma_0\omega_0^3 - S_{22}} \tag{27}$$

Finally, the stationary correlation matrix may be expressed as

$$\mathbf{R}(\tau) = A_p \mathbf{R}^{(0)}(\tau) \tag{28}$$

where A_p, refered to as the *Parametric Amplification Factor*, is defined by

$$A_p = \frac{1}{1-\lambda} \tag{29}$$

with

$$\lambda = \frac{S_{22}}{4\varsigma_0 \omega_0^3} \tag{30}$$

Eqs. (23)-(24) and (28)-(30) complete the solution for the stationary correlation matrix of the response.

Eq. (28) indicates that the stationary covariance solution under the combined external and parametric excitations is proportional to the solution under the external load alone. It follows from inequality (26) that $0 \leq \lambda < 1$ and, therefore,

$$A_p \geq 1 \tag{31}$$

which implies that the existance of parametric excitation will magnify the correlation matrix including the mean square response of the system. Note that A_p is a function of the system parameters and the intensity of the parametric excitation, and is independent of the intensity of the external excitation. For a given S_{22}, a larger value of damping ratio or natural frequency will reduce A_p, implying that a system with larger damping and stiffness has better dynamic performance under the combined excitations. The above conclusions hold for the spectral densities of the response due to the well-known Wienner-Khintchine relationship.

Note that the inequality (26) gives the stability criterion for the stationary solution

$$S_{22} < 4\varsigma_0 \omega_0^3 \tag{32}$$

which agree with the result from the previous studies such as Ibrahim [1] and Benaroya & Rehak [5].

Nonstationary Second Moment Response

The second moment response may also be called as covariance response in this case because the mean of the response is zero. Recall Eq. (14) and, without loss of generality, let $t_1 = t_2 = t$. Then

$$\mathbf{Q}(t) = \mathbf{Q}^{(0)}(t) + S_{22}\mathbf{L} \int_0^t \mathbf{P}(t-\tau)\mathbf{P}^T(t-\tau) Q_{11}(\tau) d\tau \mathbf{L}^T \tag{33}$$

which is the Volterra Integral Equation of the Second Type. Eq. (33) implies that the covariance response for uncorrelated combined loads is a superposition of the response for the external excitation alone and a correction term. Performing a Laplace transform of Eq. (33) yields

$$\mathbf{Q}(s) = Q^{(0)}(s) + S_{22}Q_{11}(s)\mathbf{L}\mathbf{I}(s)\mathbf{L}^T \tag{34}$$

where $\mathbf{L}$ is defined in Eq. (16), and $\mathbf{I}(s)$ is the Laplace tranform of $\mathbf{P}(t)\mathbf{P}^T(t)$ whose components can be expressed as

$$\begin{aligned}
I_{11}(s) &= \frac{(s+2\varsigma_0\omega_0)^2 + 2\omega_d^2}{(s+2\varsigma_0\omega_0)\left[(s+2\varsigma_0\omega_0)^2 + (2\omega_{0d})^2\right]} \\
I_{12}(s) = I_{21}(s) &= \frac{\omega_{0d}}{(s+2\varsigma_0\omega_0)^2 + (2\omega_{0d})^2} \\
I_{22}(s) &= \frac{2\omega_{0d}^2}{(s+2\varsigma_0\omega_0)\left[(s+2\varsigma_0\omega_0)^2 + (2\omega_{0d})^2\right]}
\end{aligned} \tag{35}$$

Note that Eq. (35) is a set of algebraic equations which can be solved by first solving for $Q_{11}(s)$.

$Q_{11}(s)$ can be expressed as

$$Q_{11}(s) = \frac{2S_{11}}{s(s+a)\left[(s+b)^2 + c^2\right]} \tag{36}$$

where a and $b \pm ic$ are the three roots of the cubic equation:

$$(s+2\varsigma_0\omega_0)^3 + 4\omega_{0d}^2(s+2\varsigma_0\omega_0) - 2S_{22} = 0 \tag{37}$$

It can be shown that there exists one real root and two complex conjugate roots of Eq. (37). The stability criterion is obtained by requiring that the real parts of these three roots be negative. This gives

$$S_{22} < 4\varsigma_0\omega_0^3 \tag{38}$$

which agrees with the stability criterion (32) for the stationary case.

Rearrange equation (34) as follows

$$\mathbf{Q}(s) = \mathbf{L}\mathbf{J}(s)\mathbf{L}^T \tag{39}$$

where

$$\mathbf{J}(s) = \left(\frac{S_{11}}{s} + S_{22}Q_{11}(s)\right)\mathbf{I}(s) \tag{40}$$

The components of $\mathbf{J}(s)$ can be expressed in the form of partial fraction as

$$J_{mn}(s) = \frac{f_{mn}(s)}{g(s)}$$
$$= \frac{\alpha_{mn}}{s+a} + \frac{\delta_{mn}}{s} + \frac{\beta_{mn}(s+b) + \gamma_{mn}c}{(s+b)^2 + c^2}$$
$$(m,n = 1,2) \tag{41}$$

where

$$\begin{aligned} g(s) &= s(s+a)\left[(s+b)^2 + c^2\right] \\ f_{11}(s) &= S_{11}\left[(s + 2\zeta_0\omega_0)^2 + 2\omega_{0d}^2\right] \\ f_{12}(s) &= f_{21}(s) = S_{11}\omega_{0d}(s + 2\zeta_0\omega_0) \\ f_{22}(s) &= 2S_{11}\omega_{0d}^2 \end{aligned} \tag{42}$$

and

$$\alpha_{mn} = \left.\frac{f_{mn}(s)}{s((s+b)^2+c^2)}\right|_{s=-a}$$

$$i\beta_{mn} + \gamma_{mn} = \left.\frac{1}{c}\frac{f_{mn}(s)}{s(s+a)}\right|_{s=-b+ic} \tag{43}$$

$$\delta_{mn} = \left.\frac{f_{mn}(s)}{(s+a)((s+b)^2+c^2)}\right|_{s=0}$$
$$(m,n = 1,2)$$

where i represents the imaginary unit.

The inverse Laplace transform gives the final nonstationary second moment response as follows:

$$\mathbf{Q}(t) = \mathbf{L}\mathbf{J}(t)\mathbf{L}^T \tag{44}$$

in which the components of $\mathbf{J}(t)$ are:

$$J_{mn}(t) = \alpha_{mn}e^{-at} + e^{-bt}(\beta_{mn}\cos ct + \gamma_{mn}\sin ct) + \delta_{mn}$$
$$(m,n = 1,2) \tag{45}$$

Evolutionary Spectral Density Matrix

Assume

$$\mathbf{Q}^{(0)}(t) = \int_0^\infty \mathbf{G}^{(0)}(t,\omega)d\omega$$
$$\mathbf{Q}(t) = \int_0^\infty \mathbf{G}(t,\omega)d\omega \tag{46}$$

where $\mathbf{G}^{(0)}(t,\omega)$ and $\mathbf{G}(t,\omega)$ are the one-sided evolutionary spectral density matrices associated with the response for the external excitation alone and that for the conbined excitations respectively. For simplicity, only the result for $G_{11}(t,\omega)$ is presented herein. Substituting Eq. (46) into Eq. (33) yields

$$\mathbf{G}(t,\omega) = \mathbf{G}^{(0)}(t,\omega) + S_{22}\mathbf{L}\int_0^t \mathbf{P}(t-\tau)\mathbf{P}^T(t-\tau)G_{11}(\tau,\omega)d\tau\mathbf{L}^T \tag{47}$$

Performing a Laplace transform on Eq. (47) with respect to t yields

$$\mathbf{G}(s,\omega) = \mathbf{G}^{(0)}(s,\omega) + S_{22}G_{11}(s,\omega)\mathbf{L}\mathbf{I}(s)\mathbf{L}^T \tag{48}$$

where $\mathbf{I}(s)$ is defined by Eq. (35).

$G_{11}^{(0)}(s,\omega)$ may be obtained by performing the Laplace transform on $G_{11}^{(0)}(t,\omega)$ whose explicit expression may be found in [7,11]. In the partial fraction form, $G_{11}^{(0)}(s,\omega)$ may be expressed as

$$\begin{aligned} G_{11}^{(0)}(s,\omega) = &\frac{\alpha_1}{s} + \frac{\alpha_2}{s+2\varsigma_0\omega_0} + \frac{\alpha_3(s+2\varsigma_0\omega_0)+2\omega_{0d}\alpha_4}{(s+2\varsigma_0\omega_0)^2+(2\omega_{0d})^2} + \\ &\frac{\alpha_5(s+\varsigma_0\omega_0)+\alpha_6(\omega-\omega_{0d})}{(s+\varsigma_0\omega_0)^2+(\omega-\omega_{0d})^2} + \frac{\alpha_7(s+\varsigma_0\omega_0)+\alpha_6(\omega+\omega_{0d})}{(s+\varsigma_0\omega_0)^2+(\omega+\omega_{0d})^2} \end{aligned} \tag{49}$$

where

$$\begin{aligned} &\alpha_1 = A \qquad \alpha_2 = \frac{\varsigma_0^2\omega_0^2+\omega_{0d}^2+\omega^2}{2\omega_{0d}^2}A \\ &\alpha_3 = -\frac{\varsigma_0^2\omega_0^2-\omega_{0d}^2+\omega^2}{2\omega_{0d}^2}A \qquad \alpha_4 = \frac{\varsigma_0\omega_0}{\omega_{0d}}A \\ &\alpha_5 = -(1+\frac{\omega}{\omega_{0d}})A \qquad \alpha_6 = -\frac{\varsigma_0\omega_0}{\omega_{0d}}A \\ &\alpha_7 = -(1-\frac{\omega}{\omega_{0d}})A \qquad \alpha_8 = -\frac{\varsigma_0\omega_0}{\omega_{0d}}A \end{aligned} \tag{50}$$

in which

$$A = \frac{S_{11}}{\pi}|H(i\omega)|^2 \tag{51}$$

and $|H(i\omega)|^2$ is the frequency response function of the linear system, namely,

$$|H(i\omega)|^2 = \frac{1}{(\omega_0^2-\omega^2)^2+(2\varsigma_0\omega_0\omega)^2} \tag{52}$$

It can be shown that

$$G_{11}(s,\omega) = G_{11}^{(0)}(s,\omega) + G^{(1)}(s,\omega) \tag{53}$$

where

$$G_{11}^{(1)}(s,\omega) = \frac{2S_{22}}{(s+a)[(s+b)^2+c^2]} G_{11}^{(0)}(s,\omega) \tag{54}$$

The constants a, b, and c are determined by Eq. (37).

In the partial fraction form,

$$\begin{aligned} G_{11}^{(1)}(s,\omega) = & \frac{\beta_1}{s} + \frac{\beta_2}{s+2\varsigma_0\omega_0} + \frac{\beta_3(s+2\varsigma_0\omega_0)+\beta_4(2\omega_{0d})}{(s+2\varsigma_0\omega_0)^2+(2\omega_{0d})^2} \\ & + \frac{\beta_5(s+\varsigma_0\omega_0)+\beta_6(\omega-\omega_{0d})}{(s+\varsigma_0\omega_0)^2+(\omega-\omega_{0d})^2} + \frac{\beta_5(s+\varsigma_0\omega_0)+\beta_6(\omega+\omega_{0d})}{(s+\varsigma_0\omega_0)^2+(\omega+\omega_{0d})^2} \\ & + \frac{\beta_9}{s+a} + \frac{\beta_{10}(s+b)+\beta_{11}c}{(s+b)^2+c^2} \end{aligned} \tag{55}$$

where

$$\begin{aligned} & \beta_1 = \frac{2S_{22}}{a(b^2+c^2)}\alpha_1 \qquad \beta_2 = \frac{2S_{22}}{(a-2\varsigma_0\omega_0)[(b-2\varsigma_0\omega_0)^2+c^2]}\alpha_2 \\ & \beta_3 i + \beta_4 = \left.\frac{2S_{22}(\alpha_3 i+\alpha_4)}{(s+a)[(s+b)^2+c^2]}\right|_{s=-2\varsigma_0\omega_0+i2\omega_{0d}} \\ & \beta_5 i + \beta_6 = \left.\frac{2S_{22}(\alpha_5 i+\alpha_6)}{(s+a)[(s+b)^2+c^2]}\right|_{s=-\varsigma_0\omega_0+i(\omega_{0d}-\omega)} \\ & \beta_7 i + \beta_8 = \left.\frac{2S_{22}(\alpha_7 i+\alpha_8)}{(s+a)[(s+b)^2+c^2]}\right|_{s=-\varsigma_0\omega_0+i(\omega_{0d}+\omega)} \\ & \beta_9 = \left.\frac{2S_{22}}{(s+b)^2+c^2}G_{11}^{(0)}(s,\omega)\right|_{s=-a} \\ & \beta_{10} i + \beta_{11} = \left.\frac{2S_{22}}{c(s+a)}G_{11}^{(0)}(s)\right|_{s=-b+ic} \end{aligned} \tag{56}$$

The final solution for the evolutionary spectral density of the response can be written as

$$\begin{aligned} G_{11}(t,\omega) = & \alpha_1 + \beta_1 + e^{-2\varsigma_0\omega_0 t}(\alpha_2+\beta_2+(\alpha_3+\beta_3)\cos 2\omega_{0d}t + (\alpha_4+\beta_4)\sin 2\omega_d t) + \\ & e^{-\varsigma_0\omega_0 t}((\alpha_5+\beta_5)\cos(\omega_{0d}-\omega)t + (\alpha_6+\beta_6)\sin(\omega_{0d}-\omega)t + (\alpha_7+\beta_7)\cos(\omega_{0d}-\omega)t \\ & + (\alpha_8+\beta_8)\sin(\omega_{0d}-\omega)t) + \beta_9 e^{-at} + e^{-bt}(\beta_{10}\cos ct + \beta_{11}\sin ct) \end{aligned} \tag{57}$$

It can be shown that

$$G_{11}(\infty,\omega) = A_p G_{11}^{(0)}(\infty,\omega) \tag{58}$$

Stability Analysis

The mean square stability of the response is studied in this section. Consider the following envelope function

$$\eta(t) = at^b e^{-ct} U(t) \tag{59}$$

where $U(t)$ is the unit step function, a is positive, and b and c are non-negative. Note that Eq. (59) includes many envelopes prevailing in engineering as special cases.

For the case where $b = c = 0$, ths stability criterion is given by Eq. (32) for both stationary and nonstationary second moment response. It is noted that the stability condition depends only on the damping ratio ζ_0, the natural frequency ω_0 of the system, and the intensity of the parametric excitation S_{22}. A significant increase in the mean square response of the system may be observed if these parameters are close to the region of instability. In this case, the effect of the parametric excitation may not be neglected even if its magnitude is comparatively smaller than that of the external counterpart. Systems with higher damping and frequency will exhibit better resistance under combined excitations. These observations have practical implications for the aseismic design of column structures under combined horizontal and vertical ground motions.

For the general case where $c > 0$, it has been shown [7] that the second moment response for the excitations with an exponentially decaying envelope is exponentially stable based on the following theorem.

Theorem: The solution of the nonautonomous system

$$\frac{d}{dt}\mathbf{Y}(t) = \mathbf{A}\mathbf{Y}(t) + \mathbf{F}(t) + \mathbf{B}(t)\mathbf{Y}(t) \tag{60}$$

is stable if the following conditions are satisfied:

(i) The solution of $\frac{d}{dt}\mathbf{Y}(t) = \mathbf{A}\mathbf{Y}(t)$ is stable;

(ii) $\mathbf{B}(t)$ is impulsively small, i.e. $\exists K > 0$ such that

$$\int_0^{+\infty} \|\mathbf{B}(t)\| dt < K \tag{61}$$

where $\|\cdot\|$ is any matrix norm;

(iii) $\mathbf{F}(t)$ is sufficiently small as t approaches zero, i.e. $\exists$ $a > 0$ and $b > 0$ such that

$$|\mathbf{F}(t)| < ae^{(-bt)} \tag{62}$$

where $|\cdot|$ stands for the associated vector norm. A proof of this theorem is given in [7].

Examples

Some numerical results for the stationary solution are presented in Figs. 1 and 2. Figure 1 gives the variation of the Parametric Amplification Factor A_p versus the nondimensional parameter λ which is a function of damping ratio, natural frequency, and the intensity of the parametric excitation. Note that $A_p \geq 1$. This figure shows that the parametric excitation can have a considerable effect on the correlation matrix of the response, especially near the region of instability.

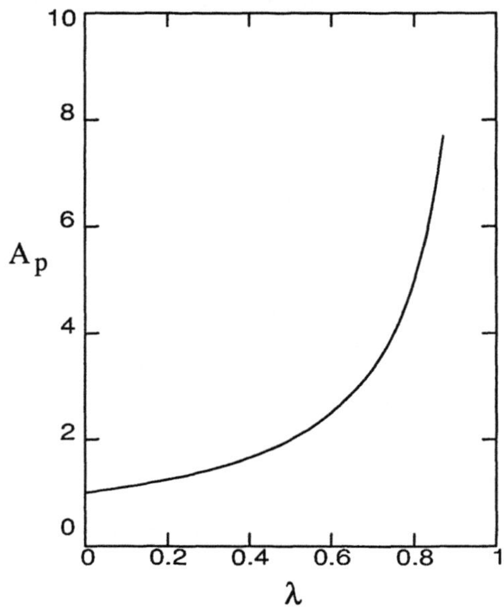

Figure 1. Parametric Amplification Factor A_p

Figure 2 shows the stationary auto- and cross-correlation of the response for three different values of the intensity of the parametric excitation, namely, $S_{22} = 0, 0.12$, and 0.24. The system parameters are chosen such that $\varsigma = 0.1$ and $\omega = 1.0$. The intensity of the external excitation is assumed to be 1.0. It is clear that the larger the parametric excitation, the greater will be the auto- and cross-correlation response of the structure. For the autocorrelation response, the curves start from the stationary values of the mean square response corresponding to $\tau = 0$ which are maxima, and then exhibit an oscillatory decrease to zero as τ approaches infinity. The cross-correlation response has the same trend as the autocorrelation response except starting from zero. Note that $R_{xv}(\tau) = -R_{vx}(\tau)$.

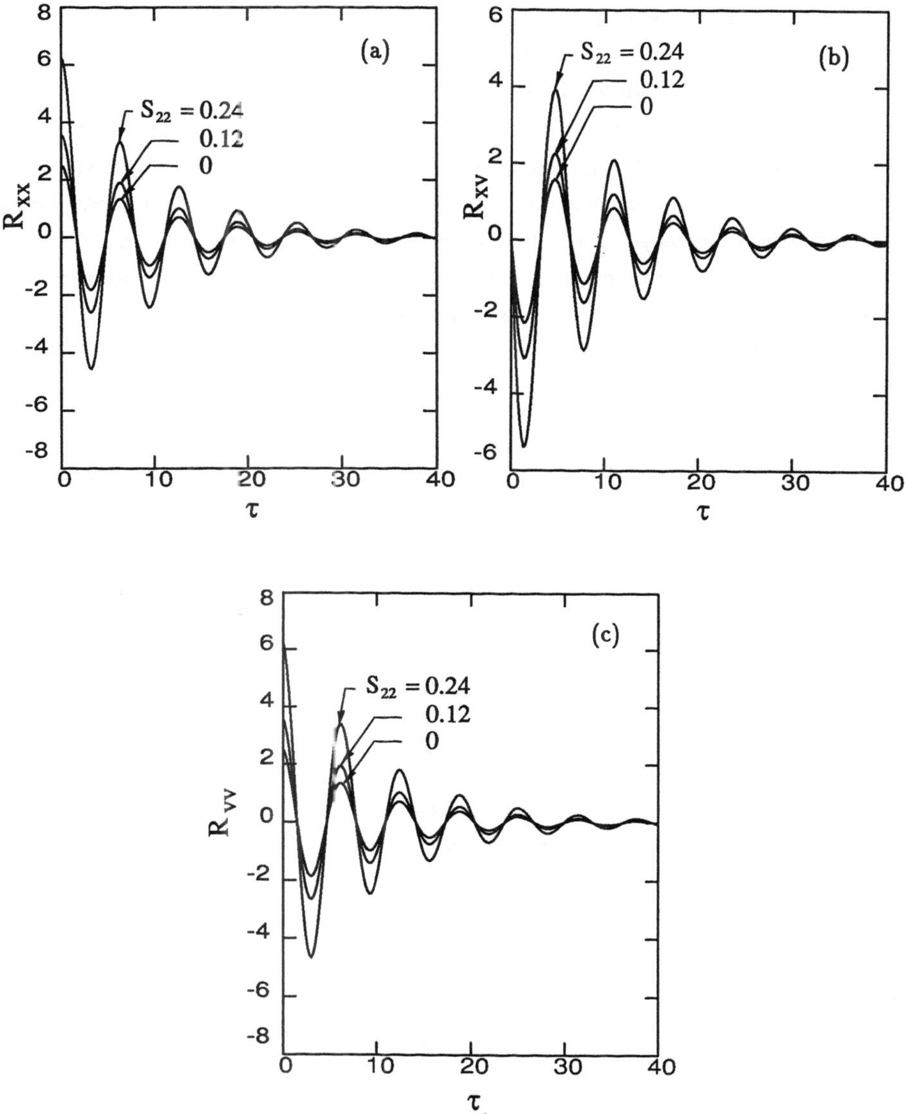

Figure 2. Stationary correlation of the structural response under uncorrelated parametric and external excitations. Structural parameters $\omega = 1.0$ and $\varsigma = 0.1$. Excitation parameters $S_{11} = 1.0$, $S_{12} = 0$, and $S_{22} = 0, 0.12$, and 0.24

Figure 3 presents some numerical results for the nonstationary second moment response. All the parameters have the same values as those for the stationary solution. $S_{22} = 0$ corresponds the case where the system is subjected to an external excitation alone. It may be observed that the covariance solution for the combined loads exhibits a similar trend as that for the external excitation alone. These responses all approach their stationary values as t approaches $+\infty$. The time required to achive stationarity depends on the damping, frequency of the system as well as the intensity of the parmetric excitation. The existance of the parametric excitation generally magnifies the covariance response. The larger the intensity of the parametric excitation, the greater is the response. The proportionality between the responses due to the pure external load and the combined loads generally does not hold during the transient stage of the solution but is still valid for the stationary values.

The evolutionary spectral density of the response at $t = 2.5, 5, 10, 20, 30, 40$, and 45 second are presented in Fig. 4. All the parameters are the same as in Figs. 2 and 3. The results show changes with time in the amplitude and frequency content of the spectral density of the response. It is observed that the results approach their stationary counterparts for sufficiently large time. The larger the intensity of the parametric excitation, the more time is required to achive the stationary value.

Conclusions

Explicit solutions for the stationary correlation matrix, the nonstationary secondary moment response, and the evolutionary spetral density of the response of a linear system subjected to suddenly applied, uncorrelated parametric and external excitationas are obtained by the simplified state-variable approach. It has been shown that the solutions are the superposition of the corresponding solution for the external excitation alone and a correction term resulting from the parametric excitation.

In the stationary case, the correlation response under combined loads is simply equal to that for the external solution alone multiplied by the Parametric Amplification Factor, which depends only on the system parameters and the intensity of parametric excitation. In the nonstationary case, this proportionality between the solutions for the external load alone and the combined excitations generally does not hold during the nonstationary stage of the solutions but may still be valid for the stationary values. The existance of the parametric excitation generally magnifies the response.

It has been shown that for the step envelope function, the solution exhibits conditional stability. The stability region is determined by the system parameters and the intensity of the parametric excitation. A system with large damping and natural frequency

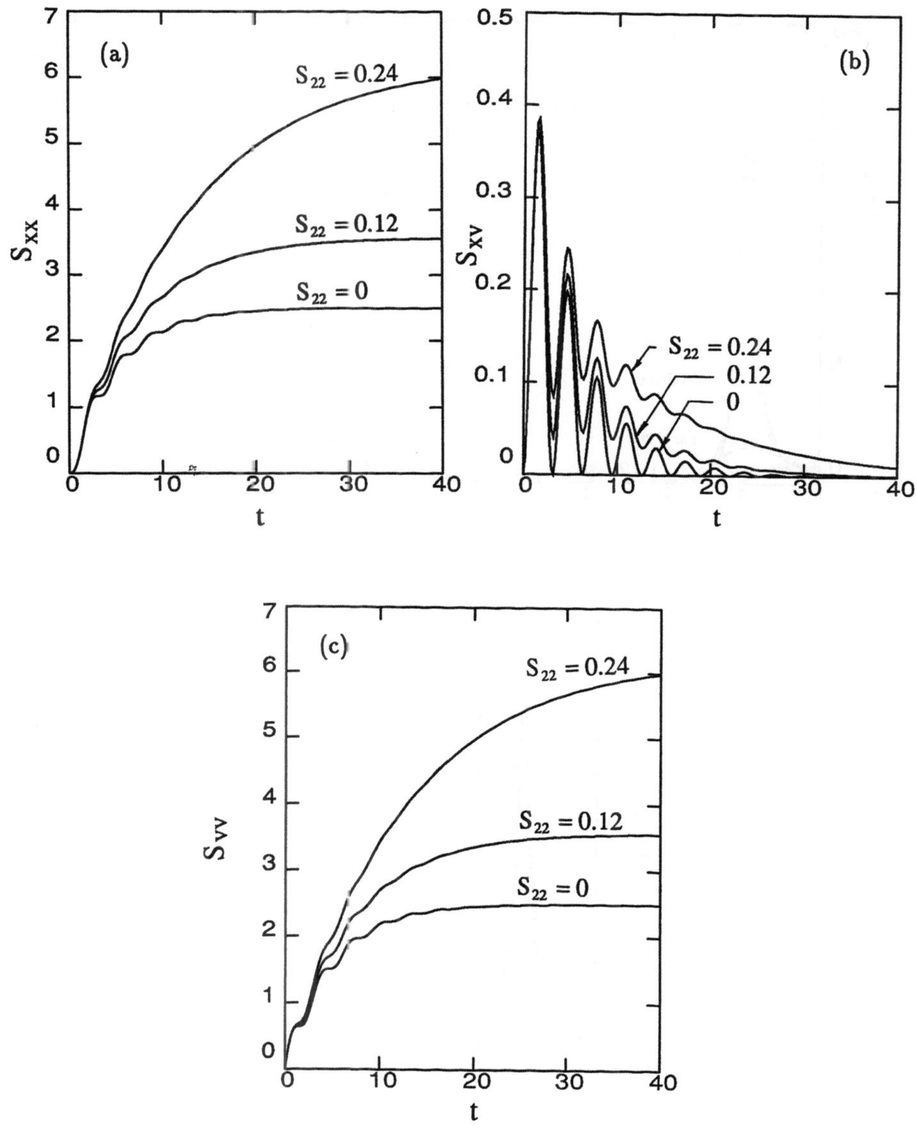

Figure 3. Nonstationary second moment response under uncorrelated parametric and external excitations using the unit step envelope. Structural parameters $\omega = 1.0$ and $\varsigma = 0.1$. Excitation parameters $S_{11} = 1.0$, $S_{12} = 0$, and $S_{22} = 0, 0.12$, and 0.24

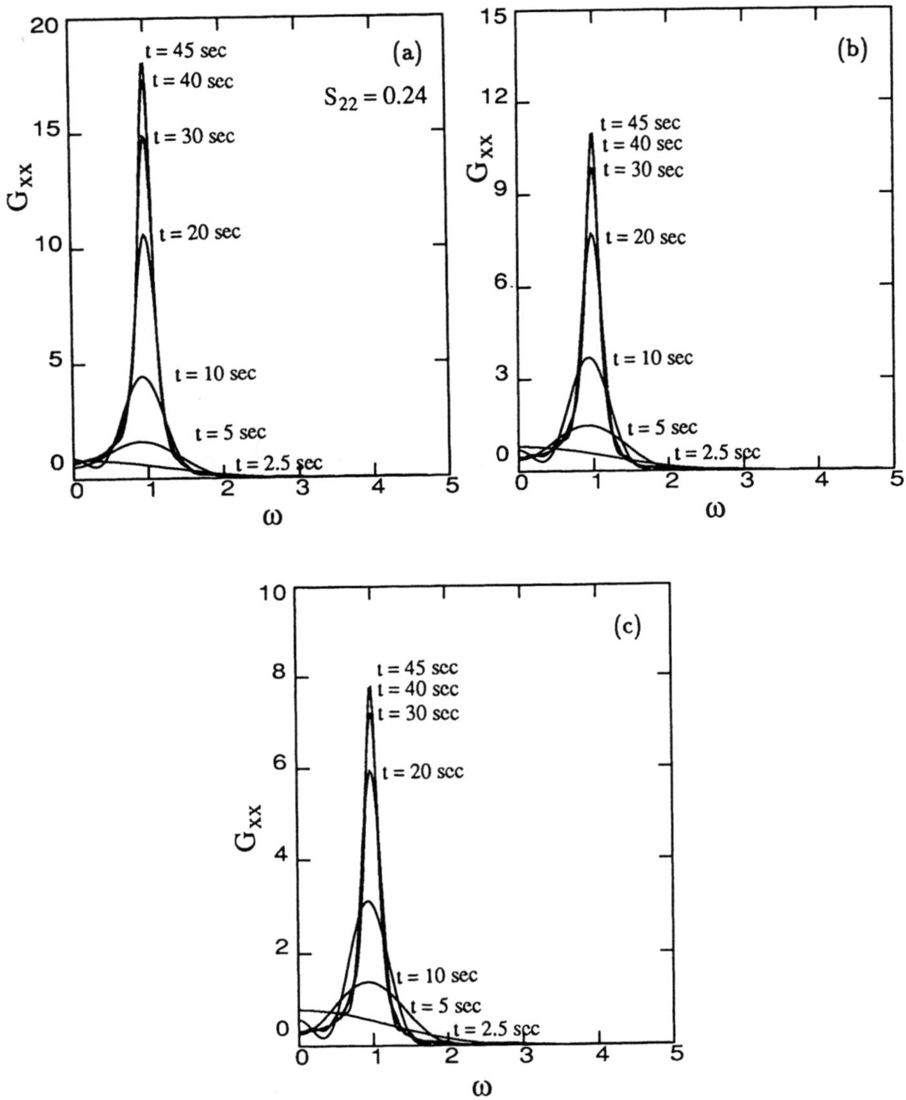

Figure 4. Evolutionary spectral densities of the response under uncorrelated parametric and external excitations using the unit step envelope. $t = 2.5, 5, 10, 20, 30, 40$ and 45 second. Structural parameters $\omega = 1.0$ and $\varsigma = 0.1$. Excitation parameters $S_{11} = 1.0$, $S_{12} = 0$, and $S_{22} = 0, 0.12$, and 0.24

has better dynamic performance under the combined excitations. The same stability criterion applies for both the stationary and nonstationary solutions. It is also found that the second moment responses are exponentially stable for a class of envelopes which includes the most commonly used envelopes in engineering as special cases.

The results obtained give guidelines to account for the effect of the vertical component of ground motion in earthquake resistant design. The conclusions for uncorrelated excitations apply for the case when the excitations are correlated provided the mean value of the displacement remains zero [4,7]. A detailed discussion for the correlated excitations may be found in [7,8].

Acknowledgement

A version of this paper has appeared in the Proceedings of the Second International Conference on Stochastic Structural Dynamics. The authors acknowledge the permission from Springer-Verlag, publisher of the proceedings, to publish this article in the present journal.

References

1. Ibrahim, R.A., Parametric Random Vibration, Research Studies Press Ltd. and John Wiley & Sons Inc., 1985.

2. Kozin, F.,"Some Results of Stability of Stochastic Dynamic Systems," in Random Vibration—Status and Recent Developments, Edited by Elishakoff, I. and Lyon, R.H., Elsevier, 1986, pp. 163-191.

3. To, C.W.S., "Vibration Analysis of System with Random Parametric Excitations," *The Vibration Digest*, No. 2, Vol. 21, 1989, pp. 2-12.

4. Lin, Y.K., and Shih, T.Y., "Column Response to Horizontal-Vertical Earthquakes," *Journal of Engineering Mechanics Division*, ASCE, No. EM6, Vol. 106, 1980. pp.1099-1109.

5. Benaroya, H. and Rehak, M., "Response and Stability of a Random Differential Equation: Part I—Moment Equation Method," *Journal of Applied Mechanics*, ASME, Vol. 56, March 1989, pp. 192-195.

6. Iwan, W.D. and Hou, Z.K., "Explicit Solutions for the Response of Simple Structures Subjected to Nonstationary Random Excitation," *Structural Safety*, 6(1989), pp. 77-86.

7. Hou, Z.K., "Nonstationary Solution for Dynamic Response of Linear MDOF Systems and Its Applications in Earthquake Engineering," Ph.D. Thesis, California Institute of Technology, 1990.

8. Iwan, W.D. and Hou, Z.K., "Exact Solutions for the Response of Systems under Uncorrelated Parametric and External Excitations," Accepted by the Second International Conference on Vibration Problems in Engineering, China, June 1990.

9. Samuels, J.C., "On the Stability of Random Systems and the Stabilization of Deterministic Systems with Random Noise," *Journal of the Acoustic Society of America*, No. 5, Vol. 32, 1960, pp. 594-601.

10. Benaroya, H. and Rehak, M., "Response and Stability of a Random Differential Equation: Part II—Expansion Method," *Journal of Applied Mechanics*, ASME, Vol. 56, March 1989, pp. 196-201.

11. Corotis, R.B. and Vanmarche, E.H., "Time-Dependant Spectral Content of System Response," *Journal of Engineering Mechanics Division*, ASCE, Vol. 103, No. EM4, 1977, pp. 501-513.

Appendix

A brief description of the simplified state-variable method for nonstationary solution of linear MDOF systems is given in the following. A detailed derivation and applications of the method can be found in [7].

The governing equation of motion of a linear MDOF system be written in a state-variable form as

$$\frac{d}{dt}\mathbf{Y}(t) = \mathbf{A}\mathbf{Y}(t) + \mathbf{F}(t) \qquad \mathbf{Y}(0) = \mathbf{Y}_0 \tag{A-1}$$

where

$$\mathbf{Y} = \begin{pmatrix} \mathbf{x}(t) \\ \dot{\mathbf{x}}(t) \end{pmatrix} \quad \mathbf{Y}_0 = \begin{pmatrix} \mathbf{x}_0 \\ \mathbf{v}_0 \end{pmatrix} \quad \mathbf{A} = \begin{pmatrix} 0 & \mathbf{I} \\ -\mathbf{M}^{-1}\mathbf{K} & -\mathbf{M}^{-1}\mathbf{C} \end{pmatrix} \quad \mathbf{F} = \mathbf{F}_0\eta(t)n(t)$$

in which $\mathbf{M}, \mathbf{C}$ and $\mathbf{K}$ represent mass, damping and stiffness properties of the system $\mathbf{I}$ is an identity matrix, $\mathbf{F}_0$ is a constant vector, and $n(t)$ is a stationary white noise modulated by a deterministic envelope $\eta(t)$.

The fundamental matrix of the system can be written as

$$\mathbf{\Phi}(t) = \mathbf{U}\mathbf{\Lambda}(t)\mathbf{U}^{-1} \tag{A-2}$$

where

$$\mathbf{U} = (\mathbf{u}_1, \mathbf{u}_2, \ldots, \mathbf{u}_{2k-1}, \mathbf{u}_{2k}, \ldots, \mathbf{u}_{2N-1}, \mathbf{u}_{2N})$$

$$\mathbf{\Lambda}(t) = \begin{pmatrix} \mathbf{\Lambda}_{11}(t) & 0 & \ldots & 0 \\ 0 & \mathbf{\Lambda}_{22}(t) & \ldots & 0 \\ \vdots & \vdots & \ddots & \vdots \\ 0 & 0 & \ldots & \mathbf{\Lambda}_{NN}(t) \end{pmatrix} \qquad (A-3)$$

The column components of $\mathbf{U}$ and the submatrice of $\mathbf{\Lambda}(t)$ are:

$$\mathbf{u}_{2k-1} = Re(\mathbf{v}^{(2k-1)}) \qquad \mathbf{u}_{2k} = Im(\mathbf{v}^{(2k-1)})$$

$$\mathbf{\Lambda}_k(t) = e^{-\varsigma_k \omega_k t} \begin{pmatrix} \cos \omega_{dk} t & \sin \omega_{dk} t \\ -\sin \omega_{dk} t & \cos \omega_{dk} t \end{pmatrix} \qquad k = 1, \ldots, N$$

where

$$-\varsigma_k \omega_k = Re(\lambda_k)$$

$$\omega_{dk} = \omega_k \sqrt{1 - \varsigma_k^2} = Im(\lambda_k) \qquad k = 1, \ldots, N \qquad (A-4)$$

The general solution can be expressed as

$$\begin{aligned} \mathbf{Y}(t) &= \mathbf{\Phi}(t)\mathbf{Y}_0 + \int_0^t \mathbf{\Phi}(t-\tau)\mathbf{F}(\tau)d\tau \\ &= \mathbf{\Phi}(t)\mathbf{Y}_0 + \mathbf{L}\int_0^t \mathbf{P}(t-\tau)\eta(\tau)n(\tau)d\tau \end{aligned} \qquad (A-5)$$

where

$$\mathbf{P}(t) = \begin{pmatrix} \mathbf{P}_1(t) \\ \mathbf{P}_2(t) \\ \cdots\cdot \\ \mathbf{P}_N(t) \end{pmatrix} \qquad \mathbf{L} = \mathbf{U}\mathbf{V} \quad \mathbf{V} = diag(\mathbf{V}_1, \mathbf{V}_2, \ldots, \mathbf{V}_N) \qquad (A-6)$$

and the submatrice $\mathbf{P}_k(t)$ and $\mathbf{V}_k, k = 1, ..., N$ are defined as

$$\mathbf{P}_k(t) = e^{-\varsigma \omega_k t} \begin{pmatrix} \cos \omega_{dk} t \\ \sin \omega_{dk} t \end{pmatrix} \quad \mathbf{V}_k = \begin{pmatrix} D_{2k-1} & D_{2k} \\ D_{2k} & -D_{2k-1} \end{pmatrix} \qquad (A-7)$$

where

$$\mathbf{D} = \mathbf{U}^{-1}\mathbf{F}_0 \qquad (A-8)$$

In the special case of SDOF systems, $\mathbf{P}(t)$ and $\mathbf{L}$ are given by Eq. (16).

For zero-mean response, the nonstationary correlation matrix of the response can be expressed a compact matrix form as

$$\mathbf{Q}(t_1, t_2) = S_0 \mathbf{L} \int_0^{min(t_1, t_2)} \eta^2(\tau)\mathbf{P}(t_1 - \tau)\mathbf{P}^T(t_2 - \tau)d\tau \mathbf{L}^T \qquad (A-9)$$

For most envelopes used in engineering, Eq. (A-9) can be explicitly integrated.

Wide-Band Response of Multiple Subsystems with High Modal Density

Takeru Igusa[1] and Kangming Xu[2]

Department of Civil Engineering
Northwestern University
Evanston, Ill 60208

SUMMARY

The basic system under consideration is a single-degree-of-freedom system connected to a large number of oscillators. The oscillators are assumed to have natural frequencies which densely cover a range of frequencies which includes that of the main system. The goal of this paper is to determine how the oscillators affect the response characteristics of the main system subjected to harmonic and wide-band base acceleration. By making certain assumptions on the properties of the oscillators, and using asymptotic techniques and contour integration methods it is possible to obtain surprisingly simple closed-form analytical expressions describing the dynamic effect of the multiple subsystems. These expressions, as well as the analysis approach, are novel and provide new understanding of coupled subsystems.

INTRODUCTION

The dynamic characteristics of coupled linear subsystems are of considerable interest in civil, mechanical, and aerospace engineering problems. For cases where the natural frequencies of each subsystem are well spaced, there exist a number of analysis approaches, such as mode synthesis [1] and subsystem receptance methods [2] which are the basis of many computational techniques; perturbation methods [3,4] which produce analytical results that provide insight into the coupling problem; and Green's function techniques [5] and Lagrange formulations [6,7] which are applicable to continuous subsystems. For cases where one or more of the coupled subsystems contain large numbers of closely-spaced natural frequencies, statistical energy analysis [8], asymptotic modal analysis [9], and power flow methods [10] are used for dynamic response analysis.

---

[1] Assoc. Prof., Dept. of Civ. Engrg., Northwestern Univ., Evanston, Ill. 60208.
[2] Grad. student, Dept. of Civ. Engrg., Northwestern Univ., Evanston, Ill. 60208.

Y. K. Lin · I. Elishakoff (Eds.)
Stochastic Structural Dynamics 1
New Theoretical Developments
2nd International Conference on Stochastic Structural Dynamics
May 9–11, 1990, Boca Raton, Florida

In the present paper, the influence of a collection of subsystems with closely-spaced natural frequencies on a single mode of another subsystem is of interest. The primary goal is to develop an understanding of the dynamic characteristics of such collections of coupled subsystems. This is achieved by examining in detail a fundamental system configuration and obtaining simple, yet physically meaningful closed-form expressions for response to harmonic and wide-band excitations.

The system configuration analyzed in this paper consists of a collection of oscillators (subsystems) mounted on a main oscillator as shown in Fig. 1. The response of the main oscillator to a base acceleration is of interest. The analysis begins with the development of a series expansion form of the frequency response function. Then, the asymptotic behavior of the frequency response function for increasingly large numbers of subsystems is determined by using contour integration techniques. The results for the frequency response function and mean-square response to wide-band input are of surprisingly simple form, considering the complexity of the coupled system. It is found that the multiple subsystems have an equivalent damping effect, resulting in a reduction of the response of the main oscillator.

SERIES EXPANSION FORM FOR THE FREQUENCY RESPONSE FUNCTION

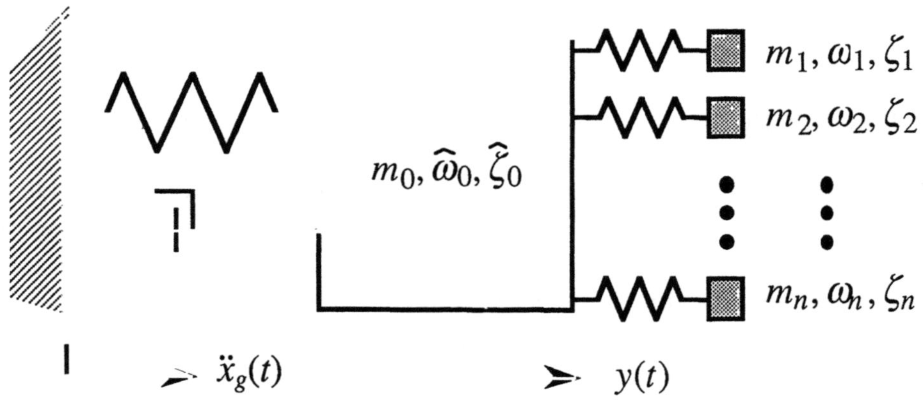

Fig. 1. System Configuration.

Consider the system of coupled linear oscillators in Fig. 1. The main oscillator, which is mounted on the base of the system, has natural frequency, $\hat{\omega}_0$, damping ratio, $\hat{\zeta}_0$, and mass, m_0. Connected to the main oscillator are a set of n oscillator subsystems in which the j-th subsystem has natural frequency, ω_j, damping ratio, ζ_j, and mass, m_j, for $j = 1, \ldots, n$. The system is subjected to a harmonic base acceleration and the relative displacement response of the main oscillator mass with respect to the base is of interest.

To simplify notation, the following expressions are used

$$\omega_0 \equiv \sqrt{\hat{\omega}_0^2 + \sum_{j=1}^{n} \mu_j \, \omega_j^2} \tag{1a}$$

$$\zeta_0 \equiv \hat{\zeta}_0 \frac{\hat{\omega}_0}{\omega_0} + \sum_{j=1}^{n} \mu_j \, \zeta_j \frac{\omega_j}{\omega_0} \tag{1b}$$

where $\mu_j \equiv m_j / m_0$ is the ratio of masses of the j-th subsystem and the main oscillator. Herein, the subscript j ranges from 1, ... , n unless noted otherwise. Physically, ω_0 and ζ_0 represent the natural frequency and damping of the main oscillator if the masses of all of the subsystems are fixed.

The system response to harmonic base acceleration is [11]

$$\left[-\omega^2 \mathbf{M} - \mathrm{i}\, \omega \mathbf{C} + \mathbf{K} \right] \mathbf{X}(\omega)\, e^{-\mathrm{i}\omega t} = -\mathbf{M}\, \mathbf{r}\, e^{-\mathrm{i}\omega t} \tag{2}$$

where $\mathbf{X}(\omega)\, e^{-i\omega t}$ is the vector of displacement responses where the first component corresponds to the main oscillator and the j+1 component corresponds to subsystem oscillator j. The vector $\mathbf{r} = \{ 1, 1, \cdots, 1 \}$ is the influence vector relating the base motion to the motion of each degree of freedom [12], and **M**, **C**, and **K** are the mass, damping, and stiffness matrices, respectively, which are given by

$$\mathbf{M} \equiv \mathbf{diag}\, \{ m_0, m_1 \cdots, m_n \} \tag{3a}$$

$$\mathbf{C} \equiv \begin{bmatrix} 2m_0\omega_0\zeta_0 & -2m_1\omega_1\zeta_1 & \cdots & -2m_n\omega_n\zeta_n \\ & 2m_1\omega_1\zeta_1 & \cdots & 0 \\ & & \ddots & \vdots \\ (sym) & & & 2m_n\omega_n\zeta_n \end{bmatrix} \tag{3b}$$

$$\mathbf{K} \equiv \begin{bmatrix} m_0\omega_0^2 & -m_1\omega_1^2 & \cdots & -m_n\omega_n^2 \\ & m_1\omega_1^2 & \cdots & 0 \\ & & \ddots & \vdots \\ (sym) & & & m_n\omega_n^2 \end{bmatrix} \tag{3c}$$

where **diag**{ . } represents the diagonal matrix with diagonal elements given in braces and (*sym*) is used to denote symmetrical matrices.

Pre-multiplying Eq. 2 by $\Gamma = \mathbf{diag}\left\{ m_0^{-1/2}, \cdots, m_n^{-1/2} \right\}$ and dividing by $e^{-i\omega t}$ yields

$$\mathbf{H}^{-1}(\omega)\, \Gamma^{-1} \mathbf{X}(\omega) = \Gamma\, \mathbf{M}\, \mathbf{r} \tag{4}$$

The matrix $\mathbf{H}(\omega)$ is the frequency response matrix

$$\mathbf{H}(\omega) \equiv \begin{bmatrix} g_0(\omega) & q_1(\omega) & \cdots & q_n(\omega) \\ & g_1(\omega) & \cdots & 0 \\ & & \ddots & \vdots \\ (sym) & & & g_n(\omega) \end{bmatrix}^{-1} \tag{5}$$

in which

$$g_j(\omega) \equiv \omega^2 - \omega_j^2 + 2\,\mathrm{i}\,\omega\,\omega_j\,\zeta_j \qquad (j = 0, 1, \cdots, n) \tag{6a}$$

$$q_j(\omega) \equiv \sqrt{\mu_j}\,(-2\,\mathrm{i}\,\omega\,\omega_j\,\zeta_j + \omega_j^2) \qquad (j = 1, 2, \cdots, n) \tag{6b}$$

The function $g_j(\omega)$ is the reciprocal of the frequency response function of oscillator j subjected to a base harmonic acceleration, and $q_j(\omega)$ represents the dynamic coupling between the main oscillator and subsystem j.

The harmonic response $y(\omega)$ of the main oscillator is found by solving Eq. 4

$$y(\omega) = \frac{1}{\sqrt{m_0}} \mathbf{H}_1(\omega)\, \Gamma\, \mathbf{M}\, \mathbf{r} \tag{7}$$

where $\mathbf{H}_1(\omega)$ is the first row of $\mathbf{H}(\omega)$. The terms of $\mathbf{H}_1(\omega)$ are obtained by inverting the matrix given in Eq. 5. This inverse matrix has been analytically determined for a similar system [13] and it can be verified by substitution that the first row is

$$\mathbf{H}_1(\omega) \equiv \frac{1}{h_0(\omega)} \left[1 \quad -\frac{q_1(\omega)}{g_1(\omega)} \quad \cdots \quad -\frac{q_n(\omega)}{g_n(\omega)} \right] \tag{8}$$

in which $h_0(\omega)$ is the determinant of $\mathbf{H}(\omega)$

$$h_0(\omega) \equiv g_0(\omega) - \sum_{j=1}^{n} \frac{q_j^2(\omega)}{g_j(\omega)} \tag{9}$$

Substituting Eq. 8 into Eq. 7 yields the exact expression for the frequency response function

$$y(\omega) = \frac{1}{h_0(\omega)}\left[1 - \sum_{j=1}^{n} \frac{q_j(\omega)\sqrt{\mu_j}}{g_j(\omega)}\right] \tag{10}$$

For the special case $n = 1$, Eq. 10 is the frequency response response function for a two-degree-of-freedom system. Such systems have been studied extensively by Den Hartog [14], Crandall and Mark [15], Sackman and Kelly [3], and Igusa and Der Kiureghian [4]. In this paper, interest is on large values of n. For such cases, Eq. 10 is a high-order rational expression in ω which, in general, is too complicated for mathematical analysis. Although it is always possible to analyze the system using numerical techniques, the results of such studies do not provide insight into the physics of the problem.

In this paper, a special, yet fundamental class of systems with the configuration in Fig. 1 is examined. The analysis leads to surprisingly simple results which provide a clear understanding of the dynamic effect of multiple subsystems.

ASYMPTOTIC RESULTS FOR A SPECIAL CLASS OF SYSTEMS

Consider the class of systems with configuration in Fig. 1 specified by the following three properties:

1. The subsystem natural frequencies are in an arithmatic series with an average close to the main oscillator natural frequency. Mathematically, this is represented as

$$\omega_j = \omega_0 [\, 1 + (j - \tfrac{n}{2})\, \beta + \beta_0 \,] \tag{11}$$

where β is a dimensionless measure of the spacing of the natural frequencies, and $\beta_0 = (\omega_a - \omega_0)/\omega_0$ in which ω_a is the average subsystem natural frequency. It is assumed that β and $\beta_0 << 1$ so that the natural frequencies of the subsystems are close to that of the main oscillator:

$$\omega_j \approx \omega_0 \tag{12}$$

These concepts are illustrated in Fig. 2 where the oscillator frequencies are shown on a frequency axis.

It is also assumed, for mathematical convenience, that the number of subsystems is odd, i.e., $n = 2m + 1$. Later in this section the odd assumption for n will be lifted and a more rigorous criteria for β will be given.

2. The damping ratios of all of the oscillator subsystems are equal and small in relation to unity:

$$\zeta_j = \zeta << 1 \tag{13}$$

3. The masses of all of the oscillator subsystems are nearly equal and are specified by the mass ratios

$$\mu_j = \left(\frac{\omega_0}{\omega_j}\right)^4 \mu \tag{14}$$

There is no restriction on the magnitude of μ; thus, the coupling between the subsystems and the main oscillator can be arbitrarily large.

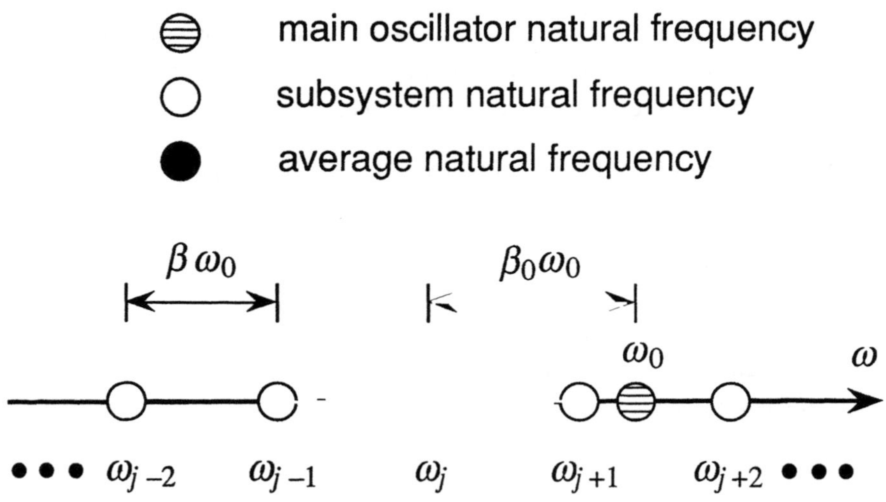

Fig. 2. Distribution of Subsystem Natural Frequencies on the Frequencies Axis.

For this special and fundamental class of systems, several simplifications and an asymptotic approximation can be developed to reduce the original expression for the frequency response function in Eq. 10 into a simple, physically meaningful result.

The first approximation involves the excitation frequency ω. Since the response of the main oscillator is of interest, the range of values for ω is restricted to a neighborhood of the main oscillator natural frequency ω_0,

$$\omega \approx \omega_0 \tag{15}$$

The highest response values are expected for such excitation frequencies; this is confirmed at the conclusion of the analysis.

By virtue of the fact that the natural frequencies of the subsystems ω_j and the excitation frequency ω are approximately equal to ω_0 (Eqs. 12 and 15), the function $g_j(\omega)$ in Eq. 6a can be simplified to

$$g_j(\omega) = (\omega - \omega_j)(\omega + \omega_j) + 2\,i\,\zeta_j\,\omega_j\,\omega \approx 2\,\omega_0\,(\omega - \omega_j + i\,\zeta\,\omega_0) = 2\,\omega_0^2\,\gamma_j(\omega) \quad (16)$$

where $\gamma_j(\omega) \equiv (\omega - \omega_j)/\omega_0 + i\,\zeta$ for $j = 0, 1, \ldots, n$. Also, the coupling functions $q_j(\omega)$ in Eq. 6b can be simplified by neglecting the small imaginary damping component (see Eq. 13) and using Eq. 14

$$q_j(\omega) \approx \sqrt{\mu_j}\,\omega_j^2 = \sqrt{\mu}\,\omega_0^2 \quad (17)$$

Substituting Eqs. 16 and 17 into Eq. 10 yields

$$y(\omega) \approx \frac{1 - \dfrac{\mu}{2}\displaystyle\sum_{j=1}^{n}\frac{1}{\gamma_j(\omega)}}{2\,\omega_0^2\,\gamma_0(\omega) - \dfrac{\mu\omega_0^2}{2}\displaystyle\sum_{j=1}^{n}\frac{1}{\gamma_j(\omega)}} = \frac{1 - \dfrac{\mu}{2\,\beta}\,\gamma(u)}{2\,\omega_0^2\left[\gamma_0(\omega) - \dfrac{\mu}{4\,\beta}\,\gamma(u)\right]} \quad (18)$$

where $u = (\omega_a - \omega_0)/\omega_0\beta$ is a normalized difference between the excitation frequency ω and the average subsystem natural frequency ω_a, and $\gamma(u)$ represents the contribution of the subsystems to the frequency response function

$$\gamma(u) \equiv \sum_{k=-m}^{m}\frac{1}{u - k + i\,\xi} \quad (19)$$

in which $\xi = \zeta/\beta$. Under the restriction in Eq. 15, it can be assumed that $u << m$.

The summation of Eq. 19 can be further evaluated using contour integration [16]. The final result for $\gamma(u)$ is approximately

$$\gamma(u) \approx i\,\pi \quad (20)$$

This approximate result is notably simple, especially when compared with the original rational polynomial expression in Eq. 19. It is independent of the excitation frequency u, subsystem damping ζ, frequency spacing β, and number of subsystems n. This independence property has important physical implications which is examined in the following discussion.

Substituting Eq. 20 into Eq. 18 and rearranging terms yields

$$y(\omega) \approx \frac{1}{2\omega_0} \frac{1 - 2\,i\,\zeta_e}{\omega - \omega_0 - i\,\omega_0\,(\,\zeta_0 + \zeta_e\,)} \tag{21}$$

where

$$\zeta_e \equiv \frac{\pi\,\mu}{4\,\beta} \tag{22}$$

At this point, the approximation used in Eqs. 16 and 17 are applied in reverse order to obtain a frequency response function that is in standard form:

$$y(\omega) \approx \frac{1}{\omega^2 - \hat{\omega}_0^2 - 2\,i\,\omega\,\hat{\omega}_0\,(\,\zeta_0 + \zeta_e\,)} \tag{23}$$

This result corresponds to the response of a single oscillator with the same natural frequency as the main oscillator in Fig. 1, but with a damping ratio that is increased by the amount ζ_e. The physical interpretation is that the cumulative dynamic effect of the collection of n subsystems on the main system response is completely described by a simple increase in damping. This additional damping value, ζ_e, which is defined in Eq. 22, is called the equivalent damping ratio.

There are several remarkable properties of the equivalent damping ratio.

1. *Independence with respect to the excitation frequency ω.* Actually, if ω is sufficiently far from the main oscillator natural frequency $\hat{\omega}_0$, the approximation in Eqs. 16 and 20 is no longer valid. However, this behavior is unimportant because the frequency response function is sensitive to the subsystem properties only for excitation frequencies near $\hat{\omega}_0$.

2. *Independence with respect to the frequency offset parameter β_0.* In other words, the effect of the subsystems is insensitive to the relative spacing between the average subsystem natural frequency ω_a and $\hat{\omega}_0$.

   The first two properties of ζ_e are mathematically explained by the fact that $\gamma(u)$ is independent of the frequency parameter u.

3. *Independence with respect to the number of subsystems n.* The physical explanation is that the subsystem oscillators with natural frequencies close to $\hat{\omega}_0$ have the dominant effect on the main oscillator response. The other subsystems have a weak effect even when taken collectively for increasing

n [16]. A conclusion is that the odd-valued assumption for n made earlier in this section can be lifted.

4. *Independence with respect to the subsystems damping* ζ. This may appear counter-intuitive since the damping of a system is usually dependent upon the damping of its components. The phenomena is mathematically described by Eq. 20 which shows that $\gamma(u)$ is independent of the parameter ξ.

5. *Proportionality to* μ/β. The ratio can be interpreted as a non-dimensional mass density along the frequency axis, i.e., it is the amount of (normalized) mass corresponding to the subsystems with natural frequencies lying within a (normalized) unit frequency increment. A higher μ/β ratio can arise from two equivalent arrangements: a larger μ, i.e., subsystems with larger masses, or a smaller β, i.e., subsystems with closer frequency spacing.

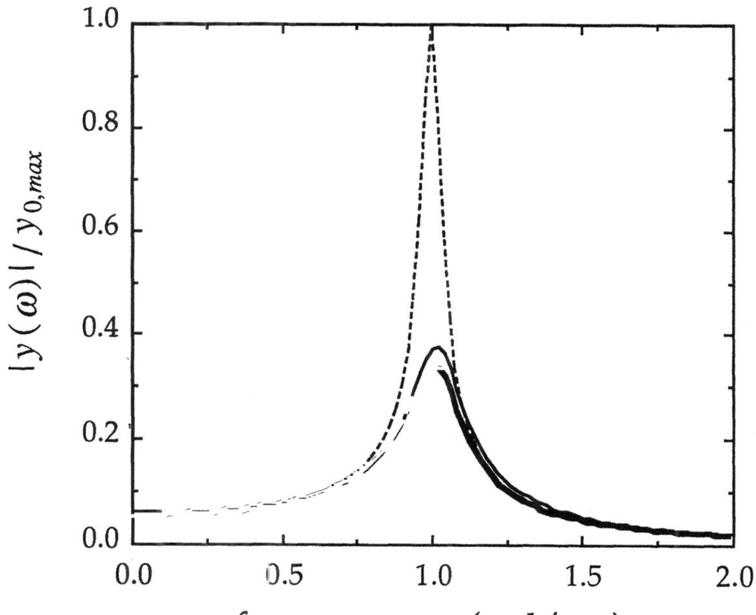

Fig. 3. Normalized Frequency Response Amplitude. ——— Exact (Eq. 10); - - - - - Exact (without Subsystems); ——— Approximate (Eq. 23).

The characteristics of the frequency response function is illustrated by a numerical comparison of the exact result in Eq. 10, which is for an n+1-degree-of-freedom system, and the approximate result in Eq. 21 in which the n subsystem oscillators are represented by the constant equivalent damping ζ_e. The parameters used are $\mu = 0.0025$, $\beta = 0.04$, $\zeta = 0.1$, $\hat{\zeta}_0 = 0.03$, and $n = 20$; the corresponding equivalent damping ratio is $\zeta_e = 0.049$. The modulus $|y(\omega)|$ is normalized with respect to the maximum frequency response for the main oscillator without subsystems, $y_{0,max}$, and the results are plotted in Fig. 3. It can be seen that the exact solution, represented by the thin curve, closely follows the approximate solution, represented by the bold curve.

The frequency response function of the main oscillator without subsystems is also plotted in Fig. 3 by a dashed curve. The results show that oscillator subsystems with closely-spaced natural frequencies have significant vibration reduction capabilities. Based on the fundamental results derived herein, the authors are currently investigating the use of multiple subsystems as a passive vibration control device [17].

RESPONSE TO RANDOM EXCITATION

The root-mean-square of the response of the main oscillator, σ_y, for stationary, wide-band base acceleration is given by [11]

$$\sigma_y^2 = \int_0^\infty G(\omega)\,|y(\omega)|^2\,d\omega \tag{24}$$

where $G(\omega)$ is the one-sided power spectral density function. This integral can be computed using the exact expression for the frequency response function in Eq. 10 and any form of the spectral density using standard numerical techniques. For some algebraic forms of $G(\omega)$, the integral may also be evaluated analytically using contour integration methods. However, neither numerical nor complicated analytical results would readily provide insight into the dynamic characteristics of the system. In the following, the theoretical results of the previous section are used to develop simple expressions for σ_y which provide a physical understanding of the dynamic effects of multiple subsystem oscillators.

Although the analysis can proceed with any power spectral density, a white-noise model, specified by $G(\omega) = G_0$, is used since it leads to the simplest and clearest results. It has been shown that the model is a good approximation to general, smoothly-varying power spectral densities for structures with natural frequencies within the most significant band of frequencies of the input excitation [11,18].

Substituting the approximate expression for the frequency response function in Eq. 23 into the integral in Eq. 24 yields

$$\sigma_y^2 \approx \frac{\pi G_0}{4\,\hat{\omega}_0^3\left(\hat{\zeta}_0 + \zeta_e\right)} \tag{25}$$

As expected, this is the mean-square response of the main oscillator modified by the additional damping, ζ_e.

The main features of this result are illustrated by several representative parameter studies. To maintain non-dimensional results, the root-mean-square of the response, σ_y, is divided by the same quantity for the main oscillator without subsystems, $\sigma_{y0} = \pi G_0/(4\,\hat{\omega}_0^3\,\hat{\zeta})$. The approximate value of this ratio is determined by substitution into Eq. 25 to yield

$$\frac{\sigma_y}{\sigma_{y0}} \approx \sqrt{\frac{\hat{\zeta}_0}{\hat{\zeta}_0 + \zeta_e}} \tag{26}$$

which is a simple ratio of damping coefficients.

The base system for comparison is the example used in Fig. 3, i.e., a main oscillator with damping ratio $\hat{\zeta}_0 = 0.03$, and a set of twenty subsystem oscillators with damping ratios $\zeta = 0.10$, frequency spacing $\beta = 0.04$, frequency offset $\beta_0 = 0$, and nominal mass ratio $\mu = 0.0025$. By simple substitution of the parameter values into Eqs. 22 and 26, it can be seen that the equivalent damping for the base example system is $\zeta_e = 0.049$ and the normalized response is $\sigma_y/\sigma_{y0} = 0.617$.

First, the effect of mass variation on the root-mean-square response is examined. The actual total mass of the mounted subsystems is given by the sum

$$\mu_{\text{total}} \equiv \frac{1}{m_0}\sum_{j=1}^{n} m_j = \sum_{j=1}^{n} \mu_j = \mu \sum_{j=1}^{n}\left(\frac{\omega_0}{\omega_j}\right)^4 \tag{27}$$

which, for the base system, is $\mu_{total} = 0.080$. It can be seen from Eq. 22 that the additional damping ζ_e is proportional to μ_{total}. In particular, if $\mu_{\text{total}} = 0$ (i.e., there are no mounted subsystems), then $\zeta_e = 0$ and the normalized response is unity, as expected. As μ_{total} increases, the influence of the mounted subsystems increases, resulting in a proportional increase in ζ_e and a corresponding decrease in the normalized response according to Eq. 26. This is illustrated in Fig. 4a, where the normalized response σ_y/σ_{y0} is plotted with respect to the normalized total mass, μ_{total}. The figure also show that the exact results, obtained by substituting Eq. 10 into Eq. 24 and performing repeated numerical integrations, is closely approximated by the simple results in Eqs. 22 and 26 for small mass ratios.

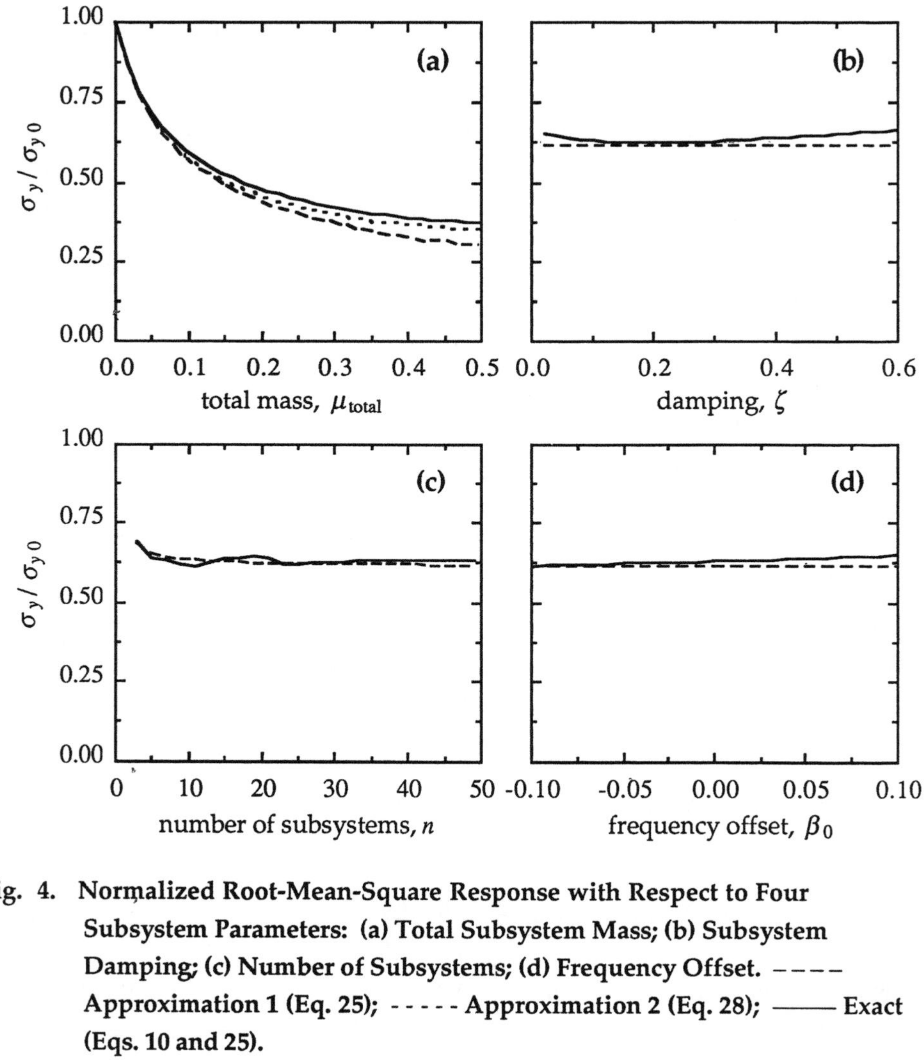

Fig. 4. Normalized Root-Mean-Square Response with Respect to Four Subsystem Parameters: (a) Total Subsystem Mass; (b) Subsystem Damping; (c) Number of Subsystems; (d) Frequency Offset. ---- Approximation 1 (Eq. 25); - - - - - Approximation 2 (Eq. 28); ——— Exact (Eqs. 10 and 25).

For large mass ratios, the equivalent damping ζ_e becomes of order unity (e.g., $\zeta_e = 0.3125$ when $\mu_{total} = 0.5$), thus the damping term in the numerator of Eq. 21 becomes significant. By including this term in the evaluation of the mean-square response, an improved approximation is obtained

$$\sigma_y^2 \approx \frac{\pi G_0 (1 + 4 \zeta_e^2)}{4 \hat{\omega}_0^3 (\hat{\zeta}_0 + \zeta_e)} \tag{28}$$

This expression, valid for large subsystem masses, differs from Eq. 25 only in the ζ_e^2 term in the numerator. The accuracy of this result for a wide range of subsystem masses is shown in Fig. 4a, where the normalized response using Eq. 28 is plotted by a dotted line. This demonstrates that the results of this paper are not restricted to light subsystems.

The effect of the damping ζ of each oscillator subsystem on the main system response is shown in Fig. 4b. As predicted by the results in this paper, the response is nearly independent of the damping ratio, even for damping ratios above 50%.

Next, the effect of the number of subsystems, n, and the frequency spacing, β, is examined. To preserve symmetry in the system, odd values of n are used. The nominal mass ratio and normalized frequency spacing are given by

$$\mu = \frac{0.05}{n} \quad \text{and} \quad \beta = \frac{0.80}{n-1} \tag{29}$$

so that the equivalent damping ζ_e is nearly constant and approximately equal to the value obtained for the base system except for small values of n. It can be seen that the frequency spacing becomes increasingly larger as n becomes smaller. This is illustrated in Fig. 4c, where the normalized response is plotted versus the number of subsystems, n.

Finally, the effect of the frequency offset β_0 on the response is investigated. The theory predicts that the response is independent of β_0, as indicated by the absence of this parameter in the approximate response expressions. This is illustrated in Fig. 4d, where the base system is examined for $-0.1 < \beta_0 < 0.1$.

CONCLUSIONS

The dynamic effect of a large number of subsystems on a single main oscillator is investigated. A special and fundamental class of configurations is examined in detail, and an analysis procedure is developed which uses contour integration and asymptotic approximations. The final results are remarkably simple and provide insight into the subsystem dynamic effect.

It is found that a set of oscillator subsystems with closely-spaced natural frequencies can be effectively replaced by a single viscous damper whose damping constant is proportional to the mass of each oscillator, inversely proportional to the spacing of the oscillator natural frequencies, and independent of the damping constants of the individual oscillators. The third property is physically counter-intuitive; however, its explanation is shown in clear mathematical terms.

Comparisons of the root-mean-square response calculations between the exact analysis, which involves lengthy numerical integration, and the simple approximate analysis shows good agreement even for large subsystem masses. However, the value in the theoretical results is not simply in the reduction in calculation time on a computer. Rather, it is in the insight gained in the dynamic behavior of the complicated system.

The theory developed in this paper is applicable to more general system configurations and other related dynamics problems. For instance, the equivalent damping concept can be used in optimal design of multiple tuned mass dampers as a passive vibration control device [17]. Furthermore, the analysis approach for subsystems with closely spaced natural frequencies can be directly applied to the study of periodic systems [19].

ACKNOWLEDGMENTS

The research work in this paper was partially supported by the National Science Foundation under Grant No. BCS-8858549 and by the Ohsaki Research Institute of Shimizu Corporation. This support is gratefully acknowledged.

APPENDIX III – REFERENCES

1. Hurty, W. C. and Rubinstein, M. E. (1964). *Dynamics of Structures,* Prentice-Hall, Inc., Englewood Cliffs, N. J.
2. Gladwell, G. M. L., and Bishop, R. E. D. (1960). "Interior Receptances of Beams." *Journal of Mechanical Engineering Science,* 2(1), 1–15.
3. Sackman, J. L. and Kelly, J. M. (1979). "Seismic Analysis of Internal Equipment and Components in Structures." *Engineering Structures,* 1(4), 179–190.
4. Igusa, T. and Der Kiureghian, A. (1985). "Dynamic Characterization of Two-Degree-of-Freedom Equipment-Structure Systems." *J. Engrg. Mech.,* ASCE, 111(1), 1–19.

5. Nicholson J. W. and Bergman L. A. (1985). "Vibration of a Class of Complex Discrete-Distributed Systems." *Proceedings of the 29th AIAA structures, Structural Dynamics and Materials Conference*, Williamsburg, VA, USA.
6. Dowell, E. H. (1979). "On Some General Properties of Combined Dynamical Systems." *J. Appl. Mech.*, ASME, 46(3), 206–209.
7. Igusa, T., Achenbach, J. D., and Min, K.-W. "Resonance Characteristics of Connected Subsystems: Theory and Simple Configurations." *J. Sound Vib.*, (accepted for publication).
8. Lyon R. H. (1975). *Statistical Energy Analysis of Dynamical Systems: Theory and Application*, MIT Press, Inc., Cambridge, Mass.
9. Dowell, E. H. and Kubota, Y. (1985). "Asymptotic Modal Analysis and Statistical Energy Analysis of Dynamical Systems." *J. Appl. Mech.*, ASME, 52(12), 949–957.
10. Pinnington, R. J. and White, R. G. (1981). "Power Flow Through Machine Isolations to Resonant and Non-resonant Beams". *J. Sound Vib.*, 75(1), 179–197.
11. Lin, Y. K. (1967). *Probabilistic Theory of Structural Dynamics*, Krieger Publishing, Inc., Huntington, New York.
12. Clough, R. W. and Penzien, J. (1975). *Dynamics of Structures*, McGraw-Hill, Inc., New York.
13. Igusa, T. and Der Kiureghian, A. (1983). *Dynamics Analysis of Multiply Tuned and Arbitrarily Supported Secondary Systems*, Report No. UCB/EERC-83/07, University of California, Berkeley.
14. Den Hartog J. P. (1956). *Mechanical Vibration*, McGraw-Hill, Inc., New York.
15. Crandall, S. H., and Mark, W. D. (1963). *Random Vibration in Mechanical Engineering*, Academic Press, Inc., New York.
16. Igusa, T. and Xu, K. (1991). "Vibration Reduction Characteristics of Distributed Tuned Mass Dampers." submitted to *J. Engrg. Mech.*, ASCE.
17. Igusa, T. and Xu, K. (1991). "Dynamic Characteristics of Coupled Multiple Subsystems with High Modal Density." *Proceedings of the Fourth International Conference on Recent Advances in Structural Dynamics*, Southampton, England.
18. Der Kiureghian, A. (1980). "Structural Response to Stationary Excitation." *J. Engrg. Mech.*, ASCE, 106(6), 1195–1213.
19. Igusa, T. and Tang, Y. (1991). "Mobilities of Periodic Structures in Terms of Asymptotic Modal Properties." *Proceedings of the 32nd AIAA structures, Structural Dynamics and Materials Conference*, Baltimore, Maryland (submitted paper).

On Almost Sure Sample Stability of Nonlinear Ito Differential Equations

Frank Kozin and Zhi Yu Zhang

Department of Electrical Engineering
Polytechnic University
333 Jay Street, New York, NY11202

Abstract

The classic stability studies for linear Ito differential equations were developed by Khasminskii in the 1960's. The main concept is to norm the solution and study the properties of the normed vector on the surface of the unit sphere. In the 1970's many ordinary second-order dynamical systems were generated to their exact stability regions by Kozin and his students. The recent numerical method due to Wedig is the most efficient way to determine the stability regions and Lyapunov exponents for the Ito one degree of freedom equations. There has not been in the past an exact study for nonlinear Ito equations. In this paper we shall show that there is a class of homogeneous non-linear oscillators that can be transformed on the unit sphere and the exact stability regions can be determined. Two simple examples will be presented.

1. Introduction

The classical transformation methods for the linear Ito equations can be found in [1]. Consider the system

$$dx_i(t)=\sum_{j=1}^{l} b_i^j x_j dt+\sum_{r=1}^{n}\sum_{j=1}^{l} \sigma_{ir}^j x_j dB_r(t), i=1,..,l \tag{1}$$

where $x=(x_1 \cdots x_l)^T$, $b=(b_i^j)$, $B=(B_1 \cdots B_n)^T$, $[\ ]^T$ denotes the transposition. B is a vector of independent Wiener components with $E\{B_r(t)\}=0$, $E\{[B_r(t)-B_r(s)]^2\}=|t-s|$. The classical generator for the system (1) can be written as

$$Lu=(bx\,,\text{grad}\,u)+\frac{1}{2}\sum_{i,j=1}^{l} a_{ij}(x)\frac{\partial^2 u}{\partial x_i \partial x_j} \tag{2}$$

where $a_{ij}(x)=\sum_{k,s=1}^{l}\sum_{r=1}^{n} \sigma_{ir}^k \sigma_{jr}^s x_k x_s$.

* This paper was presented in the Frank Kozin Memorial Session

Y. K. Lin · I. Elishakoff (Eds.)
Stochastic Structural Dynamics 1
New Theoretical Developments
2nd International Conference on Stochastic Structural Dynamics
May 9–11, 1990, Boca Raton, Florida

The classical method for linear systems is that one may make the change into the unit vector $\lambda = x/|x|$ and study the magnitude of the solution in logarithm form $\rho = \log|x|$, $|\cdot|$ indicates the Euclidean norm in R^l. Thus the stability properties of the solution process can be determined. The following form can be obtained

$$d\rho(t) = Q(\lambda)\,dt + \sum_{r=1}^{n} \big(\sigma(r)\lambda(t), \lambda(t)\big) dB_r(t) \tag{3}$$

where $Q(\lambda) = (b\lambda, \lambda) + \frac{1}{2}\sum_{i=1}^{l} a_{ii}(\lambda) - \sum_{\substack{i,j=1 \\ i\neq j}}^{l} a_{ij}(\lambda)\lambda_i\lambda_j.$

In the limit, we obtain

$$J = \lim_{t\to\infty} \frac{1}{t}\{\log|x(t)| - \log|x_0(0)|\} = \lim_{t\to\infty} \frac{1}{t}\int_0^t Q(\lambda(s))ds. \tag{4}$$

On the unit sphere (or circle), the $\lambda(t)$ is an ergodic process, and thus the stability of the solution process is determined by the value of the average $E\{Q(\lambda)\}$. Indeed if the average is positive the system is unstable and if the average is negative the system is stable. Thus, the stability curve is obtained by the expression $E\{Q(\lambda)\} = 0$. One of the early papers to present the stability regions for a large class of stochastic mechanical oscillators is [2].

The interesting fact is that we can do this same analysis for an unusual class of non-linear oscillators as we shall see in the next section.

2. Stability Study for a Certain Class of Non-linear Ito Differential Equations

We are interested in the non-linear equations of the form

$$dx = F(x(t))\,dt + G(x(t))\,dB(t) \tag{5}$$

where x and F are l-dimensional vectors, G is an l×n matrix and B is the n-vector of Brownian motions.

The two general properties of (5) are

$$\text{(a)} \lim_{|x|\to 0} F(x(t)) = 0, \lim_{|x|\to 0} G(x(t)) = [0] \tag{6a}$$

$$\text{(b)} \left(G(x)G^T(x)\alpha, \alpha\right) \geq m\,|x|^2|\alpha|^2 \tag{6b}$$

or arbitrary vector α and a certain scalar $m > 0$.

We now assume the simple homogeneous property

$$\text{(c) } k\,F(x) = F(kx),\ k\,G(x) = G(kx) \tag{7}$$

holds for any k>0. The interesting surprise is that Eqn. (7) allow us to determine the same functions as Eqn. (3) and (4) for the linear systems. The exact function $Q(\lambda)$ can be written as

$$Q(\lambda) = \lambda^T F(\lambda) + \frac{1}{2}\sum_{i=1}^{n} (G(\lambda)G(\lambda)^T)_{ii}\left(\sum_{j=1}^{n} \lambda_j^2 - 2\lambda_i^2\right) - \sum_{\substack{i,j=1\\ i\neq j}}^{n} (G(\lambda)G(\lambda)^T)_{ij}\lambda_i\lambda_j$$

The exact stability region is determined simply by $E\{Q(\lambda)\} = 0$. What is interesting is that one can now look at various homogeneous oscillators that could possess unusual non-linear functions of $(x,\dot{x})$. Furthermore, the numerical procedure $\int_0^{2\pi} Q(\varphi)p(\varphi)d\varphi = E\{Q(\varphi)\}$ on the unit circle is obtained by Wedig's method for computing $p(\varphi)$ [3].

In order to show this method for a second-order system, let us consider the equation

$$dx_1 = x_2 dt,\ dx_2 = -f(x_1,x_2)dt - g(x_1,x_2)dB. \tag{9a}$$

We recall from Section 1 that

$$d\begin{pmatrix} x_1 \\ x_2 \end{pmatrix} = \begin{pmatrix} x_2 \\ -f(x_1,x_2) \end{pmatrix} dt + \begin{pmatrix} 0 & 0 \\ 0 & -g(x_1,x_2) \end{pmatrix} \begin{pmatrix} 0 \\ dB \end{pmatrix} \tag{9b}$$

Further, we know that $\lambda = (\lambda_1,\lambda_2)$ where $\lambda_1 = x_1/\sqrt{x_1^2+x_2^2}$, $\lambda_2 = x_2/\sqrt{x_1^2+x_2^2}$ and $\rho(x_1(t),x_2(t)) = \log\sqrt{x_1^2(t)+x_2^2(t)}$.

The backward operator L must be applied on $\rho(x_1,x_2)$ to determine the function $Q(\lambda_1,\lambda_2)$ to generate the almost sure stability region. From (9a), the operator, L, is easily obtained as,

$$L = x_2\frac{\partial}{\partial x_1} - f(x_1,x_2)\frac{\partial}{\partial x_2} + \frac{1}{2}\sigma^2 g^2 \frac{\partial^2}{\partial x_2^2}. \tag{10}$$

We obtain in an exact fashion yielding

$$L\log\sqrt{x_1^2+x_2^2} = \frac{x_1x_2}{x_1^2+x_2^2} - \frac{f(x_1,x_2)x_2}{x_1^2+x_2^2} + \frac{\sigma^2}{2}g^2(x_1,x_2)\cdot\frac{x_1^2-x_2^2}{[x_1^2+x_2^2]^2}. \tag{11}$$

Now recalling from (7), we find that

$$\frac{f(x_1,x_2)}{\sqrt{x_1^2+x_2^2}} = f\left(\frac{x_1}{\sqrt{x_1^2+x_2^2}}, \frac{x_1}{\sqrt{x_1^2+x_2^2}}\right) = f(\lambda_1,\lambda_2) \tag{12a}$$

$$\frac{g^2(x_1,x_2)}{x_1^2+x_2^2} = g^2\left(\frac{x_1}{\sqrt{x_1^2+x_2^2}}, \frac{x_1}{\sqrt{x_1^2+x_2^2}}\right) = g^2(\lambda_1,\lambda_2) \tag{12b}$$

Thus, Eqn.(11) becomes, exactly

$$Q(\lambda) = L\rho = \lambda_1\lambda_2 - f(\lambda_1,\lambda_2)\lambda_2 + \frac{1}{2}\sigma^2 g^2(\lambda_1,\lambda_2)(\lambda_1^2 - \lambda_2^2) \tag{13}$$

which is a simple direct form for a non-linear oscillating system.

We will now show two examples.

3. Examples

Example 1. Consider the homogeneous oscillator,

$$dx_1 = x_2\, dt\,, \quad dx_2 = -(x_1+2\zeta x_2)\, dt - \sigma\sqrt{x_1^2+x_2^2}\, dB(t). \tag{14}$$

Since $\lambda = (\cos\varphi, \sin\varphi)$ on the unit circle we find that the φ-equaton becomes

$$d\varphi(t) = -\left[1 + (2\zeta+\sigma^2)\sin\varphi\cos\varphi\right] dt - \sigma\cos\varphi\, dB(t). \tag{15}$$

The generator of L of the φ-process becomes

$$L = -\left[1 + (2\zeta + \sigma^2)\sin\varphi\cos\varphi\right]\frac{d}{d\varphi} + \frac{1}{2}\sigma^2\cos^2\varphi\frac{d^2}{d\varphi^2} = \Phi(\varphi)\frac{d}{d\varphi} + \frac{1}{2}\Psi^2(\varphi)\frac{d^2}{d\varphi^2}. \tag{16}$$

On the unit circle, the singular points that are left shunts [2] occur only at $\varphi = \pm\pi/2$. It is easy to show that the $Q(\varphi)$ function as from (8) is

$$Q(\varphi) = -\frac{1}{2}\sigma^2 - (2\zeta+\sigma^2)\sin^2\varphi. \tag{17}$$

To determine the stability region of the form

$$\int_0^{2\pi} Q(\varphi)P(\varphi)\, d\varphi = 2\int_{-\frac{\pi}{2}}^{\frac{\pi}{2}} Q(\varphi)P(\varphi)\, d\varphi$$

the method of [3] is used to write the Fokker-Planck equation

$$\frac{1}{2}\frac{d^2}{d\varphi^2}\left[\Psi^2(\varphi)P(\varphi)\right] - \frac{d}{d\varphi}\left[\Phi(\varphi)P(\varphi)\right] = 0$$

into the first-order equation

$$\frac{1}{2}\Psi^2(\varphi)\frac{dp(\varphi)}{d\varphi} + \left(\Psi(\varphi)\frac{d\Psi(\varphi)}{d\varphi} - \Phi(\varphi)\right)p(\varphi) = c \tag{18}$$

where c is the integral constant.

The sequential fraction to obtain the discrete values of p_n, which is a backward difference form [3], can be written as

$$p_n = \frac{2\Delta\varphi p_1 + \Psi_n^2 p_n}{\Psi_n^2 + 2\Delta\varphi(\Psi_n\Psi'_n - \Phi_n)}, \quad n=1,2,\cdots,N \tag{19}$$

where $\Delta\varphi$ is a small increment in the unit circle angle φ, and Ψ'_n is the value of the derivative of Ψ at the nth φ-value. Figure 1 shows us the probability function for values of σ^2, and $\zeta=0.1$. Figure 2 yields the curve for the exact stability region based upon the function $J = E\{Q(\lambda)\}$.

Example 2. Let us now consider another unusual non-linear oscillator, which is homogeneous,

$$dx_1 = x_2\,dt, \quad dx_2 = -(x_1+2\zeta x_2)\,dt - \sigma x_1^{4/5}\sqrt[5]{x_1 - x_2}\,dB(t). \tag{20}$$

The transformation onto the unit circle yields the φ-equation,

$$\begin{aligned} d\varphi = &\left[-1 - 2\zeta\cos\varphi\sin\varphi - \sigma^2\cos^{13/5}\varphi\sin\varphi(\mathrm{con}\varphi - \sin\varphi)^{2/5}\right]dt \\ &+\sigma\cos^{9/5}\varphi(\mathrm{con}\varphi - \sin\varphi)^{1/5}dB(t). \end{aligned} \tag{21}$$

The generator L for the φ-process is given as

$$\begin{aligned} L = &\left[-1 - 2\zeta\cos\varphi\sin\varphi - \sigma^2\cos^{13/5}\varphi\sin\varphi(\mathrm{con}\varphi - \sin\varphi)^{2/5}\right]\frac{d}{d\varphi} \\ &+\frac{1}{2}\sigma^2\cos^{18/5}\varphi(\mathrm{con}\varphi - \sin\varphi)^{2/5}\frac{d^2}{d\varphi^2}. \end{aligned} \tag{22}$$

Recall, $L = \Phi(\varphi)\,(d/d\varphi) + \frac{1}{2}\Psi^2(\varphi)(d^2/d\varphi^2)$. For this nonlinear oscillator, we find the singularity, $\Psi^2(\varphi) = 0$, occurs at $\varphi_{1,2} = \pm\pi/2$, and $\varphi_{3,4} = \pi/4, 5\pi/4$. The points $\varphi_{1,2}$ are usual left shunts and $\varphi_{3,4}$ are regular shunts and regular boundaries.

We evaluate the stability region, as above for

$$J = E\{Q(\varphi)\} = \int_0^{2\pi} Q(\varphi)p(\varphi)d\varphi$$

where

$$Q(\varphi) = \left[- 2\zeta\sin^2\varphi + \frac{\sigma^2}{2}\cos^{8/5}\varphi(\cos\varphi + \sin\varphi)(\cos\varphi - \sin\varphi)^{7/5}\right]. \tag{23}$$

Upon solving the as in (19a) above, we find the exact stability region for J<0. Figure 3 shows $p(\varphi)$ for $\zeta=1$, $\sigma^2=4$ which contains unusual discontinuity. Figure 4 yields the exact stability region.

4. Conclusions

We find this interesting accurate method for homogeneous non-linear oscillator, can allow us to determine that stability regions. A very large class of oscillators can be studied, for interesting applications. A whole new set of the criteria for the classification of singular boundaries have been developed for these engineering applications [4].

References

[1]. R.Z. Khasminskii, "Necessary and Sufficient Conditions for the Asymptotic Stability of Linear Stochastic Systems," Th. Prob. Appl. 1, pp. 144-147, 1967.

[2] R.R. Mitchell and F. Kozin, "Sample Stability of Second Order Linear Differential Equations with Wide Band Noise Coefficients," SIAM J. Appl. Math. Vol. 22, No.4, Dec. 1974.

[3] W. Wedig, "Pitchfork and bifurcations in stochastic systems - Effective Methods to Calculate Lyapunov exponents," in: P. Kree and W. Wedig (Eds.), "Effective Stochastic Analysis," Springer-Verlag, Berlin, 1989.

[4] Z.Y. Zhang, "New Developments in Almost Sure Sample Stability of Nonlinear Stochastic Dynamic Systems", Ph.D. Thesis, Dept. of Electrical Engineering, Polytechnic University, 1990.

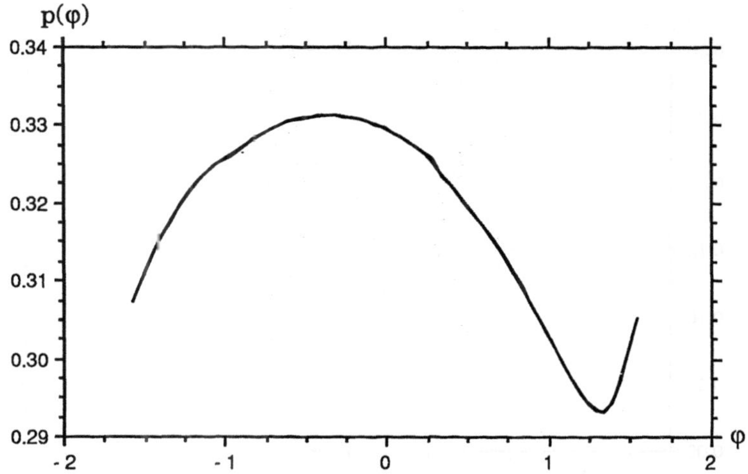

Fig. 1. Plot of p.d.f. of zeta=0.1 and sigsq=4.

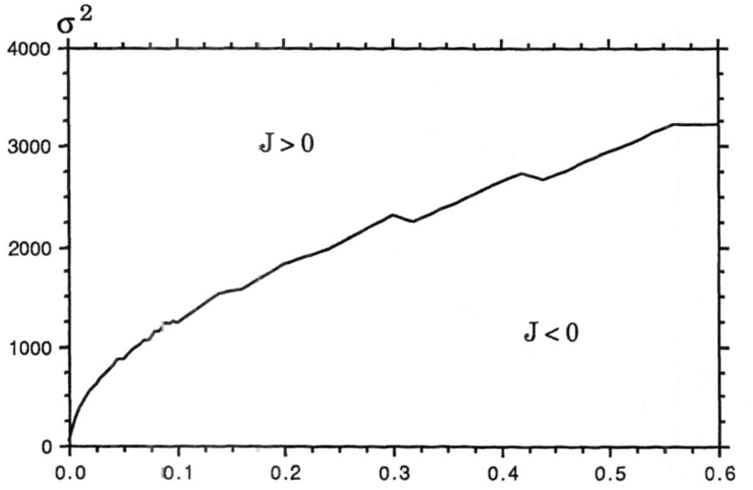

Fig. 2. Plot of the stable region of Example 1.

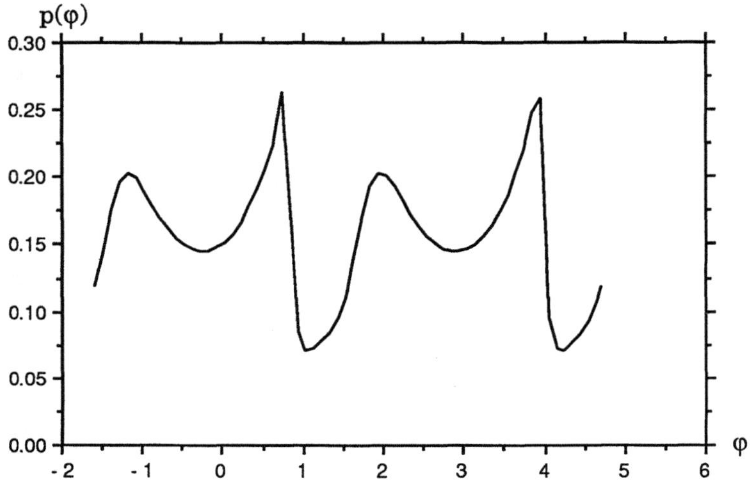

Fig. 3. Plot of p.d.f. for zeta=1 and sigsq=4.

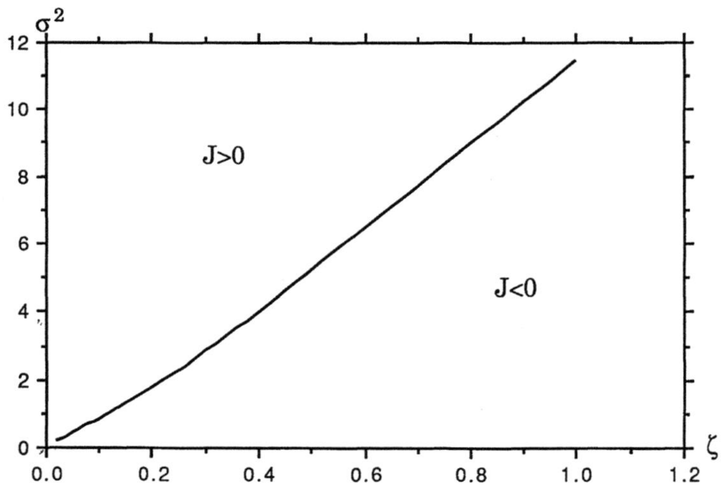

Fig. 4. Plot of the stable region of Example 2.

Padé-Type Approach to Nonlinear Random Vibration Analysis

R. Valéry Roy

Department of Mechanical Engineering, University of Delaware

Newark, Delaware 19716

Pol D. Spanos

L. B. Ryon Endowed Chair in Engineering, Rice University

P.O. Box 1892, Houston, Texas 77251

Summary

A new approach is proposed for the prediction of the response statistics of nonlinear systems under additive and multiplicative white noise excitation. The prime concept of this approach is to use ordinary perturbation expansion techniques, not for representing the random system response itself, but its statistical moments. Specifically, it is shown that the infinite hierarchy of stationary moments equations may be solved in a closed form by expressing each unknown moment in the form of a perturbation expansion in powers of the parameter quantifying the system nonlinearity. Then, by recasting the series solutions by means of Padé-type transformations, quite reliable approximations are found for even strongly nonlinear systems. The method is applied on various nonlinear systems. The derived results are validated by comparison with pertinent Monte Carlo simulation data.

Introduction

In view of the fact that Fokker-Planck-Kolmogorov (FPK) equations are rarely amenable to exact solutions, and that approximate solutions are difficult to obtain, considerable effort has been devoted in stochastic dynamics to the determination of response statistical functions such as means, mean-squares or higher-order moments. It was recognized long ago (see Caughey and Dienes,[1] 1962) that with the knowledge of the FPK equation corresponding to a given nonlinear stochastic system, a sequence of coupled differential equations (or algebraic equations, in the stationary case) governing the response statistical moments can be readily generated. However, nonlinearities in the system lead to an infinite hierarchy of equations. Indeed, any equation derived for moments of a given order involves moments of higher order. A number of closure schemes have therefore been devised in order to truncate these infinite hierarchies to finite sets of equations. A class of such schemes consists of approximating higher-order moments by lower-order ones, such as in Gaussian closure, cumulant-neglect closure (see Wu and Lin,[2] 1984, or Ibrahim *et al.*,[3] 1985) or quasi-moment-neglect closure (see Bover,[4] 1978). Another class of approximations consists of assuming a non-Gaussian response probability density function

Y. K. Lin · I. Elishakoff (Eds.)
Stochastic Structural Dynamics 1
New Theoretical Developments
2nd International Conference on Stochastic Structural Dynamics
May 9–11, 1990, Boca Raton, Florida

(p.d.f.), usually in the form of an asymptotic expansion whose unknown parameters are determined from an appropriate set of moments equations. Clearly, these approximations have greatly contributed to understanding nonlinear stochastic dynamics. However, they seem to lack a rigorous mathematical basis. For example, in a recent study, Sun and Hsu[5] (1987) have examined the validity of cumulant-neglect closure for a single-degree-of-freedom (s.d.o.f.) oscillator whose exact stationary p.d.f. is available. They have shown that the resulting approximations can lead to erroneous predictions in certain ranges of the system's parameters. Crandall[6] (1980) also pointed out possible problems in the use of Gram-Charlier expansions in closure schemes.

Formulation

In the present article, a new approach is proposed for addressing the "infinite hierarchy problem" in connection with the determination of stationary response moments. Consider the following nonlinear multi-degree-of-freedom (m.d.o.f.) system with white Gaussian noise excitation $\xi(t)$

$$\ddot{\mathbf{X}} + \mathbf{C}\dot{\mathbf{X}} + \mathbf{K}\mathbf{X} + \bar{\epsilon}\,\mathbf{f}(\mathbf{X},\dot{\mathbf{X}}) = \left(\mathbf{G}_0 + \bar{\epsilon}\,\mathbf{G}_1(\mathbf{X},\dot{\mathbf{X}})\right)\xi(t) \quad . \tag{1}$$

In this equation, a dot denotes time differentiation, $\mathbf{X}$ is the response vector, $\mathbf{f}(\mathbf{X},\dot{\mathbf{X}})$ and $\mathbf{G}_1(\mathbf{X},\dot{\mathbf{X}})$ are the system additive and multiplicative nonlinearities respectively, and $\mathbf{C}$, $\mathbf{K}$, $\mathbf{G}_0$ are constant matrices. Note that the adopted format for equation (1) is dictated by the intention to pursue a perturbation solution about parameter $\bar{\epsilon} = 0$. The method of perturbation was one of the very first techniques proposed for the determination of solution process $\mathbf{X}(t)$ (see Crandall[7] , 1963). Yet continued interest in this technique has been severely limited for two reasons. First, only a few terms (at most up to $\bar{\epsilon}^2$) of the expansion $\mathbf{X} = \mathbf{X}_0 + \mathbf{X}_1\bar{\epsilon} + \mathbf{X}_2\bar{\epsilon}^2 + \cdots$ can be realistically calculated, due to increasing algebraic difficulty. Second, such series can be divergent for all values of $\bar{\epsilon}$, even though they seem to possess asymptotic properties. Note that the determination of statistics, such as moments or correlation functions, from the obtained solution involves tedious operations.

However, perturbation solutions can be found in a straightforward manner for the response stationary moments. Denote by $\mathbf{m}^{[p]}$ the vector defined by the listing of all joint moments of order $p_1 + p_2 + \cdots + p_n = p$

$$m_{p_1,p_2,\ldots,p_n} = \langle Y_1^{p_1} Y_2^{p_2} \cdots Y_n^{p_n} \rangle \tag{2}$$

of the state response vector $\mathbf{Y} = (\mathbf{X},\dot{\mathbf{X}})^T$ of dimension n. Then, with the use of the FPK equation associated with the system or with Itô stochastic differential calculus, after first converting equation (1) to a stochastic differential equation in the Itô sense, an infinite sequence of linear equations between the unknowns $\mathbf{m}^{[p]}$ can be written in the following form:

$$(\mathbf{A}_0 + \bar{\epsilon}\,\mathbf{A}_1)\begin{bmatrix} \mathbf{m}^{[0]} \\ \mathbf{m}^{[1]} \\ \vdots \\ \mathbf{m}^{[p]} \\ \vdots \end{bmatrix} = \mathbf{0}. \tag{3}$$

If the system's nonlinearities are of the polynomial kind, it is possible to write the infinite hierarchy of moments in the form of equation (3). In equation (3), $\mathbf{A}_0$ is a matrix of infinite size, whose elements are computed from the linear part of the system, that is, $\bar{\epsilon} = 0$ in equation (1). Further, matrix $\mathbf{A}_1$ is computed from the nonlinear terms of the system, and causes the equations of system (3) to be coupled. A solution of system (3) is sought by extracting from it the unknowns in the form of expansions in successive powers of $\bar{\epsilon}$. That is, for $p = 0, 1, 2, \ldots$

$$\mathbf{m}^{[p]} = \mathbf{m}_0^{[p]} + \mathbf{m}_1^{[p]}\,\bar{\epsilon} + \cdots + \mathbf{m}_k^{[p]}\,\bar{\epsilon}^k + \cdots. \tag{4}$$

This will have the effect of uncoupling the infinite sequence of equations. Starting from the value $\mathbf{m}^{[0]} = 1$, the coefficients of expansion (4) can be obtained successively by simple recursion relations. This will be demonstrated in the following sections. Note that these coefficients are not the result of an approximation, but rather are exact. Yet perturbation expansions are useless as they stand in equation (4). Nevertheless, it has been known for quite some time to mathematicians that the information contained in these coefficients may be used successfully in a variety of ways to obtain meaningful quantitative or qualitative results. In fact, the use of series transformations, such as Padé approximants, the Borel summation process, or the u-transform, lead to various sequences of approximations, named herein Padé-type solutions. The definition of these transforms is briefly discussed in the Appendix of this paper. In this context, it is noted that in the last thirty years an extensive amount of literature has appeared on the mathematical properties, the numerical computation, and the applications of Padé approximants, and other related methods. For references see Baker and Graves-Morris[8] (1981), and Brezinski[9] (1985).

In the following sections, Padé-type solutions are presented for the stationary response moments of a variety of nonlinear oscillators. First, the method is used for a system with both additive and multiplicative excitations. Then, the classical Duffing's oscillator is studied: the first few approximants are given in a closed form, and the results are compared with those resulting from a Galerkin-type solution of the corresponding FPK solution. Finally, results are shown for the Van der Pol oscillator and a system with parametric excitation.

A Scalar Nonlinear System with Additive and Multiplicative Excitations

Consider the scalar nonlinear system governed by the following differential equation:

$$\dot{X}(t) + (1 + \xi_2(t))\,X(t) + \epsilon X^3(t) = \xi_1(t), \tag{5}$$

where the equation is meant in the sense of Stratonovich. The symbols ξ_1 and ξ_2 denote two independent white Gaussian noises with autocorrelations

$$\langle \xi_1(t)\xi_1(t+\tau)\rangle = 2D\delta_0(\tau)\,, \quad \langle \xi_2(t)\xi_2(t+\tau)\rangle = 2\rho\delta_0(\tau)\,, \tag{6}$$

where $\delta_0(\tau)$ represents the Dirac delta function. Equation (5) can be transformed into the following Itô stochastic differential equation:

$$dX(t) = -\left((1-\rho)X + \epsilon X^3\right)dt + (2D + 2\rho X^2)^{1/2} dW(t), \tag{7}$$

in which a drift correction term $-\rho X$ has been included. The stationary p.d.f. of solution process $X(t)$ exists and can be written as

$$p(x) = N(D + \rho x^2)^{\alpha} \exp(-\frac{\epsilon}{2\rho}x^2)\ , \tag{8}$$

where $\alpha = -1/2 - 1/2\rho + \epsilon D/2\rho^2$, and N is a corresponding normalization constant. This exact result is used to assess the reliability of the ensuing approximations. Application of Itô stochastic differential calculus yields the following moments equations:

$$\dot{m}^{[p]} = p(p\rho - 1)m^{[p]} - p\epsilon m^{[p+2]} + p(p-1)Dm^{[p-2]} \tag{9}$$

for $p = 1, 2, \ldots$. These equations are solved by perturbation about the linear solution of (3), that is, about $\epsilon = \rho = 0$. Thus, after introducing a parameter $\bar{\epsilon}$, as in equation (1), the stationary response moments of the system

$$\dot{X}(t) + X(t) + \bar{\epsilon}X(t)\left(\xi_2(t) + \epsilon X^2(t)\right) = \xi_1(t), \tag{10}$$

are sought in form

$$m^{[p]} = m_0^{[p]} + m_1^{[p]}\,\bar{\epsilon} + \cdots + m_k^{[p]}\,\bar{\epsilon}^k + \cdots\ , \tag{11}$$

where the solution of (5) is recovered for $\bar{\epsilon} = 1$. Upon substituting equation (11) in (9), the following recursive relations are obtained:

$$\begin{aligned}
&m_0^{[0]} = 1 \\
&for\ p \geq 1\,, && m_0^{[p]} = (p-1)Dm_0^{[p-2]} \\
&for\ k \geq 1\,,\ p \geq 1\,, && m_k^{[0]} = 0 \\
& && m_k^{[p]} = (p-1)Dm_k^{[p-2]} - \epsilon m_{k-1}^{[p+2]} + p\rho m_{k-1}^{[p]}\ .
\end{aligned} \tag{12}$$

Thus, any finite set $\{m_k^{[p]}\}_{p,k\geq 0}$ of moment perturbation coefficients can be determined exactly by simple recursions. In particular, these equations lead to the following coefficients for $m^{[2]} = \langle X^2 \rangle$:

$$\begin{aligned} m_0^{[2]} &= D \\ m_1^{[2]} &= D(2\rho - 3\epsilon D) \\ m_2^{[2]} &= D(4\rho^2 - 24\epsilon\rho D + 24\epsilon^2 D^2) \\ &\vdots \end{aligned} \tag{13}$$

Results are restricted herein to the determination of $\langle X^2 \rangle$ for the values $\epsilon = D = \rho = 1.0$. Table 1 shows that meaningless results are obtained by direct summation of the perturbation coefficients as given in (13). This will generally be true of all perturbation solutions, unless the system in question is very slightly nonlinear, in which case direct summation up to a given order may give reliable results. This is characteristic of asymptotic series with zero radius of convergence. In general, the perturbation coefficients tend to grow as fast as $(n!)$, as their order n is increased. Thus, the next necessary step of the proposed method consists in transforming the perturbation expansions into Padé-type approximations. See Appendix for more details. First, ordinary Padé approximants are used, yielding rational expressions in the variable $\bar{\epsilon}$. It is seen in Table 1 that convergence is achieved, yet at a rate not as fast as desired. Indeed, it is critical to use as few coefficients as possible, since the latter will be more expensive to determine for systems of higher dimension than equation (5). Furthermore, the stronger the nonlinearities in the system, the more coefficients will be needed. This higher rate of convergence is achieved by Padé-Borel approximants and by the u-transform. The former are especially suitable since they take advantage of the asymptotic properties of the series, and they seem to make better use of their coefficients. The success of the u-transform is attributed to the alternating character in the signs of the coefficients. For irregular sign patterns, the u-transform is known to fail. Clearly, the relative errors shown in Table 1 are smaller than those expected for practical purposes. Further, by the addition of higher-order coefficients, higher-order approximations can be obtained without difficulty. In this manner, the reliability of the method can be easily assessed. This feature of the proposed approach makes it more advantageous than closure schemes, whose order of approximation has been found laborious to increase. In this regard it is noted that the determination of the perturbation coefficients, even for systems of higher dimension, requires only linear equations to be solved consecutively. Closure schemes invariably involve nonlinear equations between the unknowns, thus often leading to potential numerical complications, such as multiple solutions.

n	% mean-square error			DIRECT
	PADE	PADE-BOREL	U-TRANSFORM	SUMMATION
2	30%	16.6%	6.8%	0
3	11.8%	2.1%	14%	4
4	11.0%	0.5%	1.4%	-21
5	4.8%	1.9%	0.08%	187
6	4.7%	0.6%	0.03%	-1959
7	2.3%	0.09%	0.04%	24409
8	2.3%	0.04%	0.007%	-351324
9	1.2%	0.002%	0.004%	5740708

Table 1: Relative error for $\langle X^2 \rangle$ obtained by 3 summation techniques versus the number n of coefficients used in perturbation expansion; also shown is $\langle X^2 \rangle$ obtained by direct summation of perturbation coefficients ($\epsilon = D = \rho = 1.0$).

The Duffing Oscillator

The Duffing oscillator, with linear damping and cubic restoring force, has been extensively studied for its capacity to model many physical systems. When excited by a Gaussian white noise, the corresponding stationary FPK equation can be solved exactly, and its solution can be used for comparison with the results of the proposed approach. In this context consider the system

$$\ddot{X}(t) + 2\eta\dot{X}(t) + X(t) + \epsilon\, X^3(t) = \xi(t) \ , \tag{14}$$

where $\xi(t)$ is a white Gaussian noise with autocorrelation $\langle \xi(t)\xi(t+\tau)\rangle = 2D\delta_0(\tau)$. Setting $m_{i,j} = \langle X^i \dot{X}^j \rangle$, the stationary moments equations

$$i m_{i-1,j+1} - 2\eta j m_{i,j} - j m_{i+1,j-1} - \epsilon\, j m_{i+3,j-1} + Dj(j-1) m_{i,j-2} = 0 \tag{15}$$

are readily derived by using the Itô stochastic differential formula. These equations can be regrouped for each order $p = i + j$ ($p \geq 1$) in the vector equation

$$\mathbf{A}_p \mathbf{m}^{[p]} + \epsilon\, \mathbf{B}_p \mathbf{m}^{[p+2]} + \mathbf{C}_p \mathbf{m}^{[p-2]} = \mathbf{0} \ , \tag{16}$$

where $\mathbf{m}^{[p]} = (\langle X^p \rangle, \langle X^{p-1}\dot{X}\rangle, \ldots, \langle \dot{X}^p \rangle)^T$. See Roy and Spanos[10] (1989) for the details of matrices $\mathbf{A}_p$, $\mathbf{B}_p$, and $\mathbf{C}_p$. The first step of the proposed technique is to express each unknown $\mathbf{m}^{[p]}$ in perturbation series of the parameter ϵ. After identifying terms of equal powers of ϵ in equation (16), one finds the following recursive relations:

$$\begin{array}{ll} \mathbf{m}_0^{[0]} = 1 & \\ for\ p \geq 1\ , & \mathbf{A}_p \mathbf{m}_0^{[p]} = -\mathbf{C}_p \mathbf{m}_0^{[p-2]} \\ for\ k \geq 1\ ,\ p \geq 1\ , & \mathbf{m}_k^{[0]} = 0 \\ & \mathbf{A}_p \mathbf{m}_k^{[p]} = -\mathbf{C}_p \mathbf{m}_k^{[p-2]} - \mathbf{B}_p \mathbf{m}_{k-1}^{[p+2]}\ . \end{array} \tag{17}$$

It is seen from equation (17) that the determination of the moments perturbation coefficients is essentially similar to that of the system previously studied. At each step of the recursion, a sparse linear system must be solved. Note that the perturbation coefficients must be determined with sufficient numerical precision. Indeed, accuracy in the coefficients is essential for the numerical computation of sequence or series transformations (see Baker and Graves-Morris,[8] 1981). As expected, all velocity moments $m_{0,i}$ are determined by their first perturbation coefficient. Furthermore, all odd response moments $\mathrm{m}^{[2p+1]}$ are zero. With the help of the symbolic manipulation program Maple,[11] the following expansions can be determined from algorithm (17):

$$\begin{aligned} \langle X^2\rangle/\langle X^2\rangle_0 &= 1 - 3\bar{\epsilon} + 24\bar{\epsilon}^2 - 297\bar{\epsilon}^3 + 4896\bar{\epsilon}^4 - \cdots \\ \langle X\dot{X}\rangle &= 0 \\ \langle X^4\rangle/\langle X^4\rangle_0 &= 1 - 8\bar{\epsilon} + 99\bar{\epsilon}^2 - 1632\bar{\epsilon}^3 + 33426\bar{\epsilon}^4 - \cdots \quad , \end{aligned} \tag{18}$$

where $\langle X^2\rangle_0 = d/2\eta$ and $\langle X^4\rangle_0 = 3(d/2\eta)^2$ are the second and fourth moments for the associated linear system (that is, $\epsilon = 0$ in equation (14)). The parameter $\bar{\epsilon} = d\epsilon/2\eta$ measures the effective nonlinearity of the system.

First, examine the summation of normalized mean-square $\sigma_X^2 = \langle X^2\rangle/\langle X^2\rangle_0$, in terms of parameter $\bar{\epsilon}$ with the three transformations mentioned above. By using 5, 6, and 7 coefficients, the following Padé approximants can be found:

$$\begin{aligned} [2/2](\bar{\epsilon}) &= \frac{1 + 21\bar{\epsilon} + 45\bar{\epsilon}^2}{1 + 24\bar{\epsilon} + 93\bar{\epsilon}^2} \\ [2/3](\bar{\epsilon}) &= \frac{1 + 32\bar{\epsilon} + 177\bar{\epsilon}^2}{1 + 35\bar{\epsilon} + 258\bar{\epsilon}^2 + 231\bar{\epsilon}^3} \\ [3/3](\bar{\epsilon}) &= \frac{1 + 45\bar{\epsilon} + 450\bar{\epsilon}^2 + 585\bar{\epsilon}^3}{1 + 48\bar{\epsilon} + 570\bar{\epsilon}^2 + 1440\bar{\epsilon}^3} \quad . \end{aligned} \tag{19}$$

Similarly, Padé-Borel approximants take the form

$$\begin{aligned} f_B^{[1/2]}(\bar{\epsilon}) &= \frac{1}{\bar{\epsilon}}\int_0^{+\infty} \exp(-t/\bar{\epsilon})\frac{2 + 3t}{2 + 9t + 3t^2}dt \\ f_B^{[2/2]}(\bar{\epsilon}) &= \frac{1}{\bar{\epsilon}}\int_0^{+\infty} \exp(-t/\bar{\epsilon})\frac{2 + 2t - t^2}{2 + 8t - t^2}dt \\ f_B^{[2/3]}(\bar{\epsilon}) &= \frac{1}{\bar{\epsilon}}\int_0^{+\infty} \exp(-t/\bar{\epsilon})\frac{10 + 78t + 97t^2}{10 + 108t + 301t^2 + 102t^3}dt \quad , \end{aligned} \tag{20}$$

with the use of 4, 5, and 6 coefficients respectively. Finally, the following approximants resulting from the u-transform of 3 and 4 perturbation coefficients are given:

$$\begin{aligned} l_2(\bar{\epsilon}) &= \frac{1 + 13\bar{\epsilon}}{1 + 16\bar{\epsilon} + 24\bar{\epsilon}^2} \\ l_3(\bar{\epsilon}) &= \frac{32 + 795\bar{\epsilon} + 2847\bar{\epsilon}^2}{32 + 891\bar{\epsilon} + 4752\bar{\epsilon}^2 + 2376\bar{\epsilon}^3} \quad . \end{aligned} \tag{21}$$

Numerical evaluation of these groups of approximants are compiled in Table 2 for the value $\bar{\epsilon} = 1.0$ and in Table 3 for $\bar{\epsilon} = 10.0$. These results exhibit a trend similar to that of Table 1. All three methods are seen to converge, the rate of convergence being slower for $\bar{\epsilon} = 10.0$ than for $\bar{\epsilon} = 1.0$. The convergence of Padé approximants is seen to be substantially improved by the use of the Borel summation procedure. The u-transform seems to make the best use of the perturbation coefficients. Figure 1 shows the resulting approximations of the u-transform with the use of 5, 6, and 7 coefficients compared to the exact result. Results can also be given for response moments of higher-order: Figure 2 shows the results of Padé approximants and the u-transform on the perturbation expansion of the kurtosis number, defined as the ratio $\langle X^4\rangle/\langle X^2\rangle^2$. Also shown in Tables 2 and 3, are the results derived by applying a Galerkin solution technique to the Duffing oscillator's FPK equation. This solution method is obviously not related to Padé-type approximations. However, it yields very similar expressions for the response moments. By seeking an approximate solution to the stationary FPK equation in the form of a linear combination of trial functions, scaled Hermite functions, with unknown coefficients, the Galerkin technique leads to an infinite system of coupled linear equations between these coefficients; for more details, see Wen,[12] 1975. By direct truncation of this system, approximations of increasing order can be obtained for the response p.d.f., and thus for σ_X^2. The resulting normalized mean-squares are rational expressions of $\bar{\epsilon}$, as shown by the first 3 approximations:

$$\sigma_X^2 = \frac{1+3\bar{\epsilon}}{1+6\bar{\epsilon}} \tag{22}$$

$$\sigma_X^2 = \frac{1+15\bar{\epsilon}+15\bar{\epsilon}^2}{1+18\bar{\epsilon}+45\bar{\epsilon}^2} \tag{23}$$

$$\sigma_X^2 = \frac{1+33\bar{\epsilon}+210\bar{\epsilon}^2+105\bar{\epsilon}^3}{1+36\bar{\epsilon}+294\bar{\epsilon}^2+420\bar{\epsilon}^3} \ . \tag{24}$$

The Van der Pol Oscillator

The Van der Pol oscillator is governed by the following stochastic differential equation:

$$\ddot{X}(t) + 2\eta\left(-1+\epsilon\, X^2(t)\right)\dot{X}(t) + X(t) = \xi(t) \ , \tag{25}$$

where the excitation $\xi(t)$ is defined as in the case of the Duffing oscillator. Equation (25) is representative of many self-excited systems encountered in electrical, mechanical and biochemical oscillations, and its study has attracted considerable interest. The exact solution for the probability density of the stationary response of this system is not known. However, several attempts have been made to find approximate solutions. Here, the proposed method of Padé-type solutions is applied to determine the mean-square response

n	% error σ_X^2 , $\bar{\epsilon} = 1.0$			
	PADE	PADE-BOREL	U-TRANSFORM	GALERKIN
2	46%	17.5%	22%	22%
3	42%	7%	27%	3.5%
4	25%	1%	2.5%	0.7%
5	21%	*	1%	1.5%
6	14.5%	1%	0.4%	1%
7	12%	0.5%	0.2%	0.6%
8	9%	0.2%	0.01%	0.3%
9	7.5%	0.003%	0.05%	0.2%

Table 2: Relative error for σ_X^2 obtained by 3 summation techniques versus the number n of coefficients used in perturbation expansion; A '*' indicates that the corresponding approximant does not exist; also indicated is σ_X^2 obtained by a Galerkin solution of the FPK equation for $\bar{\epsilon} = 1.0$.

n	% error σ_X^2 , $\bar{\epsilon} = 10.0$			
	PADE	PADE-BOREL	U-TRANSFORM	GALERKIN
2	83%	48%	170%	170%
3	240%	64%	73%	87%
4	71%	15%	46%	49%
5	160%	*	18%	28%
6	63%	16%	1%	16%
7	120%	14%	5%	9%
8	56%	6.5%	2.5%	4.5%
9	98%	2.5%	0.5%	2%

Table 3: Relative error for σ_X^2 obtained by 3 summation techniques versus the number n of coefficients used in perturbation expansion; also indicated is σ_X^2 obtained by a Galerkin solution of the FPK equation for $\bar{\epsilon} = 10.0$.

of equation (25). The results are compared with several other approximations and with digital simulations.

A direct perturbation of equation (25) about $\epsilon = 0$ is not possible, since the corresponding linear system has no stationary solution. To make a perturbation meaningful, the system is replaced by the following equation:

$$\ddot{X}(t) + 2\alpha\eta\dot{X}(t) + X(t) + 2\eta\bar{\epsilon}\left(-1 - \alpha + \epsilon X^2(t)\right)\dot{X}(t) = \xi(t) \ , \tag{26}$$

where $\alpha > 0$. The solution of (25) is then recoved for $\bar{\epsilon} = 1$. The mean-square response of (26) can be found in power series of $\bar{\epsilon}$, and is subsequently summed by Padé-type transformations for $\bar{\epsilon} = 1$. The choice of α is made so that the linear system from which the perturbation is performed is close in some optimal way to the original system. The most convenient method for obtaining an initial choice for α is that of equivalent linearization, since it provides an equivalent damping coefficient

$$2\alpha\eta = 2\eta(-1 + \epsilon\langle X^2\rangle_{e.l.}) = \eta(-1 + \sqrt{1 + 2D\epsilon/\eta}) \ . \tag{27}$$

Even though the response of (25) is known to be highly non-Gaussiàn, equation (27) provides a good initial guess for the parameter α, except for $\epsilon \ll 1$. The moments equations can be easily written for system (26). Specifically, it is found

$$im_{i-1,j+1} - 2\alpha\eta j m_{i,j} - j m_{i+1,j-1} + Dj(j-1)m_{i,j-2} = 2\eta\bar{\epsilon}j\left((-1-\alpha)m_{i,j} + \epsilon m_{i+2,j}\right) \ . \tag{28}$$

Perturbation of these equations about $\bar{\epsilon} = 0$ is done as for the Duffing oscillator, and the coefficients are determined using an algorithm similar to the one described by equation (17). In particular, one can obtain for the second moment $m_{2,0}$ the following 2-term expansion:

$$\langle X^2\rangle = \frac{D}{2\alpha\eta} + \frac{D}{2\alpha\eta}(1 + \frac{1}{\alpha} - \frac{\epsilon D}{2\alpha^2\eta})\bar{\epsilon} + \cdots \ . \tag{29}$$

The summation of series (29) at $\bar{\epsilon} = 1.0$ is achieved by Padé-Borel approximants. The parameters ϵ and η are fixed, and the influence of the noise intensity D on $\langle X^2\rangle$ is examined. Using different approximate techniques, Caughey[13] (1959) and Piszczek[14] (1977) showed that the mean square amplitude, and thus $\langle X^2\rangle$, decrease with increasing noise intensity. However, Zhu and Yu[15] (1987) showed an opposite trend by using the method of stochastic averaging, and by subsequent digital simulations. The results obtained by Padé-Borel summation of 10 coefficients, and with values of α as given by equation (27), are compared in Table 4 and Figure 3 with those obtained by digital simulations for the values $\epsilon = 10.0$, $\eta = 0.1$, and for D varying between 0 and 2.0. In addition, the resulting approximations obtained by stochastic averaging (see Spanos,[16] 1980 and Zhu and Yu,[15] 1987) and by an approximate solution due to Ebeling *et al.*[17] (1986) are examined. The

D	$\langle X^2\rangle_{P.B.}$	$\langle X^2\rangle_{M.C.}$	$\langle X^2\rangle_{E.}$	$\langle X^2\rangle_{S.A.}$	$\langle X^2\rangle_{E.L.}$
0.0	0.2	0.2	0.2	0.2	0.1 (50%)
0.01	0.2078 (0.1%)	0.2080	0.2077 (0.2%)	0.2055 (1%)	0.137 (34%)
0.10	0.3615 (0.4%)	0.3600	0.3653 (2%)	0.3402 (6%)	0.279 (22%)
0.50	0.7344 (0.2%)	0.7325	0.8568 (17%)	0.6433 (12%)	0.552 (25%)
1.00	1.028 (0.3%)	1.031	1.454 (41%)	0.875 (15%)	0.759 (26%)
2.00	1.452 (0.1%)	1.454	2.793 (92%)	1.204 (17%)	1.051 (28%)

Table 4: The effect of the noise intensity D on the mean-square response of the Van der Pol oscillator obtained by Padé-Borel summation ($\langle X^2\rangle_{P.B.}$), Monte Carlo simulations ($\langle X^2\rangle_{M.C.}$), Ebeling *et al.* approximation ($\langle X^2\rangle_{E.}$), stochastic averaging ($\langle X^2\rangle_{S.A.}$), and equivalent linearization ($\langle X^2\rangle_{E.L.}$). $\epsilon = 10.0$, $\eta = 0.1$.

results of equivalent linearization, as given by equation (27), are also shown. First, it is seen that digital simulations lead to excellent agreement with the proposed method. When the value of D becomes too large, the approximation by Ebeling *et al.* breaks down. It is also seen that the relative error in the stochastic averaging approximation increases with the noise intensity, and that, as expected for Van der Pol oscillator, equivalent linearization leads to unacceptable results both for small, that is, in the deterministic limit, and for large values of D.

A Random Parametric Oscillator

The importance of parametric vibrations in many problems of engineering interest has been extensively discussed by Ibrahim[18] (1985). Parametric vibrations can have catastrophic consequences near the critical parameter regions of instability. Interest here is limited to the influence of nonlinearity on the mean square stability of a linear s.d.o.f. oscillator with both external excitation and parametric excitation in the stiffness term. In particular, the stationary response of the following nonlinear oscillator is examined:

$$\ddot{X}(t) + 2\eta\dot{X}(t) + \left(1 + \epsilon X^2(t) + \xi_\rho(t)\right) X(t) = \xi_D(t) \ . \tag{30}$$

In this equation, $\xi_\rho(t)$ and $\xi_D(t)$ are two independent white Gaussian noises with autocorrelations

$$\langle \xi_\rho(t)\xi_\rho(t+\tau)\rangle = 2\rho\delta_0(\tau)\,, \quad \langle \xi_D(t)\xi_D(t+\tau)\rangle = 2D\delta_0(\tau). \tag{31}$$

The question whether this system is to be interpreted in the Itô sense or the Stratonovich sense does not arise, since the Wong-Zakai correction terms happen to be zero for equation (30). For $\epsilon = 0$, the system is linear, and the second moments equations can be solved exactly to yield

$$\langle X^2\rangle = \langle \dot{X}^2\rangle = \frac{D}{2\eta - \rho} \, . \tag{32}$$

Thus, beyond the critical parametric noise intensity $\rho_{cr.}$, the system may not have bounded response samples. This phenomenon of instability is exhibited by moments of any order, when $\epsilon = 0$. In fact, the stability becomes more restrictive with increasing order of the moments.

For non-zero nonlinearity, the study of equation (30) is far more difficult. As yet, the corresponding stationary probability density of the response has not been found. Stability analysis of nonlinear parametric systems is usually limited to the asymptotic properties of the response moments which are solutions to the infinite hierarchy of equations

$$i m_{i-1,j+1} - 2\eta j m_{i,j} - j m_{i+1,j-1} - \epsilon\, j m_{i+3,j-1} + Dj(j-1) m_{i,j-2} + \rho j(j-1) m_{i+2,j-2} = 0\ . \quad (33)$$

The solution $m_{2,0} = \langle X^2 \rangle$ of this hierarchy of equations is sought in the form of Padé-type approximants, and is compared to the solution resulting from non-Gaussian closure accomplished by neglecting joint cumulants of order higher or equal to 6. A perturbation of these equations about $\epsilon = 0$ cannot be performed. It has been seen from the previous sections that the p-th moment perturbation coefficients $\mathbf{m}_k^{[p]}$ of order k are found recursively starting from the "linear" coefficients $\mathbf{m}_0^{[p]}$, which are the moments of the oscillator described by equation (30) for $\epsilon = 0$. However, depending on the values taken by the system parameters η, D, and ρ, the latter coefficients may not exist. Therefore, by perturbing the system about $\epsilon = 0$, there always exists an order beyond which the perturbation coefficients of $\langle X^2 \rangle$ are meaningless. An alternative approach is to perturb the system

$$\ddot{X}(t) + 2\eta \dot{X}(t) + X(t) + \bar{\epsilon}\left(\xi_\rho(t) X(t) + \epsilon X^3(t)\right) = \xi_D(t)\ , \quad (34)$$

about $\bar{\epsilon} = 0$. Then, by summing the resulting series for $\bar{\epsilon} = 1$, the solution of equation (30) is recovered. The following expansion for the mean-square response of (34) is, up to order $\bar{\epsilon}$,

$$\langle X^2 \rangle = \frac{D}{2\eta} - D/4 \frac{3\epsilon D - \rho}{\eta^2}\, \bar{\epsilon} + \cdots\ . \quad (35)$$

With the use of at most 10 coefficients from this expansion, the obtained series is transformed in Padé-Borel approximants. The parameters η and D are fixed to the value of 0.1. The parametric noise intensity ρ is varied from 0 to 1.0. The two cases examined involve small nonlinearity, $\epsilon = 0.1$, and large nonlinearity, $\epsilon = 1.0$. Digital simulations are also performed to assess the reliability of these approximations. The results for $\langle X^2 \rangle$ versus ρ are shown for both cases on Figure 4. The case of $\epsilon = 0$ is also shown. It is seen that, according to the Padé-type and simulation results, that the mean-square response is always physically realizable and that the system seems to be stable in the entire range $0 \leq \rho \leq 1$, for both $\epsilon = 0.1$ and $\epsilon = 1.0$. Both methods agree quite well. However, the conclusions derived from non-Gaussian closure are quite different. It is seen that, for $\epsilon = 0.1$, no physical mean-square response can be found in the approximate range $0.2 \leq \rho \leq 0.75$. In other ranges of ρ, the system would seem to be stable, yet multiple solutions coexist. This feature is not uncommon in non-Gaussian schemes, since the infinite linear system is replaced by a finite, yet nonlinear one. However, it is not exhibited by Monte Carlo simulations of the system response. For $\epsilon = 1.0$, results with similar characteristics are found. It seems that the replacement of the infinite hierarchy of equations by a system of finite dimension, as is done in closure schemes, can result in the loss of important qualitative properties of the original system. These properties seem to be properly reproduced by perturbation solutions followed by Padé-type transformations.

Concluding Remarks

A new approach has been proposed for treating nonlinear random vibration problems. The first step of the proposed approach involves a perturbation solution of the infinite hierarchy of the system response moments equations about a system nonlinearity parameter $\bar{\epsilon}$. It has been shown that it is crucial to perform this perturbation properly. That is, moments of all orders of the linear system corresponding to $\bar{\epsilon} = 0$ must exist. This might not be the case for a linear system with parametric excitations, in which case the perturbation coefficients may be meaningless. It has been seen that the method of equivalent linearization can be useful for determining a linear system about which a perturbation solution can be pursued. In this context, the higher-order perturbation coefficients represent correction terms to the equivalent linear system.

The second step of the proposed approach involves transforming the obtained perturbation series into Padé-type approximations. The use of various transformations has been demonstrated. The quality of the analytical approximations of the unknown statistical functions improves with the number of perturbation coefficients used. The technique has proven to be very reliable for a variety of nonlinear s.d.o.f. systems.

Finally, it is emphasized that the method is limited herein to first-order statistics of systems with polynomial nonlinearities. However, it appears that no serious difficulties will be encountered in extending its applicability to other important statistics, such as response correlation functions or spectra, and for systems with non-analytical nonlinearities.

APPENDIX: Padé-type Transformations

Padé approximants (P.A.). Briefly stated, a P.A. is a rational approximation of a function $f(z)$ such that its power series expansion matches the formal power series of $f(z)$ as far as possible. More explicitly, the $[L/M]_f$ P.A. associated with the formal power series $f(z) = \sum_{n=0}^{\infty} c_n z^n$ is the rational function

$$[L/M](z) = \frac{a_0 + a_1 z + \cdots + a_L z^L}{b_0 + b_1 z + \cdots + b_M z^M} , \tag{A.1}$$

such that

$$f(z) = [L/M]_f(z) + O(z^{L+M+1}) , \tag{A.2}$$

where '$O(z^n)$' means 'of order z^n as $z \to 0$'. Upon matching the coefficients of equal powers of z in (A.2) up to order $L+M$, the denominator and numerator coefficients of $[L/M]$ are found from the first $L+M+1$ coefficients of the power series of $f(z)$ by solving two linear systems. By varying the orders M and N, sequences of P.A. can be found, and are arranged in a table, called the Padé table. Note that the construction of P.A. is an ill-conditioned problem. Specifically, rounding errors in the coefficients c_n will be amplified in the P.A.'s numerator and denominator coefficients. Therefore, the higher the order of the Padé approximant to be constructed, the more precision is required (see Baker and Graves-Morris,[8] 1981). The numerical computations of P.A. can also be done by using a variety of algorithms which make use of identities between neighboring approximants in the Padé table. Further, it is worth mentioning the well known ϵ-algorithm due to Wynn,[19] (1956), which can be stated as follows:

$$\begin{aligned} \epsilon_{-1}^{(n)} = 0 \ , \ \epsilon_0^{(n)} = S_n \ , n \geq 0 \\ \epsilon_{k+1}^{(n)} = \epsilon_{k-1}^{(n+1)} + 1/\left(\epsilon_k^{(n+1)} - \epsilon_k^{(n)}\right) \ , \ n, \ k \geq 0 \ , \end{aligned} \tag{A.3}$$

where it is seen that the generated quantities $\epsilon_k^{(n)}$ are computed recursively, starting with the terms of a sequence $(S_n)_{n \geq 0}$. The fundamental property of the ϵ-algorithm is that it is closely related to P.A.: when it is initialized with the partial sums of power series $f(z)$, i.e., $\epsilon_0^{(n)} = S_n = \sum_{k=0}^{n-1} c_k z^k$, then

$$\epsilon_{2k}^{(n)} = [n + k/k]_f(z) \ , \ n, \ k \geq 0 \ . \tag{A.4}$$

Thus, the ϵ-algorithm allows the numerical value of P.A. recursively for a given point z, but does not give the coefficients of P.A. The main use of P.A. is to provide rational approximations of functions known only in terms of power series, and can either accelerate slowly converging series or provide analytical continuation beyond their domain of convergence. The capacity of P.A. can often be quite impressive, especially for series of Stieltjes type.[8] However, it should be noted that P.A. do not necessarily work well for all problems, especially far away from the domain of convergence of series.

Padé-Borel approximants. Under certain conditions, asymptotic series can have a unique sum given by the Borel summation process. Briefly stated, if the asymptotic series $\sum_{n=0}^{\infty} c_n z^n$ is Borel summable, then its sum $f(z)$ can be found from the following integral:

$$f(z) = 1/z \int_0^{\infty} g(t) \exp(-t/z) dt , \tag{A.5}$$

where $g(t) = \sum_{n=0}^{\infty} c_n z^n / n!$ is the Borel transform series. Note that the sum of Borel transform series is not known over the entire positive real axis. Thus, an analytical continuation is necessary for the evaluation equation (A.5), and can be performed by P.A. For example, from a given sequence of $[L/M]_g$ P.A. applied to Borel series $g(t)$, the following sequence of approximations known as Padé-Borel approximants can be derived:

$$f_B^{[L/M]}(z) = 1/z \int_0^{\infty} [L/M]_g(z) \exp(-t/z) dt \ . \tag{A.6}$$

The existence of (A.6) is guaranteed as long as the $[L/M]$ approximant has no pole on the positive real axis. The method of Borel summation has proved quite useful for the summation of series in physics for which other techniques converge too slowly or simply fail.

U-transform. Smith and Ford[20] (1982) reported in their review of convergence accelerators, that the u-transform can be quite powerful in accelerating the convergence of a broad range of test series. The u-transform applied on the summation of $\sum_{n=0}^{\infty} c_n z^n$ can be shown to give the following approximations in a closed form, for $n \geq 1$:

$$l_n(z) = \frac{\sum_{i=1}^{n} \alpha_{n,i} z^{i-1}}{\sum_{i=1}^{n+1} \beta_{n,i} z^{i-1}} , \tag{A.7}$$

with

$$\alpha_{n,i} = \sum_{j=1}^{i} (-1)^{n-j} \frac{c_{i-j}}{c_{n-j+1}} \binom{n}{n-j+2} \left(\frac{n-j+2}{n+1}\right)^{n-2} , \tag{A.8}$$

and

$$\beta_{n,i} = (-1)^{n-i} \frac{1}{c_{n-i+1}} \binom{n}{n-i+2} \left(\frac{n-i+2}{n+1}\right)^{n-2} . \tag{A.9}$$

The u-transform can dramatically improve the rate of convergence of P.A., especially for alternating series, whether asymptotic or not. However, it often fails to sum series with irregular sign patterns.[20]

References

[1] Caughey, T. K., and Dienes, J. K., "The behavior of linear systems with random parametric excitation," J. Math. Phys., **41**, 300-318, 1962.

[2] Wu, W. F., and Lin, Y. K., "Cumulant-neglect closure for non-linear oscillators under parametric and external excitations," *Int. J. Non-Linear Mech.*, **19 (4)**, 349-362, 1984.

[3] Ibrahim, R. A., Soundararajan, A., and Heo, H., "Stochastic response of nonlinear systems based on a non-Gaussian closure," *J. Applied Mech.*, **52**, 965-970, 1985.

[4] Bover, D. C. C., "Moment equation method for nonlinear stochastic systems," *J. Math. Anal. Applic.*, **65**, 306-320, 1978.

[5] Sun, J. Q., and Hsu, C. S., "Cumulant-neglect closure for nonlinear systems under random excitations," *J. Applied Mech.*, **54**, 649-655, 1987.

[6] Crandall, S. H., "Non-Gaussian closure for random vibration of non-linear oscillators," *Int. J. Non-Linear Mechanics*, **15**, 303-313, 1980.

[7] Crandall, S. H., "Perturbation technique for random vibration of non-linear systems," *J. Acoust. Soc. Am.*, **35 (11)**, 1700-1705, 1963.

[8] Baker, G. A., and Graves-Morris, P., *Padé Approximants*, Addison-Wesley, 1981.

[9] Brezinski, C., "Convergence acceleration methods: the past decade," *J. Comput. Appl. Math.*, **12**, 19-36, 1985.

[10] Roy, R. V., and Spanos, P. D., "Wiener-Hermite functional representation of nonlinear stochastic systems," *Structural Safety*, **6**, 187-202, 1989.

[11] Char, B. W., Geddes, K. O., Gonnet, G. H., Monagan, M. B., and Watt, S. M., *Maple Reference Manual*, 5th edition, Watcom, 1988.

[12] Wen, Y. K., "Approximate method for nonlinear random vibration," *J. Eng. Mech. Div., Proc. ASCE*, **101**, 389-401, 1975.

[13] Caughey, T. K., "Response of Van der Pol's oscillator to random excitation," *J. Applied Mech.*, **26**, 345-348, 1959.

[14] Piszczek, K., "Influence of Random disturbances on determined nonlinear vibration," in *Stochastic Problems in Dynamics* (B. L. Clarkson, editor), Pitman, London, 1977.

[15] Zhu, W.-Q., and Yu, J.-S., "On the response of the Van der Pol oscillator to white noise excitation," *J. Sound Vib.*, **117**, 421-431, 1977.

[16] Spanos, P.-T. D., "Numerical simulations of a Van der Pol oscillator," *Comp. & Maths. with Appls.*, **6**, 135-145, 1980.

[17] Ebeling, W., Herzel, H., Richert, W., and Schimansky-Geier, L., "Influence of noise on Duffing-Van der Pol oscillator," *Z. Angew. Math. Mech.*, **66**, 141-146, 1986.

[18] Ibrahim, R. A., *Parametric Random Vibration*, Research Studies Press, John Wiley, 1985.

[19] Wynn, P., "On a device for calculating the $e_m(S_n)$ transformation," *Math. Tables and A. C.*, **10**, 91-96, 1956.

[20] Smith, D. A., and Ford, W. F., "Numerical comparisons of nonlinear convergence accelerators," *Math. Comp.*, **38**, 481-499, 1982.

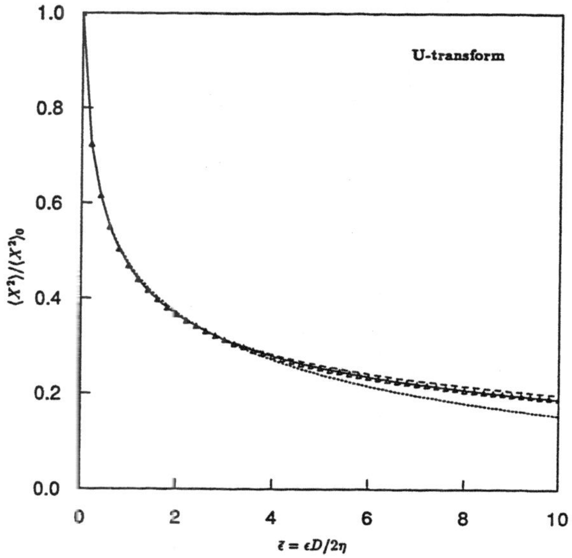

Fig.1. U-transform approximations applied to the mean-square response $\langle X^2\rangle/\langle X^2\rangle_0$ versus $\bar{\epsilon} = \epsilon D/2\eta$ for the Duffing oscillator (— exact solution; ······ $n = 5$ coefficients, △△△ $n = 6$ coefficients, - - - - $n = 7$ coefficients).

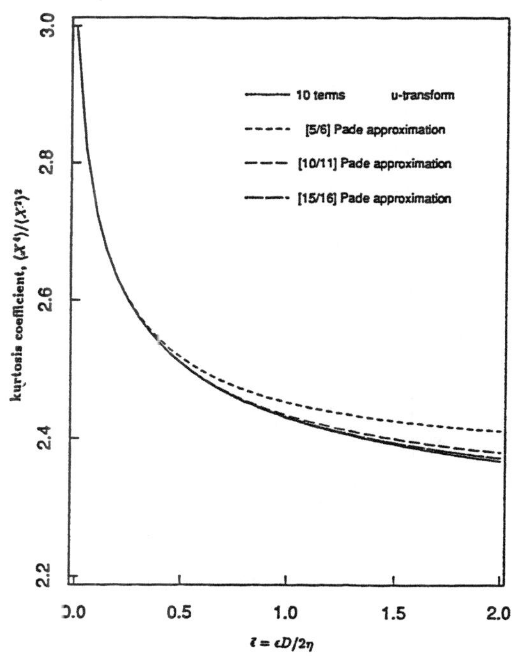

Fig. 2. Padé-type approximations applied to the kurtosis coefficient $\langle X^4\rangle/\langle X^2\rangle^2$ of the Duffing oscillator's response, versus $\bar{\epsilon} = \epsilon D/2\eta$ (— exact solution).

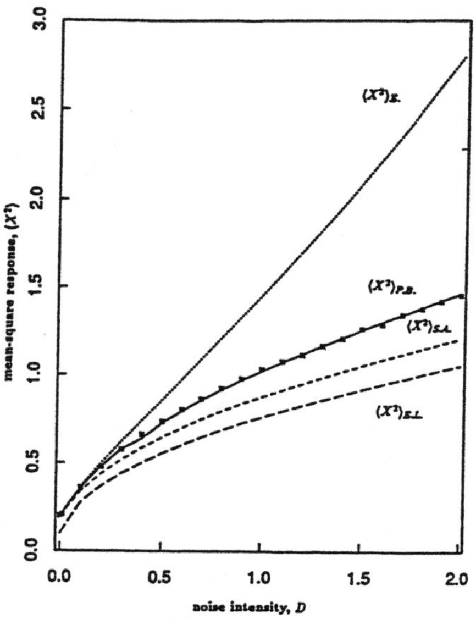

Fig. 3. Mean-square response of $\ddot{X}(t) + 2\eta\,(-1 + \epsilon\, X^2)\,\dot{X}(t) + X(t) = \xi(t)$ (Van der Pol oscillator), versus the noise intensity D. Comparison of Padé-type approximation (Borel summation) with other analytical approximations, and with digital simulations $(* * *)$. $\epsilon = 10.0$, $\eta = 0$.

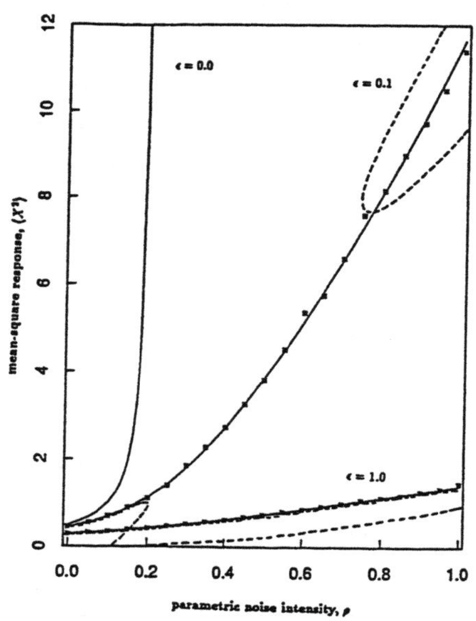

Fig. 4. Mean-square response of $\ddot{X}(t) + 2\eta\dot{X}(t) + (1 + \epsilon X^2(t) + \xi_\rho(t))\,X(t) = \xi_D(t)$ versus parametric noise intensity ρ. $D = \eta = 0.1$. Comparison with digital simulation $(* * *)$, and non-Gaussian closure $(- - -)$.

Random Vibrations of Timoshenko Beams with Generalized Boundary Conditions

M. P. SINGH and A. S. ABDELNASER

Department of Engineering Science and Mechanics
Virginia Polytechnic Institute and State University
Blacksburg, VA 24061 USA

Summary

A new method which includes the effect of shear deformations and rotatory inertia is presented to obtain the response of beams. The equations of motion are written in the state vector form. The response is expressed as a linear combination of the eigenfunctions of the homogeneous boundary value problem which can be solved for any arbitrary boundary conditions. The eigenfunctions are obtained as a solution of a nested eigenvalue problem. To define the time dependent initial value problem for the principal coordinates, the adjoint eigenvalue problem is used. Numerical results for the natural frequencies, bending moment response variance and zero crossing rates are obtained for beams with several boundary conditions and the effect of shear deformation and rotatory inertia on these responses is evaluated.

Introduction

Analysis of beam vibrations including the effect of shear deformations and rotatory inertia have been of active interest for some time. Such beams have been called Timoshenko beams, in honor of Timoshenko [1,2] who was the first one to develop the equations of motion of such beams to study the effect of shear deformation and rotatory inertia on the natural frequencies. Several other authors have examined the frequency behavior of these beams. A most comprehensive account of the frequency analysis of such beams is provided by Traill-Nash and Collar [3].

The forced vibration analysis of a simply supported Timoshenko beam is also fairly straight forward as the normal modes are the same as the modes of a simply supported Euler-Bernoulli beam. Since the algebra associated with simply supported Timoshenko beams is reasonably simple and as there are many practical situations involving simply supported beams, the literature on the subject is replete with the numerical examples of beams with simply supported boundary conditions. Among

Y. K. Lin · I. Elishakoff (Eds.)
Stochastic Structural Dynamics 1
New Theoretical Developments
2nd International Conference on Stochastic Structural Dynamics
May 9–11, 1990, Boca Raton, Florida

others, Samuels and Eringen [4], Crandall and Yildiz [5] and Elishakoff and Livshits [17] have reported results on the random vibration analysis of simply supported Timoshenko beams with different damping models. Similar analysis of beams with generalized boundary conditions are, however, more involved. The writers are aware of the publications by Hermann [6], Dolph [7,8] and Anderson [9] in which the solutions for undamped Timoshenko beams with general boundary conditions have been developed. Hermann [6] and Dolph [7] were probably the first ones to develop the orthogonality conditions of the normal modes which are essential relationships required for the solution of coupled differential equations of Timoshenko beam problem with general boundary conditions. Huang [10] has provided explicit expressions for the frequency equations and the normal modes of beams with various standard boundary conditions. Pan [11] utilized the normal mode approach along with the orthogonality relationships of Hermann and Dolph to solve the vibration problem of a viscoelastic beam. However, for illustration purposes, only the results for a simply supported beam are provided. Plass [12], on the other hand, employed the method of characteristics to transform the partial differential equations into ordinary differential equations which were subsequently solved by a finite difference approach.

A crucial step in the solution of a forced vibration problem with coupled partial differential equations by normal mode approach is the explicit formulation of the orthogonality conditions. This explicit formulation, however, can be avoided quite conveniently by adopting the state vector approach. Herein, the concept of adjoint is utilized to solve the state vector equations associated with the coupled partial differential equations of a Timoshenko beam. The deterministic solution is then used to obtain the random response characteristics of the beam with shear deformation and rotatory inertia effects. In the current microcomputer age, this solution is quite easy to implement. Furthermore, in situations where it may be quite involved to develop the orthogonality conditions explicitly, such as in the case of vibration of a beam problem formulated with higher order shear deformation theory [13], especially for layered composite beams, or in the case where the coupled torsional and lateral vibrations with shear deformations of a beam are of interest, or in situations were the exact solution of a damped shear deformable vibrating beams is desired, the approach proposed here may prove to be quite convenient. These further applications of this solution are being investigated currently.

Analytical Formulation

The equations of motion of a beam which includes shear deformation and rotatory inertia can be developed either by a direct application of the Newton's second law or by a variational method. In general, one obtains the following set of coupled linear partial differential equations:

$$m\frac{\partial^2 y}{\partial t^2}+\frac{\partial}{\partial x}\left[C\left(\theta-\frac{\partial y}{\partial x}\right)\right]=p(x,t) \tag{1}$$

$$mr^2\frac{\partial^2\theta}{\partial t^2}-\frac{\partial}{\partial x}\left(EI\frac{\partial\theta}{\partial x}\right)+C\left(\theta-\frac{\partial y}{\partial x}\right)=0 \tag{2}$$

The first equation is a shear equation whereas the second one is a moment equation. In these equations, y =lateral displacement; θ =slope due to bending only; $p(x,t)$ =lateral distributed force; x =spatial coordinate along the length; t =time variable; m =mass of the beam/unit length; r = radius of gyration of the cross section; EI =bending rigidity; $C = K'AG$, a constant; K' =shear correction factor for the cross section; a =area of the cross section and G =shear modulus of the beam material. We are interested in the solution of equations (1) and (2) for random $p(x,t)$ for arbitrary boundary conditions of the beam.

Equations (1) and (2) have to satisfy the boundary conditions which depend on the type of support at the end. For the classical cases of simply supported, fixed and free ends, these boundary conditions are:

Simply supported: $y=0\ ,\quad M=EI\dfrac{\partial\theta}{\partial x}=0$

Fixed end: $y=0\ ,\quad \theta=0$ (3)

Free end: $M=EI\dfrac{\partial\theta}{\partial x}=0\ ,\quad V=K'AG\left(\theta-\dfrac{\partial y}{\partial x}\right)=0$

where M and V, respectively, are the bending moment and shear force in the beam. For other fixity conditions at the end, the moment and shear boundary conditions are appropriately modified.

For calculating the frequencies of vibration of the beam, equations (1) and (2) are usually combined to obtain a fourth order equation in deflection $u(x,t)$. A large number of investigators have followed this route, with the primary purpose of evaluating the effects of shear deformation and rotatory inertia on the beam frequencies. Such a fourth order equation is, however, not suitable for calculating the forced response of the beam.

In this paper, we present a generalized solution approach through the application of the method of adjoint. To obtain a solution for a generalized loading function $p(x,t)$, we first develop the eigenvalue problem for the homogeneous set of equations. The solution of this eigenvalue problem provides the natural frequencies and eigenvalues of the system. These eigenproperties are then used to obtain the solution for a general loading pattern.

For the homogeneous set of equations corresponding to equations (1) and (2), it is indeed possible to obtain a solution separable in time and space. Let such a solution be:

$$\begin{Bmatrix} y \\ \theta \end{Bmatrix} = q(t)\begin{Bmatrix} u_1 \\ u_2 \end{Bmatrix} = q(t)\{u\} \tag{4}$$

Substitution of equation (4) into (1) shows that $q(t)$ must satisfy the following equation

$$\ddot{q} + \omega^2 q(t) = 0 \tag{5}$$

That is, the time variation should be harmonic. The parameter ω is the frequency parameter. For the case of a uniform cross-section beam, the spatially dependent variables, u_1 and u_2 have to satisfy the following coupled linear differential equations:

$$\begin{aligned} C\frac{d^2u_1}{dx^2} - C\frac{du_2}{dx} + \omega^2 m u_1 = 0 \\ EI\frac{d^2u_2}{dx^2} + C\frac{du_1}{dx} + \left(\omega^2 m r^2 - C\right)u_2 = 0 \end{aligned} \tag{6}$$

To obtain the solution of equation (6), we rewrite them in the state vector form as follows:

$$\frac{d}{dx}\{u\} - [A]\{u\} = \{0\} \tag{7}$$

where now, the vector $\{u\}$ has four elements, u_1, u_2, u_3 and u_4, with u_3 and u_4 defined as:

$$u_3 = \frac{du_1}{dx} \;, \quad u_4 = \frac{du_2}{dx} \tag{8}$$

The coefficient matrix $[A]$ depends upon the frequency and is defined as:

$$[A] = -[B_1] - \omega^2[B_2] \tag{9}$$

where,

$$[B_1] = \begin{bmatrix} 0 & 0 & -1 & 0 \\ 0 & 0 & 0 & -1 \\ 0 & 0 & 0 & -1 \\ 0 & -\dfrac{C}{EI} & \dfrac{C}{EI} & 0 \end{bmatrix} ; \quad [B_2] = \begin{bmatrix} 0 & 0 & 0 & 0 \\ 0 & 0 & 0 & 0 \\ \dfrac{m}{C} & 0 & 0 & 0 \\ 0 & \dfrac{mr^2}{EI} & 0 & 0 \end{bmatrix} \tag{10}$$

The solution of equation (7) can be written in terms of the eigenvalues λ_j and eigenvectors $\{\phi_j\}$ of the matrix $[A]$ as follows:

$$\{u\} = \sum_{j=1}^{4} c_j\{\phi_j\}e^{\lambda_j x} \tag{11}$$

where the eigenvalues and eigenvectors are obtained from the solution of:

$$[A]\{\phi_j\} = \lambda_j\{\phi_j\} \;, \quad j = 1, ..., 4 \tag{12}$$

The expansion coefficients c_j are obtained by applying the boundary conditions. In general, the application of boundary conditions will lead to a set of homogeneous simultaneous equations as follows:

$$[\Delta]\{c\} = \{0\} \tag{13}$$

where c_j are the element of the vector $\{c\}$. The elements of $[\Delta]$, of course, depend upon the type of boundary conditions of the problem. For example, for the case of cantilever beam, fixed at $x = 0$ and free at $x = L$, the matrix $[\Delta]$ is of the following form:

$$[\Delta] = \begin{bmatrix} \phi_{11} & \phi_{12} & \phi_{13} & \phi_{14} \\ \phi_{21} & \phi_{22} & \phi_{23} & \phi_{24} \\ (\phi_{21} - \phi_{31})e^{\lambda_1 L} & (\phi_{22} - \phi_{32})e^{\lambda_2 L} & (\phi_{23} - \phi_{33})e^{\lambda_3 L} & (\phi_{24} - \phi_{34})e^{\lambda_4 L} \\ \phi_{41}e^{\lambda_1 L} & \phi_{42}e^{\lambda_2 L} & \phi_{43}e^{\lambda_3 L} & \phi_{44}e^{\lambda_4 L} \end{bmatrix} \tag{14}$$

where ϕ_{ij} is the ith element of the eigenvector $\{\phi_j\}$.

Since the elements of the matrix in equation (13) depend upon the solution of the eigenvalue problem in equation (12), equation (13) is the nested eigenvalue problem. The eigenproperties of equation (12) depend on frequency ω. In fact, the four eigenvalues λ_j are defined as:

$$\lambda^2 = -\frac{m\omega^2}{2}\left(\frac{r^2}{EI} + \frac{1}{C}\right) \pm \left[\frac{m^2\omega^4}{4}\left(\frac{r^2}{EI} + \frac{1}{C}\right)^2 - \frac{m\omega^2}{EA}\left(\frac{m\omega^2}{C} - \frac{1}{r^2}\right)\right]^{1/2} \tag{15}$$

The eigenvector corresponding to λ_j is then defined as:

$$\{\phi_j\} = \begin{bmatrix} 1 \\ \lambda_j + \dfrac{m\omega^2}{C\lambda_j} \\ \lambda_j \\ \lambda_j^2 + \dfrac{m\omega^2}{C} \end{bmatrix} \tag{16}$$

The frequency ω in equations (15) and (16) is still unknown. The frequencies are obtained when the eigenvalue problem (13) is solved. For equation (13) to have a nontrivial solution, the determinant of $[\Delta]$ must be zero. That is:

$$|\Delta| = 0 \tag{17}$$

This provides the characteristic equation for calculating the frequencies. This equation can be solved by a root finding iterative technique. Once the frequencies,

ω_m are known, the corresponding coefficients $\{c_m\}$ in equation (13) and the eigenproperties λ_{jm} and $\{\phi_{jm}\}$ of equation (12) are also known. These in turn, define the eigenfunctions $\{u_m\}$ for equation (6) corresponding to frequency ω_m as:

$$\{u_m\} = \sum_{j=1}^{4} c_{jm}\{\phi_{jm}\}e^{\lambda_{jm}x} \tag{18}$$

These eigenfunctions can be used to obtain the solution of the nonhomogeneous forced boundary value problem of equations (1) and (2). For this we re-write these equations in the state vector form as:

$$\frac{\partial}{\partial x}\{w\} + [B_1]\{w\} - [B_2]\frac{\partial^2}{\partial t^2}\{w\} = \{F(t)\} \tag{19}$$

where the vector $\{w\}$ and $\{F(t)\}$ are defined as:

$$\{w\}^T = \left(y, \theta, \frac{\partial y}{\partial x}, \frac{\partial \theta}{\partial x}\right) \tag{20}$$

$$\{F(t)\}^T = \left(0, 0, -\frac{p(x, t)}{C}, 0\right) \tag{21}$$

Since the eigenfunctions, $\{u_m\}$, of equation (6) form a complete set, we can use them with expansion theorem to obtain the solution $\{w\}$ as:

$$\{w\} = \sum_{m=1}^{\infty} q_m(t)\{u_m(x)\} \tag{22}$$

where $q_m(t)$, $m = 1, \ldots, \infty$ are the time dependent principal coordinates. To obtain $q_m(t)$, we substitute equation (22) into equation (19). However, we immediately run into a problem when we try to decouple the resulting equations by using the eigenfunction $\{u_n(x)\}$, as the two eigenfunctions are not orthogonal for $m \neq n$. This is so, because equation (19) is not a self-adjoint equation. To be able to decouple these equations to obtain $q_m(t)$, we also need the eigenfunctions of the adjoint boundary value problem corresponding to equation (6) or the state vector equation (7).

Adjoint Boundary Value Problem

The procedure to construct the adjoint boundary value problem is described by Nyfeh [14] and Ince [15]. Let $\{v\}$ be the adjoint eigenfunction. Premultiplying equation (7) by $\{v\}^T$ and integrating it by parts over the domain of the problem, one obtains:

$$\int_0^L \left(\frac{d}{dx} \{v\}^T + \{v\}^T [A] \right) \{u\} dx - \{v\}^T \{u\} \Big|_0^L = 0 \tag{23}$$

This equation defines the adjoint equation as well as its boundary conditions. The adjoint equation is:

$$\frac{d}{dx} \{v\} + [A]^T \{v\} = 0 \tag{24}$$

The boundary conditions on $\{v\}$ which are consistant with the boundary conditions on $\{u\}$ should be such that:

$$\sum_{r=1}^{4} [u_r(L) v_r(L) - u_r(0) v_r(0)] = 0 \tag{25}$$

This identifies the required boundary conditions on $\{v\}$ corresponding to the boundary conditions on $\{u\}$. For example, for the case of a cantilever beam, these these boundary conditions are:

$$v_1(L) = 0 \ , \ \ v_2(L) = - v_3(L) \ , \ \ v_3(0) = 0 \ , \ \ v_4(0) = 0 \tag{26}$$

The solution of the boundary value problem (24) provides the adjoint eigenfunctions of the problem. These eigenfunctions can now be written in terms of the eigenproperties of matrix $[A]^T$, obtained from

$$[A]^T \{\psi_j\} = \lambda_j \{\psi_j\} \tag{27}$$

Of course, λ_j's in equation (27) and (12) are the same, but $\{\psi_j\}$'s are different from $\{\phi_j\}$'s. In terms of λ_j, $\{\psi_j\}$ is defined as:

$$\{\psi_j\} = \begin{bmatrix} 1 \\ \left(\frac{C - mr^2\omega^2}{C}\right)\left(\frac{1}{\lambda_j} + \frac{C\lambda_j}{m\omega^2}\right) \\ -\frac{C\lambda_j}{m\omega^2} \\ -\frac{EI}{C}\left(1 + \frac{c\lambda_j^2}{m\omega^2}\right) \end{bmatrix} \quad j = 1, \dots, 4 \tag{28}$$

and the adjoint eigenfunctions defined as:

$$\{v_m\} = \sum_{j=1}^{4} d_{jm}\{\psi_{jm}\}e^{-\lambda_{jm}x} \;, \quad m = 1, \dots, \infty \tag{29}$$

where $\{\psi_{jm}\}$ is the eigenvector obtained from equation (28) for $\omega = \omega_m$. The coefficients d_{jm} are such that $\{v_m\}$ satisfies the boundary conditions implied by equation (25). The application of the boundary conditions leads to the following set of homogeneous equations in $\{d_m\}$

$$[\Gamma_m]\{d_m\} = \{0\} \;, \quad m = 1, \dots, \infty \tag{30}$$

Of course, the matrix $[\Gamma_m]$ depends upon the boundary conditions. For the case of the cantilever beam, this matrix is defined as:

$$[\Gamma] = \begin{bmatrix} \psi_{31} & \psi_{42} & \psi_{43} & \psi_{44} \\ \psi_{41} & \psi_{32} & \psi_{33} & \psi_{34} \\ (\psi_{21} + \psi_{31})e^{-\lambda_1 L} & (\psi_{22} + \psi_{32})e^{-\lambda_2 L} & (\psi_{23} + \psi_{33})e^{-\lambda_3 L} & (\psi_{24} + \psi_{34})e^{-\lambda_4 L} \\ \psi_{11}e^{-\lambda_1 L} & \psi_{12}e^{-\lambda_2 L} & \psi_{13}e^{-\lambda_3 L} & \psi_{14}e^{-\lambda_4 L} \end{bmatrix} \tag{31}$$

It can be shown that the characteristic equation obtained by setting the determinant of $[\Gamma_m]$ equal to zero will provide the same frequencies as those obtained from equation (17). It can also be shown that the eigenfunctions $\{u_m\}$ and $\{v_n\}$ are orthogonal. To show this, we consider equation (7) and its adjoint equation (24). We re-write these equations in terms of matrices $[B_1]$ and $[B_2]$ for eigenfunctions $\{u_m\}$ and $\{v_n\}$ as follows:

$$\frac{d}{dx}\{u_m\} + [B_1]\{u_m\} + \omega_m^2[b_2]\{u_m\} = 0 \tag{32}$$

$$\frac{d}{dx}\{v_n\} - [B_1]^T\{v_n\} - \omega_n^2[B_2]^T\{v_n\} = 0 \tag{33}$$

We premultiply (32) by $\{v_n\}^T$, integrate it by parts over the domain of x and then transpose the entire equation:

$$\{u_m\}^T\{v_n\}\Big|_0^L - \int_0^L \{u_m\}^T\left(\frac{d}{dx}\{v_n\} - [B_1]^T\{v_n\}\right)dx + \omega_m^2\int_0^L \{u_m\}^T[B_2]^T\{v_n\}dx = 0 \tag{34}$$

For any pair of boundary conditions, the first term of equation (34) is always zero. Next we premultiply equation (33) by $\{u_m\}^T$, integrate over the domain and then add to equation (34) to obtain:

$$\left(\omega_m^2 - \omega_n^2\right)\int_0^L \{u_m\}^T[B_2]^T\{v_n\}dx = 0 \tag{35}$$

Therefore, for any two distinct frequencies,

$$\int_0^L \{v_n\}^T[B_2]\{u_m\}dx = 0 \tag{36}$$

This provides the orthogonality statement for the eigenfunctions $\{u_m\}$ and $\{v_n\}$. For $m = n$ equation (35) can be used to normalize one of the two eigenfunctions such that:

$$\int_0^L \{v_n\}^T[B_2]\{u_n\}dx = 1 \tag{37}$$

Equations (36) and (37), in conjunction with equation (33), immediately lead to another form of orthogonality equation,

$$\int_0^L \{v_n\}^T\left(\frac{d}{dx}\{u_m\} + [B_1]\{u_m\}\right)dx = -\omega_m^2\delta_{mn} \tag{38}$$

Decoupled Equations for Principal Coordinates

Equations (36) through (38) can be directly used to decouple equation (19) to define the time dependent equations for the principal coordinates. We substitute equation (22) into equation (19), premultiply by $\{v_n\}^T$, integrate over the domain of x, and utilize the orthogonality relationships in equation (36) through (38) to obtain the following equation for the principal coordinate $q_m(t)$:

$$\begin{aligned}\ddot{q}_m(t) + 2\beta_m\omega_m\dot{q}_m(t) + \omega_m^2 q_m(t) &= -\int_0^L \{v_m\}^T\{F(t)\}dx \\ &= \frac{1}{C}\int_0^L v_{3m}(x)p(x, t)dx\end{aligned} \tag{39}$$

where $v_{3m}(x)$ is the third element of the eigenfunction $\{v_m\}$. Although there were no damping terms in the original equations, here the term $2\beta_m\omega_m\dot{q}_m$ has now been added just to include some energy dissipation in the vibration of the beam.

Random Response

The solution of equation (39) provides the time variation of the response expressed by equation (22). In general, a response quantity of interest, which is linearly related to the response vector $\{w\}$, can also be expressed in a form similar to equation (22) as:

$$R(x, t) = \sum_{m=1}^{\infty} R_m(x)q_m(t) \tag{40}$$

where R_m is the mth modal response of the response quantity of our interest. If we are interested in the beam deflection, then $R_m = u_{1m}(x)$. If we are interested in the bending slope, then $R_m = u_{2m}(x)$. Likewise for the bending moment and shear force at any location x, R_m is equal to $EIu_{4m}(x)$ and $C\{u_{2m}(x) - u_{3m}(x)\}$, respectively.

For a beam exited by a randomly varying lateral force, it is now straight forward to obtain the moments of $R(x, t)$ in terms of the moments of the principal coordinates $q_m(t)$. For example, the covariance function of $R(x, t)$ can be obtained from (Lin [15]):

$$E[R(x_1, t_1)R(x_2, t_2)] = \sum_{m=1}^{\infty}\sum_{n=1}^{\infty} R_m(x_1)R_n(x_2)E[q_m(t_1)q_n(t_2)] \tag{41}$$

The correlation function for the principal coordinates can be obtained by utilizing equation (39). That is,

$$E[q_m(t_1)q_n(t_2)] = \frac{1}{C^2}\int_0^L\int_0^L v_{3m}(x_1')v_{3n}(x_2') \int_0^{t_1}\int_0^{t_2} E[p(x_1', \tau_1)p(x_2', \tau_2)]h_m(t_1 - \tau_1)h_n(t_2 - \tau_2)d\tau_1 d\tau_2 dx_1' dx_2' \tag{42}$$

In equation (42), $h_m(\tau)$ is the impulse response function of equation (39).

For known spatial and temporal correlation characteristics of the lateral load $p(x, t)$, equation (42) can be evaluated in principle. Especially, for a temporally stationary and spatially homogeneous random process, the stationary value of the correlation function of the principal coordinates can be shown to be as follows:

$$E[q_m(t_1)q_n(t_2)] = \frac{1}{C^2}\int_{-\infty}^{\infty} H_m H_n^* \int_0^L\int_0^L \Phi_{pp}(\omega, x_1', x_2') v_{3m}(x_1')v_{3n}(x_2')e^{i\omega(t_1 - t_2)}dx_1' dx_2' d\omega \tag{43}$$

where H_m is the frequency response function of equation (39), an astrisk denotes a complex-conjugate and $\Phi_{pp}(\omega, x_1', x_2')$ is the spectral density function defined as:

$$\Phi_{pp}(\omega, x_1', x_2') = \frac{1}{2\pi}\int_{-\infty}^{\infty} e^{-i\omega\tau}E[p(x_1', t)p(x_2', t + \tau)]d\tau \tag{44}$$

Numerical Results

In this section, we present some numerical results obtained for the natural frequencies and random response of rectangular beams to demonstrate the effect of shear deformation and rotatory inertia on these quantities. Beams with different boundary conditions have been considered.

Equation (18) has been used to obtain the natural frequencies. The frequencies with and without shear deformation and rotatory inertia are compared for beams with different boundary conditions. Tables 1 and 2 show the nondimensional frequencies $\omega' = \omega L^2 \sqrt{\rho / E D^2}$ for Timoshenko and Euler-Bernoulli beams and the percent error caused by the omission of the shear deformation and rotatory inertia terms, for fixed-fixed, fixed-hinged, fixed-free (cantilever) and simply supported beams. Table 1 is for depth-to-length (D/L) ratio of 0.1 and table 2 is for D/L ratio of 0.4.

Table 1: Frequencies for Timoshenko (T) and Euler-Bernoulli (EB) beams and percent errors (% E) in the Euler-Bernoulli theory for different boundary conditions. D/L = 0.1

	Fixed - Fixed			Fixed - Hinged			Fixed - Free			Hinged - Hinged		
MODE#	T	EB	% E	T	EB	% E	T	EB	% E	T	EB	% E
1	6.063	6.459	6.5	4.286	4.451	3.8	1.007	1.015	0.8	2.803	2.849	1.6
2	15.56	17.80	14	13.10	14.42	10	6.041	6.361	6.4	10.72	11.40	6.3
3	28.16	34.90	24	25.40	30.09	18	15.91	17.81	12	22.61	25.64	13
4	42.80	57.69	35	40.08	51.46	28	28.89	34.90	21	37.27	45.59	22
5	58.82	86.19	46	56.33	78.53	39	44.02	57.69	31	53.73	71.23	33
6	75.78	120.4	59	73.61	111.3	51	60.55	86.19	42	71.35	102.6	44
7	93.39	160.3	72	91.57	149.8	64	77.99	120.4	54	89.67	139.6	56
8	111.4	205.8	85	110.0	193.9	76	96.01	160.3	67	108.4	182.3	68
9	129.8	257.1	98	128.6	243.8	89	114.4	205.8	80	127.4	230.8	81
10	148.4	314.1	112	147.5	299.3	103	132.9	257.1	93	146.6	284.9	94
11	167.0	376.8	126	166.4	360.6	117	151.5	314.1	107	165.8	344.7	108
12	185.7	445.2	140	185.4	427.5	131	169.9	376.8	122	185.0	410.3	122
13	202.0*	------	----	199.6*	------	----	187.8	445.2	137	198.7*	------	----
14	204.1*	519.2	154	204.0*	500.2	145	200.9*	------	----	202.0*	------	----
15	211.8*	------	----	206.3*	------	----	205.8*	519.3	152	204.3*	481.5	136
16	221.7*	599.0	170	217.6*	578.5	166	211.0*	------	----	211.3*	------	----
17	227.3*	------	----	223.6*	662.6	196	222.0*	599.0	170	223.5*	558.4	150
18	239.7*	684.5	186	233.8*	------	----	227.2*	684.5	201	225.4*	------	----
19	246.1*	------	----	242.6*	752.3	210	239.7*	------	----	242.6*	641.0	164
20	259.1*	775.7	199	252.9*	------	----	246.1*	775.7	215	243.1*	------	----

* Denotes second spectrum frequencies

Table 2: Frequencies for Timoshenko (T) and Euler-Bernoulli (EB) beams and percent errors (% E) in the Euler-Bernoulli theory for different boundary conditions. D/L = 0.4

	Fixed - Fixed			Fixed - Hinged			Fixed - Free			Hinged - Hinged		
MODE#	T	EB	% E	T	EB	% E	T	EB	% E	T	EB	% E
1	3.673	6.549	76	2.992	4.451	49	0.910	1.015	12	2.329	2.849	22
2	7.405	17.80	140	7.139	14.42	102	3.849	6.361	65	6.777	11.40	68
3	11.82	34.90	195	11.64	30.09	159	8.244	17.81	116	11.57	25.64	122
4	14.83*	------	----	13.23*	------	-----	12.00	34.90	191	12.42*	------	----
5	16.79*	57.69	244	16.27*	51.46	216	14.96*	------	----	15.19*	------	----
6	20.08*	------	----	17.94*	------	-----	16.70*	57.69	245	16.36*	45.59	179
7	21.89*	86.19	294	21.08*	78.53	227	19.83*	------	----	20.89*	------	----
8	25.45*	120.4	373	24.08*	------	----	22.33*	86.17	286	21.10*	71.23	238
9	27.83*	------	----	25.87*	111.3	330	25.12*	------	----	25.79*	102.6	298
10	30.34*	160.3	428	30.25*	149.8	395	27.74*	120.4	334	27.54*	------	----
11	34.65*	------	----	31.22*	------	----	31.22*	------	----	30.44*	139.6	359
12	35.06*	205.9	487	35.06*	193.9	453	32.73*	160.3	390	34.62*	------	----
13	39.72*	257.1	547	38.21*	------	----	37.31*	205.8	452	35.07*	182.3	420
14	41.81*	------	----	39.74*	243.8	514	38.26*	------	----	39.67*	230.8	482
15	44.40*	314.1	608	44.20*	299.3	577	42.12*	257.1	510	41.94*	------	----
16	48.48*	376.8	677	45.70*	------	----	45.16*	314.1	595	44.26*	284.9	544
17	49.74*	------	----	48.84*	360.6	638	47.04*	------	----	48.83*	344.7	606
18	53.33*	445.2	735	53.00*	------	----	55.87*	376.8	641	49.41*	------	----
19	56.99*	------	----	53.57*	427.5	698	53.36*	------	----	53.39*	410.3	668
20	57.94*	519.3	796	57.94*	500.2	763	55.63*	445.2	700	56.97*	------	----

* Denotes second spectrum frequencies

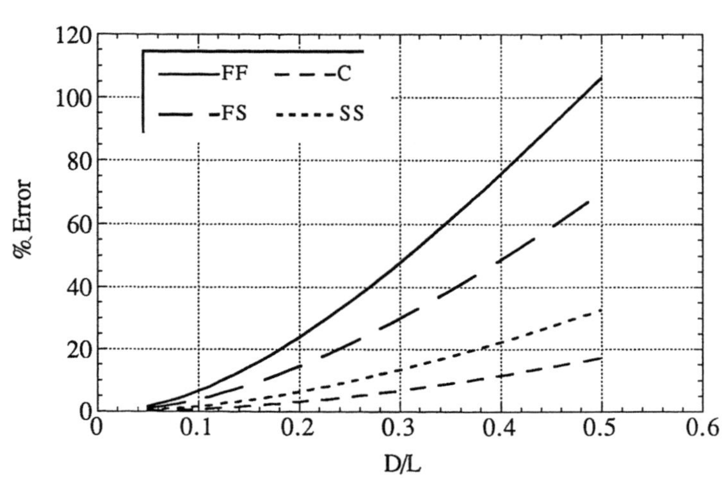

Figure 1. Percent error due to omission of shear deformation and rotatory inertia terms in the first mode frequency of simply supported (SS), cantilever (C), fixed-hinged (FS) and fixed-fixed (FF) beams.

Some values in these tables are identified by astrisks. They correspond to the frequencies in the "second spectrum". The term "second spectrum" was introduced by Traill-Nash and Collar [3]. The frequencies in the so called "first spectrum" correspond to the case when the two out of four λ's, defined by equation (15), are real and the other two are complex and conjugate. In this case, the frequency equation has only the imaginary part, with the real part being identically zero. The "second spectrum" on the other hand belong to the case when all λ's, in equation (15) are imaginary. In this case the characteristic equation has the real part, with the imaginary part being identically zero. These special characteristic, however do not pose any numerical problems in the solution of the frequency equation.

The frequencies that are not compared are those of the shear modes from the Timoshenko beams which have no counterpart in the Euler-Bernoulli beams. It is noted that the errors increase with the mode number for all cases. Also, the error is smallest for cantilever beam and highest for fixed-fixed beams for all first spectrum frequencies. In the second spectrum, however, and for high mode numbers, the error for cantilever beam becomes higher than the errors for simply supported and fixed-hinged beams.

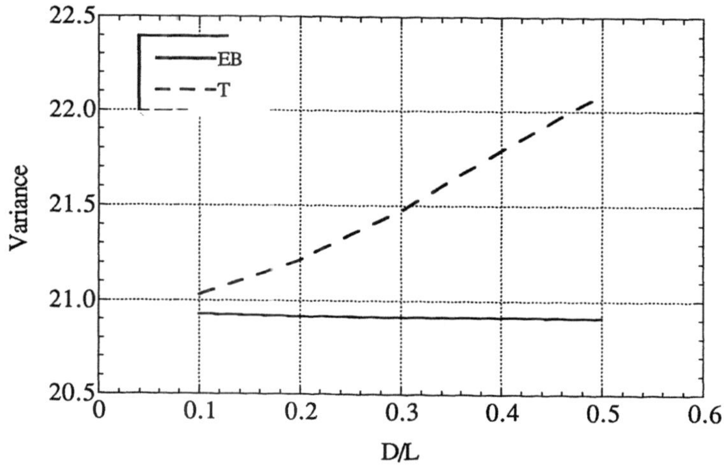

Figure 2. Bending moment response variance for Timoshenko (T) and Euler-Bernoulli (EB) cantilever beams.

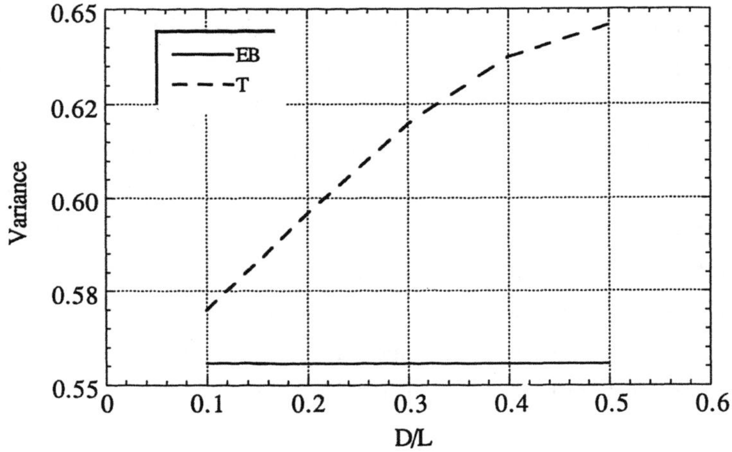

Figure 3. Bending moment response variance for Timoshenko (T) and Euler-Bernoulli (EB) fixed-fixed beams.

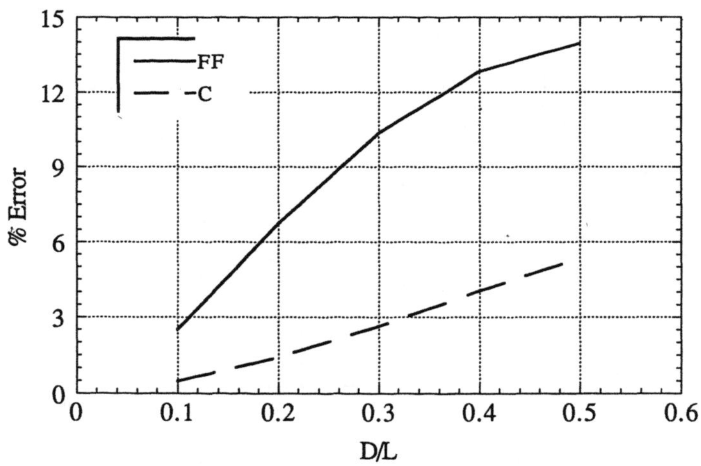

Figure 4. Percent error due to the omission of shear deformation and rotatory inertia terms in the variance of bending moment for cantilever (C) and fixed-fixed (FF) beams.

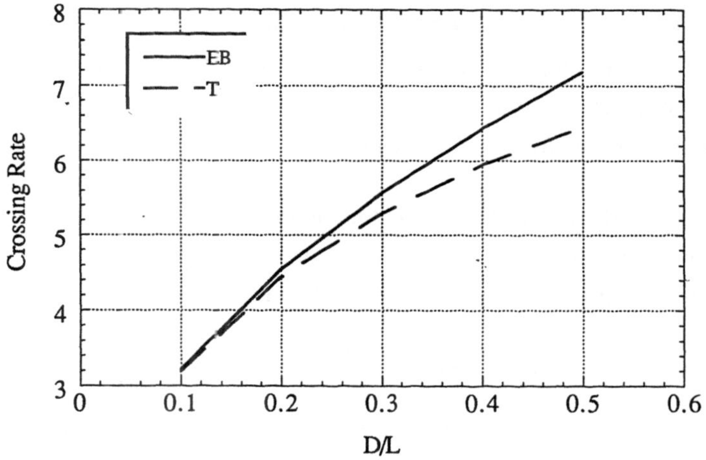

Figure 5. Zero crossing rate for bending moment of Timoshenko (T) and Euler-Bernoulli (EB) cantilever beams.

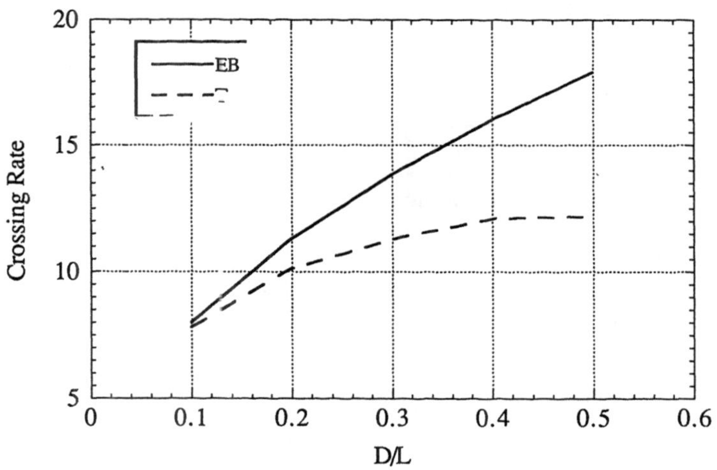

Figure 6. Zero crossing rate for bending moment of Timoshenko (T) and Euler-Bernoulli (EB) fixed-fixed beams.

Usually, the first mode frequency is the most important in the vibration of beams. In figure 1, therefore, we plot the percent error of the first mode frequencies versus the D/L ratio for different beams. The error increases as the D/L ratio is increased because the shear deformations become relatively more important for deeper beams.

Next we show the numerical results for the maximum bending moment in the beams caused by a randomly varying load. For the numerical example, the correlation characteristics of the load $p(x, t)$ is defined by the following cross correlation function:

$$E[p(x_1, t_1)p(x_2, t_2)] = \sigma^2 e^{-\alpha|x_1 - x_2| - \beta|t_1 - t_2|} \tag{45}$$

To obtain the numerical results, $\sigma = 10\ N/m$, $\alpha = 0.3$ and $\beta = 0.3$ were chosen. Figure 2 shows the mean square values of the bending moment in a cantilever beam for increasing D/L ratios, obtained with and without the shear deformation and rotatory inertia terms. Figure 3 shows similar results but for a fixed-fixed beam. As the D/l ratio increases, the difference in the two response values increases. Figure 4 shows the percent error in the bending moment variances of the cantilever and fixed-fixed beams calculated with and without shear deformation and rotatory inertia terms. Figure 5 and 6 show for the cantilever and fixed-fixed beams, respectively, the effect of the shear deformation and rotatory inertia on the zero crossing rate of the

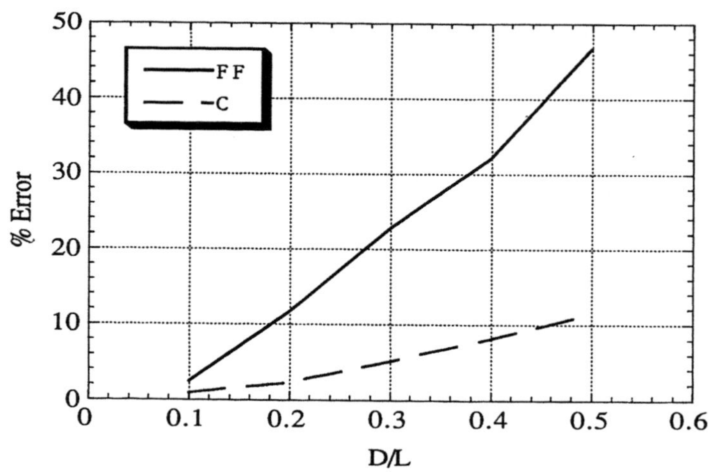

Figure 7. Percent error due to omission of shear deformation and rotatory inertia terms in the zero crossing rates of bending moment for cantilever (C) and fixed-fixed (FF) beams.

bending moment response. The crossing rates for Timoshenko beams are smaller as the shear deformations make it more flexible. In figure 7 , the error in the zero crossing rate calculated with and without shear deformation and rotatory inertia terms is plotted against the D/L ratio of the cantilever and fixed-fixed beams. The errors in figures 4 and 7 are seen to increase with the D/L ratio. This shows the relative importance of the shear deformation for deep beams. It is interesting to note that some very large errors noted in the frequencies in tables 1 and 2 do not translate into that large errors in the response. It is because the moment response calculated here is only affected by the lower modes for the type of loading considered. However it is quite possible that for other loads with different correlation characteristics, the errors could be larger or smaller than those shown herein. Also, the errors could be large in beams made of laminated fiber composites which are known to be more susceptible to shear deformation effects.

Aknowledgements

This work is partially supported by the National Science Foundation through Grant No. BSC-8822864 with Dr. S. C. Liu as its Program Director. This financial support is gratefully acknowledged.

References

1. Timoshenko, S. P. : On the Correction for Shear of the Differential Equation for Transverse Vibrations of Prismatic Bars. London Phil. Mag. series 6, 41 (1921) 744-746.

2. Timoshenko, S. P. : On the Transverse Vibrations of Bars of Uniform Cross-Sections. London Phil. Mag. series 6, 43 (1922) 125-131.

3. Traill-Nash, P. W. and Collar, A. R. : The effects of Shear Flexibility and Rotatory Inertia on the Bending Vibrations of Beams. Quart. Journal of Mech. and Applied Math., Vol VI, pt. 2 (1953) 186-222.

4. Samuels, J. C. and Eringen, A. C. : Response of Simply Supported Timoshenko Beam to a Purely Random Gaussian Process. Journal of Applied Mechanics, 25, Trans. ASME, 80 (1958) 496-500.

5. Crandall, S. H. and Yildiz, A. : Random Vibration of Beams. Journal of Applied Mechanics, 29, Trans. ASME, 84 (1962) 267-275.

6. Hermann, G. : Forced Motions of Timoshenko Beams. Journal of Applied Mechanics, 22, Trans. ASME, 77 (1955) 53-56.

7. Dolph, C. L. : Normal Modes Oscillation in Beams. Report UMM 79, Willow Run Research Center, University of Michigan, Ann Arbor, Mich., 1951.

8. Dolph, C. L. : On The Timoshenko Theory of Transverse Beam Vibrations. Quarterly Journal of Applied Math., 12 (1954) 175-187.

9. Anderson, R. A. : Flextural Vibration in Uniform Beams According to the Timoshenko Theory. Journal of applied Mechanics, 20, Trans. ASME, 75 (1953) 504-510.

10. Huang, T. C. : The Effect of Rotatort Inertia and Shear Deformation on the frequency and Normal Mode Equations of a Uniform Beam with Simple End conditions. Journal of Applied Mechanics, 28, Trans. ASME, 83 (1961) 579-584.

11. Pan, H. : Vibration of Viscoelastic Timoshenko Beam. Journal of the Engineering Mechanics Division, Proc. ASCE, 92 (1966) 213-234.

12. Plass, H. J. : Some Solutions of the Timoshenko Beam Equation for Short Pulse-Type Loading. Journal of Applied Mechanics, 25, Trans. ASME, 80 (1958) 379-385.

13. Reddy, J. N. : Energy and Variational Methods in Applied Mechanics. New York: John Wiley 1984.

14. Nayfeh, A. H. : Introduction to Perturbation Techniques. New York: John Wiley 1981.

15. Ince, E. L. : Ordinary Differential Equations. New York: Dover Publications 1956.

16. Lin, Y. K. : Probabilistic Theory of Structural Dynamics. New York: McGraw-Hill 1967.

17. Elishakoff, I. and Livshits, d. : Some Closed-Form Solutions in Random Vibration of Bresse-Timoshenko Beams. Journal of Probabilistic Engineering Mechanics, 4(1989) 49-54.

A Survey of Quantitative and Qualitative Methods of Sensitivity Analysis for Stochastic Dynamic Systems

Leslaw Socha

Department of Civil Engineering
School of Engineering and Applied Sciences
State University of New York at Buffalo
235 Ketter Hall, Buffalo, NY 14260
on leave from the Institute of Transport
Silesian Technical University
Katowice, Poland

Summary

In this paper a brief literature survey concerning sensitivity analysis of stochastic dynamic systems described by Ito equations is presented. In the first part of the paper we review the quantitative methods in the time and frequency domain. We quote the definitions of output and moment sensitivity measures and we present applications for linear systems and nonlinear oscillators with stochastic coefficients under stochastic excitations. The cases of white and coloured noise are considered. This is followed by a discussion of the application of the output sensitivity process to response approximation of nonlinear oscillators. We discuss also the application of sensitivity methods to the approximation of characteristics of the solution of stochastic differential equations. We present another approach to the study of response sensitivity of stochastic systems in the frequency-domain with applications to secondary structural systems. In the second part of the paper we review the qualitative methods. The concept of exponential and practical insensitivity for linear and a class of nonlinear stochastic dynamic systems is presented. We quote the definitions and criteria of the exponential and practical insensitivity which are very close to the definitions of stability concerning selected response variables.

1. Introduction

Over the last four decades the method of sensitivity analysis has been applied to the study of dynamic systems. However, the study of the influence of the coefficients of a differential equation on its solution started with the origins of differential equations and were merely considered by mathematicians. In 1945 Bode [1] was the first to establish the significance of sensitivity in design of feedback control systems. He has introduced a proper sensitivity definition on the basis of the frequency domain. Beginning in the period 1958-1960 the number of publications devoted to sensitivity considerations in the time domain rose considerably due to the development of the state space methods in control engineering. Both approaches have been developed by electrical and control engineers. They have also been applied successfully in a variety of disciplines such as aerospace and mechanical engineering usually for control problems and structural optimization in which gradient methods were used to find the search directions for

Y. K. Lin · I. Elishakoff (Eds.)
Stochastic Structural Dynamics 1
New Theoretical Developments
2nd International Conference on Stochastic Structural Dynamics
May 9–11, 1990, Boca Raton, Florida

optimum solution. This has been described in many papers and books, for example in survey papers written by Kokotokovic and Rutman [2], Nuguyen Thuong Ngo [3] and in the monograph by Frank [4]. Most recently, researchers in disciplines such as physiology, thermodynamics, physical chemistry and aerodynamics have been using sensitivity methodology to assess the effects of parameter variations in their analytical models and to create designs insensitive to parameter variation. This has been described in excellent survey paper by Adelman and Haftka [5]. Also some applications have been made in connection with parameter identification procedures, in the studies of dynamic response. A short review for mechanical systems can be found in the papers by Hsieh and Arora [6], Zimoch [7], Nalecz and Wicher [8], Sharp and Brooks [9,10] for linear models and in Kuo and Wada [11] for both linear and nonlinear models.

The sensitivity analysis for stochastic systems has not been developed although, in analogy with the concept of deterministic sensitivity, the answer to the question "what is the influence in some stochastic sense of parameters of the system on the solution of this system?" would seem to be very important. From numerical study of the response of stochastic dynamic systems it follows that the dependence of the solution on changes of the parameters is significant (see references in [12]).

In this paper we present a brief literature survey concerning sensitivity analysis of stochastic dynamic systems. We show also some future of the developing of this approach.

First we introduce some mathematical preliminaries using the notation from the paper written by Socha [12]

2. Mathematical Preliminaries

A class of the above mentioned systems, to be considered in what follows, can be described by the stochastic nonlinear differential equation of Ito-type

$$dx(t) = f(t, x(t), p)dt + \sum_{k=1}^{M} g_k(t, x(t), p)d\xi_k(t), \quad x(t_o) = x_o, \quad t \in [t_o, T], \tag{1}$$

where $x(t)$ is an n-dimensional vector state, $x = [x_1, \ldots, x_n]^T, p$ is a q-dimensional deterministic vector of parameters, $p = [p_1, \ldots, p_q]^T, p \in Q \subset R^q$, ($Q$ is an open set), $\xi_k(t)$ are independent standard Wiener processes, $k = 1, \ldots, M$ and $f, g_k : [t_o, T] \times R^n \times R^q \to R^n$ are nonlinear deterministic functions; initial state x_o is assumed to be deterministic, $(\cdot)^T$ denotes the transpose of matrix or vector. With the hypothesis described by Gihman and Skorohod [13] the equation (1) has a unique continuous solution.

As in the deterministic theory, there are several ways to define quantities for the characterization of the parameter sensitivity of a system. Following Gihman and Skorohod, Socha [12] proposed the following one.

Definition 1 (output sensitivity process). The limit in probability

$$y_i(t, p_{i0}) = \left.\frac{\partial x(t,p)}{\partial p_i}\right|_{p=p_0} = \lim_{\Delta p_i \to 0}\left[\frac{x(t, p_0 + \Delta p_i) - x(t, p_0)}{\Delta p_i}\right] \tag{2}$$

is defined as the output sensitivity process of the system (1), where $\Delta p_i = [0, \ldots, \Delta_i, \ldots, 0]^T$ (Δ_i is real). The output sensitivity process is a vector stochastic process, $y_i \in R^n$, and is defined if this limit in probability exists. Besides, this y_i is a process of the nominal parameter value p_0. If one assumes that p_0 is a constant and the parameter changes around p_0 are small, y_i can be regarded as a stochastic process independent of p. Gihman and Skorohod [13] showed that the output sensitivity process, under suitable assumptions, exists and satisfies the following stochastic differential equation (sensitivity equation):

$$\begin{aligned} dy_i(t) = {} & [f_{p_i}(t, x(t), p) + f_x(t, x(t), p) y_i(t)]_{p=p_o} dt + \\ & + \sum_{k=1}^{M} [g_{k_{p_i}}(t, x(t), p) + g_{k_x}(t, x(t), p) y_i(t)]_{p=p_o} d\xi_k(t), \quad y_i(t_o) = 0. \end{aligned} \tag{3}$$

where

$$\begin{aligned} f_{p_i}(t, x(t), p) &= \frac{\partial f(t, x(t), p)}{\partial p_i}, & f_x(t, x(t), p) &= \frac{\partial f(t, x(t), p)}{\partial x}, \\ g_{k_{p_i}}(t, x(t), p) &= \frac{\partial g_k(t, x(t), p)}{\partial p_i}, & g_{k_x}(t, x(t), p) &= \frac{\partial g_k(t, x(t), p)}{\partial x}. \end{aligned} \tag{4}$$

For convenience we will write the equations (1) and (3) in vector form and omit the index i remembering that all calculations are valid for every i.

$$dz(t) = F(t, z(t), p)dt + \sum_{k=1}^{M} G_k(t, z(t), p) d\xi_k(t), \quad z(t_o) = z_o, \tag{5}$$

where

$$z(t) = \begin{bmatrix} y(t) \\ x(t) \end{bmatrix}, \quad F(t, z, p) = \begin{bmatrix} f_p(t, x, p) + f_x(t, x, p) y \\ f(t, x, p), \end{bmatrix}$$

$$G_k(t, z, p) = \begin{bmatrix} g_{k_p}(t, x, p) + g_{k_x}(t, x, p) y \\ g_k(t, x, p) \end{bmatrix}, \quad z_o = \begin{bmatrix} 0 \\ x_o \end{bmatrix}, \quad k = 1, \ldots, M. \tag{6}$$

For equation (5) one can write the corresponding Fokker-Planck equation for probability density φ of the solution z

$$\frac{\partial \varphi(z,t)}{\partial t} = -\frac{\partial^T}{\partial z}[F(t, z, p)\varphi(z, t)] + \frac{1}{2} tr \left\{ \frac{\partial}{\partial z} \frac{\partial^T}{\partial z} G(t, z, p) G(t, z, p)^T \varphi(z, t) \right\}, \tag{7}$$

where the symbol "tr" denotes the trace of the matrix and, $G = [G_1, \ldots, G_M]$.

Another approach of the study of sensitivity of stochastic dynamic system is to investigate the moment equations of the system (1). In many cases one can write the moment equations using one of the closure techniques.

$$\frac{dE[x^{[1]}(t)]}{dt} = F_1(E[x^{[1]}(t)], \ldots, \quad E[x^{[N]}(t)], p, t), \quad E[x^{[1]}(t_o)] = x_o,$$

$$\vdots$$

$$\frac{dE[x^{[N]}(t)]}{dt} = F_N(E[x^{[1]}(t)], \ldots, \quad E[x^{[N]}(t)], p, t), \quad E[x^{[N]}(t_o)] = x_o^{[N]}, \tag{8}$$

where

$x^{[r]}$ denotes the list of all r-th forms of the components of vector x, i.e.,

$$x_1^{r_1} x_2^{r_2} \ldots x_n^{r_n}, \quad \sum_{j=1}^{n} r_j = r, \tag{9}$$

F_r are nonlinear functions derived by applying one of closure techniques.

Differentiating the equations (9) with respect to parameter p_k we find the following deterministic differential equations (moment sensitivity equations).

$$\frac{dv^{[1]}}{dt} = F_{1p} + F_{11}v^{[1]} + \ldots + F_{1N}v^{[N]}, \quad v^{[1]}(t_o) = 0,$$

$$\frac{dv^{[N]}}{dt} = F_{Np} + F_{N1}v^{[1]} + \ldots + F_{NN}v^{[N]}, \quad v^{[N]}(t_o) = 0, \tag{10}$$

where

$$v^{[r]} = v^{[r]}(t) = \frac{\partial x^{[r]}(t)}{\partial p_k}, \quad F_{ip} = \frac{\partial F_i}{\partial p_k}, \quad F_{ij} = \frac{\partial F_i}{\partial x^{[j]}}. \tag{11}$$

Usually the considerations are limited to first and second order moments.

3. Quantitative Methods in Time Domain

As in deterministic theory the first works in which sensitivity method were applied to the study of stochastic models appeared in the field of automatic control. Evlanov [14,15] used sensitivity methods to the study of accuracy estimation of random parameter linear and nonlinear systems respectively.

Evlanov [14] considered linear nonstationary system

$$\dot{x}_i = \sum_{k=1}^{n} a_{ik}(t,p)x_k + b_i(t,p)u_i(t), \qquad (i = 1, \ldots, n), \tag{12}$$

where

p is vector of random parameters, $p = [p_1, \ldots, p_q]$, $p \epsilon Q \subset R^q$ is an open set, $a_{ik}(t,p), b_i(t,p) : R^+ \times R^q \to R^1$, $u(t)$ is an input signal which is split into two parts $S(t)$ deterministic and $\xi(t)$ stochastic, i.e.,

$$u_i(t) = S_i(t) + \xi_i(t). \tag{13}$$

The system is intended for transformation of the useful signal $S_i(t)$ into the output useful signal

$$x_T = L_T S_i(t), \tag{14}$$

where L_T is the operator of the required transformation. The error of the system is defined as

$$e(t) = x_i(t) - x_T(t). \tag{15}$$

For the estimation of the accuracy of the operation of the nonlinear system Evlanov chose a certain probabilistic characteristic of the error. With the goal of separate calculation of the influence of the random disturbances and of the random parameters, he introduced the conditional characteristic of the error.

$$\mu = E[\psi(e)|p], \tag{16}$$

which is the conditional expectation of the loss function $\psi(e)$ for a fixed vector of the random parameters. The conditional characteristic determines only the influence on the accuracy of the input random disturbances. For the calculation of the influence of the random parameters the unconditional probability measure was introduced

$$R = E[f(\mu)], \tag{17}$$

where $f(\mu)$ is the loss function and $E[\cdot]$ is the expectation.

Evlanov estimated the accuracy of operation of the system by the mean square error. In this case the loss function must be taken as quadratic. Here the conditional and unconditional mean square errors of the system are determined by the formulas

$$\mu(p) = E[e^2|p] = m_e^2(p) + D_e(p), \tag{18}$$

$$R = E[\mu^2] = m_\mu^2 + D_\mu, \tag{19}$$

where m_e, m_μ and D_e, D_μ are the mean values and variances of variables e and μ respectively.

The problem of estimating the accuracy consists in the calculation of the quantities μ and R for given probabilistic characteristics of the input signals and of the random parameters and for the known equations of the system. To solve this problem, it is convenient to apply sensitivity method in the following way. We represent the conditional characteristic of accuracy $\mu(p)$ by a Taylor series in the random parameters p_k with a center of development corresponding to the expectations of all the parameters

$$\mu(p) = \mu(m_p) + \sum_{k=1}^{q} \left(\frac{\partial \mu}{\partial p_k}\right)_{m_p} p_k^o + \frac{1}{2} \sum_{k=1}^{q} \sum_{l=1}^{q} \left(\frac{\partial^2 \mu}{\partial p_k \partial p_l}\right)_{m_p} p_k^o p_l^o + \dots, \tag{20}$$

where $p_k^o = p_k - m_{p_k}, m_p = E[p]$ and the index m_p in the partial derivatives denotes that these derivatives are calculated at the point $p = m_p$.

For the calculation of the expectation and variance of the quantity μ it is necessary to limit the number of terms of the series and to take into account the probabilistic characteristics of the random parameters. In what follows we shall assume that all the random parameters are independent.

For the calculation of only the linear terms of the series we obtain the following formulas for the unconditional expectation and variance:

$$m_\mu^{(1)} = \mu(m_p), \qquad D_\mu^{(1)} = \sum_{k=1}^{q} \left(\frac{\partial \mu}{\partial p_k}\right)^2_{m_p} D_{p_k}. \tag{21}$$

Taking into account that the conditional measure $\mu(p)$ is a function of the expectation and variance of the error (18), the partial derivatives of $\mu(p)$ with respect to the parameters can be represented in the form

$$\frac{\partial \mu}{\partial p_k} = 2m_e \frac{\partial m_e}{\partial p_k} + \frac{\partial D_e}{\partial p_k}, \tag{22}$$

$$\frac{\partial^2 \mu}{\partial p_k \partial p_l} = 2m_e \frac{\partial m_e}{\partial p_k}\frac{\partial m_e}{\partial p_l} + 2m_e \frac{\partial^2 m_e}{\partial p_k \partial p_k} + \frac{\partial^2 D_e}{\partial p_k \partial p_l}. \tag{23}$$

Thus, to complete the calculations it is necessary to determine the partial derivatives with respect to the parameters of the expectation and of the error. We note that these derivatives must be calculated for the expectations of the parameters. To do this, Evlanov [14] used the explicit formula for solution, i.e.,

$$x(t) = \int_{t_o}^{t} g(t,\tau,p) u(\tau) d\tau \tag{24}$$

where $g(t,\tau,p) = [g_{ij}(t,\tau,p)]$ is a matrix Green function of linear system (12). Hence,

$$\left(\frac{\partial m_e}{\partial p_k}\right)_{m_p} = \sum_{j=1}^{n} \int_{t_o}^{t} \frac{\partial g_{ij}(t,\tau,m_p)}{\partial m_{p_k}} m_{u_j}(\tau) d\tau, \tag{25}$$

$$\left(\frac{\partial D_e}{\partial p_k}\right)_{m_p} = \sum_{j=1}^{n} \int_{t_o}^{t}\int_{t_o}^{t} \left\{ \frac{\partial g_{ij}(t,\tau,m_p)}{\partial m_{p_k}} g_{ij}(t,\tau',m_p) + \right.$$

$$\left. + g_{ij}(t,\tau,m_p) \frac{\partial g_{ij}(t,\tau',m_p)}{\partial m_{p_k}} \right\} K_{u_j}(\tau,\tau') d\tau d\tau', \tag{26}$$

where $m_{p_k} = E[p_k]$, $\quad m_u = E[u(\tau)]$, $\quad K_{u_j}(\tau,\tau') = E[u_j(\tau)u_j(\tau')]$.

The quantities $\frac{\partial g_{ij}}{\partial m_{p_k}}$ can be calculated in two different ways. For instance, Evlanov proposed to differentiate the equation (12) with respect to parameter p_k. Introducing the notation $y_{ik} = \frac{\partial x_i}{\partial p_k}$ one can obtain

$$\dot{y}_{ik} = \sum_{i=1}^{n} a_{ij}(t,p) y_{jk} + \sum_{j=1}^{n} \frac{\partial a_{ij}(t,p)}{\partial p_k} x_j + \frac{\partial b_j(t,p)}{\partial p_k} u_j \tag{27}$$

and then to the equations (12) and (27) to apply the method of adjoint systems. The other possibility is to find the differential equation for Green function

$$\frac{dg_{ij}(t,\tau,p)}{dt} = \sum_{l=1}^{n} a_{il}(t,p)g_{ij}(t,\tau,p) + b_j(t,p)\delta(t-\tau)\delta_{ji}, \tag{28}$$

where $\delta(\cdot)$ is impulse function, δ_{ij} are the Kronecker symbols.

Differentiating with respect to parameter p_k, changing the differentiation with respect to time and parameter p_k and next introducing the weight function $w_{ji}(t,\tau,p)$ corresponding to the single coefficient $b_i(\tau)$, i.e.,

$$g_{ij}(t,\tau,p) = b_i(\tau,p)w_{ij}(t,\tau,p)$$

and replacing the random parameters by their expectations Evlanov [14] obtained the computational formula

$$\left(\frac{\partial g_{ij}(t,\tau,p)}{\partial p_k}\right)_{m_p} = w_{ij}(t,\tau,m_p)\left(\frac{\partial b_j(\tau,p)}{\partial p_k}\right)_{m_p} +$$

$$+ b_j(\tau,m_p)\sum_{r=1}^{n}\sum_{\nu=1}^{n}\int_{\tau}^{t}\left(\frac{\partial a_{\nu r}(\sigma,p)}{\partial p_k}\right)_{m_p} w_{ir}(t,\sigma,m_p)w_{\nu j}(\sigma,\tau,m_p)d\sigma. \tag{29}$$

Evlanov considered also higher order approximation and stationary case. In the second paper Evlanov [15] extended this approach for nonlinear systems

$$\dot{x}_i = \varphi_i(x,p,t) + b_i(t,p)u_i(t), \quad x_i(0) = p_{oi}, \qquad (i = 1,\ldots n), \tag{30}$$

where φ_i are nonlinear functions, and p_{oi} are random initial conditions he applied to the system (30) statistical linearization for fixed random parameters $p = p_f$, i.e., the nonlinear functions φ_i are approximate by linear relations.

$$\varphi_i(x,p_f,t) \simeq \varphi_{io}(m_x,\Theta_x,p_f,t) + \sum_{i=1}^{n} k_{ij}(m_x,\Theta_x,p_f,t)x_j^o, \qquad (i = 1,2,\ldots n), \tag{31}$$

where m_x is the expectation vector of the variables, Θ_x is the matrix of correlation moments of the variables, $\varphi_{i0}(m_x,\Theta_x,p_f,t)$ is the statistical characteristic of the nonlinear component, and k_{ij} is the statistical coefficient of the intensification of fluctuations. In the statistical linearization we use the assumption on the normality of the distribution of the input variables of the nonlinear characteristics. In this case

$$k_{ij} = \frac{\partial \varphi_{i0}}{\partial m_j}. \tag{32}$$

Substituting the functions φ_i from (31) into Eq. (30) we obtain

$$\dot{x}_i = \varphi_{i0}(m_x,\Theta_x,p_f,t) + \sum_{j=1}^{n} k_{ij}(m_x,\Theta_x,p_f,t)x_j^o + b_i(t,p_f)u_i(t) \qquad i = 1,2,\ldots,n. \tag{33}$$

Hence one can calculate the moment equations for the system (23) replacing fixed value of parameter by its expectation, i.e.,

$$p_f = E[p] = m_p, \tag{34}$$

$$\dot{m}_{x_i} = \varphi_{i0}(m_x, \Theta_x, m_p, t) + b_i(t, m_p) m_{u_i}(t), \quad m_{x_i}(t_o) = m_{i0}, \tag{35}$$

$$\dot{\Theta}_{x_{ij}} = \sum_{l=1}^{n} [k_{il}(m_x, \Theta_x, m_p, t)\Theta_{x_{jl}} + k_{jl}(m_x, \Theta_x, m_p, t)\Theta_{x_{il}}] +$$

$$+ \frac{1}{2} b_i(t, m_p) b_j(m_p, t)(G_i + G_j), \quad \Theta_{x_{ij}}(t_o) = \Theta_{oij}, \tag{36}$$

where G_i and G_j are the intensities of the input white noise at the i and j inputs of the system and m_{io} and Θ_{oij} are the mean values and correlation moments of the random initial conditions.

For the determination of the partial derivatives of the expectations and the correlation moments with respect to the random parameters for the values of the parameters equal to their expectations, we differentiate the system of Eqs. (35) and (36) with respect to the parameter p_k and set $p_k = m_{p_k}$. Introducing the notation

$$V_{ik} = \left(\frac{\partial m_i}{\partial p_k}\right)_{m_p}, \quad V_{ij_k} = \left(\frac{\partial Q_{ij}}{\partial p_k}\right)_{m_p}, \quad k = 1, \ldots, q, \quad (i, j = 1, \ldots, n) \tag{37}$$

and changing the order of differentiation with respect to time and with respect to parameter, we write the following system of nonlinear differential equations

$$\dot{V}_{ik} = \sum_{l=1}^{n} \left(\frac{\partial \varphi_{i0}}{\partial m_{x_l}}\right)_{m_p} V_{lk} + \left(\frac{\partial \varphi_{i0}}{\partial p_k}\right)_{m_p} + \left(\frac{\partial b_i}{\partial p_k}\right)_{m_p} +$$

$$+ \sum_{l=1}^{n} \sum_{j=1}^{n} \left(\frac{\partial \varphi_{i0}}{\partial \Theta_{x_{ij}}}\right)_{m_p} V_{ljk}, \tag{38}$$

$$\dot{V}_{ijk} = \sum_{l=1}^{n} [k_{il}(m_x, \Theta_x, m_p, t) V_{jlk} + k_{jl}(m_x, \Theta_x, m_p, t) V_{ilk}] +$$

$$+ \sum_{l=1}^{n} \sum_{\nu=1}^{n} \left[\left(\frac{\partial k_{i0}}{\partial m_{x_l}}\right)_{m_p} V_{lk} \Theta_{x_{j\nu}} + \left(\frac{\partial k_{j\upsilon}}{\partial m_{x_l}}\right)_{m_p} V_{lk} \Theta_{x_{i\nu}} \right] +$$

$$+ \sum_{l=1}^{n} \sum_{\mu=1}^{n} \sum_{\nu=1}^{n} \left[\left(\frac{\partial k_{i0}}{\partial m_{x_l}}\right)_{m_p} V_{lk} \Theta_{x_{j\nu}} + \left(\frac{\partial k_{j\upsilon}}{\partial m_{x_l}}\right)_{m_p} V_{lk} \Theta_{x_{i\nu}} \right] +$$

$$+ \sum_{l=1}^{n} \sum_{\mu=1}^{n} \sum_{\nu=1}^{n} \left[\left(\frac{\partial k_{i\upsilon}}{\partial \Theta_{x_{l\mu}}}\right)_{m_p} V_{l\mu_k} \Theta_{x_{j\nu}} + \left(\frac{\partial k_{j\upsilon}}{\partial \Theta_{x_{l\mu}}}\right)_{m_p} V_{l\mu k} \Theta_{x_{i\nu}} \right] +$$

$$+ \frac{1}{2} \left[\left(\frac{\partial b_i}{\partial p_k}\right)_{m_p} b_j(t, m_p) + b_i(t, m_p) \left(\frac{\partial b_i}{\partial p_k}\right)_{m_p} \right] (G_i + G_j), \quad V_{ijk}(t_o) = 0, \tag{39}$$

$$(k = 1, \ldots, q), (i, j = 1, \ldots n).$$

The possibility of changing the order of differentiation with respect to time and the parameters is based on the assumption that the expectations and correlation moments of the variables have continuous second mixed derivatives with respect to time and to the parameters. This continuity holds in many practical problems.

Summarizing, Evlanov proposed the following procedure of estimating the accuracy of nonlinear system with random parameters.

1. By simultaneous integration of the systems of Eqs. (35-39) we determine the quantities $m_{x_i}, \Theta_{x_{ij}}, V_{ij_k}$
2. According to (15) we calculate the expectation and variance of the error $m_{e_i} = m_{x_i} - m_T, \quad D_e = \Theta_p$ and the partial derivatives of the following quantities

$$\left(\frac{\partial m_{e_i}}{\partial p_k}\right)_{m_p} = V_{i_k}, \quad \left(\frac{\partial De_i}{\partial p_k}\right)_{m_p} = V_{ii_k} \quad (k = 1, \ldots, r) \tag{40}$$

3. Using relations (22), (23) we determine the quantities $\left(\frac{\partial \mu}{\partial p_k}\right)_{m_p}$
4. Using relations (21) we calculate m^1_μ and D^1_μ
5. Using relation (19) we determine $R^1 = R$
6. The deterioration of the accuracy of the system on account of the randomness of the parameters can be estimated by the relation

$$\beta = \frac{\sqrt[4]{R^1}}{\sqrt{m^1_\mu}}. \tag{41}$$

7. The estimate of the influence of each parameter on the deterioration of the accuracy is carried out by determining the quantities

$$\beta_i = \frac{\sqrt[4]{R^1_i}}{\sqrt{m^1_\mu}} \qquad i = 1, \ldots r, \tag{42}$$

where R^1_i is the unconditional characteristic of the accuracy, calculated with consideration of the randomness of only the i-th parameter.

This procedure was applied to simple one dimensional system

$$\dot{x} = -p \cdot sgnX + g \cdot \dot{\xi}(t), \tag{43}$$

where p is a parameter random variable with known expectation m_p and variance $D_p, \dot{\xi}(t)$ is standard white noise, g is a known parameter. In this case Evlanov showed that $\beta = 1.37$, it means, the accuracy deteriorated by 37%. This approach has been also discussed and developed in the book written by Evlanov and Konstantinov [16].

The general case of nonlinear systems with stochastic coefficients and external excitations described by equation (1) has been studied by Socha [12,17]. He considered the cases of white and coloured noise by assumption that both initial conditions and parameter are constant. We present briefly this approach because it will be convenient to quote some other results using the same notation.

First we quote definitions of sensitivity measures.

3.1 Sensitivity Measures

In section 2 we discussed sensitivity processes which are functions not only of a nominal parameter vector p_0 but also functions of time and probability space. Therefore it is desirable to characterize the sensitivity of a system simply by a number, rather than by a stochastic process. Sensitivity definitions of this kind shall be termed sensitivity measures. There are various ways to exclude time and randomness from the sensitivity definition, depending on what sort of study of the dynamic system one is making. We introduce the following classification:

1. The output sensitivity measures

$$E\int_{t_o}^{T}[y_i^{[r]}(t)]^T Q_i [y_i^{[r]}(t)]dt = \delta_1^i(t_o,T) \qquad \text{integral measure} \tag{44}$$

$$\sup_{t\in[t_o,T]} |E[y_i^{[r]}(t)| = \delta_2^i(t_o,T) \qquad \text{supremum measure} \tag{45}$$

Here the Q_i are matrices of weight coefficients $i = 1,\ldots,n$, $r = 1,\ldots,N$, $|.|$ denotes the Euclidean norm.

2. The domain sensitivity measures

One can define also the domain sensitivity measures

$$P\{y_i(t) \in A\} = \delta_3^i(t_o,T,A) \tag{46}$$

In particular case when the set A is specified by Euclidean norm the following measure was proposed by Socha [18]

$$P\{|y_i(t)|^2 < \varepsilon | x(t_o) = x_o, y(t_o) = 0\} = \delta_4^i(t_o,T,\delta) \tag{47}$$

3. Moment sensitivity measures

Using the solutions of moment sensitivity equations the following measures were defined by Socha [17]

$$\int_{t_o}^{T}[v_i^{[r]}(t)]^T Q_i [v_i^{[r]}(t)]dt = \delta_5^i(t_o,T) \qquad \text{integral measure} \tag{48}$$

$$\sup_{t\in[t_o,T]} |[v_i^{[r]}(t)| = \delta_6^i(t_o,T) \qquad \text{supremum measure} \tag{49}$$

The main results for analysis of output sensitivity measures were done in [12]. In that paper the linear systems and three nonlinear oscillators with stochastic coefficients under noise excitation were analyzed. The cases of white and coloured noise were considered. For illustration we quote some of these results.

3.2 Linear Systems

First we consider linear dynamical systems with white noise coefficients and white noise excitations described by the Ito stochastic differential equation.

$$dx(t) = [A_0(t,p) + A(t,p)x(t)]dt + \sum_{k=1}^{M}[g_{k0}(t,p) + g_k(t,p)x(t)]d\xi_k(t), \qquad x(t_0) = x_0, \tag{50}$$

where $x(t) = [x_1(t),\dots,x_n(t)]^T$ is the state vector, $x_0 = [x_{10},\dots,x_{n0}]^T$ is the deterministic vector of initial conditions, $p = [p_1,\dots,p_q]^T$ is a vector of parameters, $A_0(t,p) = [a_0^1(t,p),\dots,a_0^n(t,p)]^T$ and $g_{k0}(t,p) = [\sigma_{k0}^1(t,p),\dots,\sigma_{k0}^n(t,p)]^T$ are n-dimensional deterministic vectors, $A(t,p) = [a_k^i(t,p)]$ and $g_k(t,p) = [\sigma_{km}^i(t,p)]$ are $n \times n$ dimensional deterministic matrices, and $\xi_k(t)$, $(k = 1,\dots,m)$ are independent Wiener processes. From relation (3) one has the sensitivity stochastic differential equation with respect to the parameter p_j,

$$dy(t) = \left[\frac{\partial A_0(t,p)}{\partial p_j} + \frac{\partial A(t,p)}{\partial p_j}x(t) + A(t,p)y(t)\right]dt +$$

$$+\sum_{k=1}^{M}\left[\frac{\partial g_{k0}(t,p)}{\partial p_j} + \frac{\partial g_k(t,p)}{\partial p_j}x(t) + g_k(t,p)y(t)\right]d\xi_k, \qquad y(t_0) = 0, \tag{51}$$

where $y(t) = \partial x(t)/\partial p_j$ is the output sensitivity process. To simplify the notation the arguments t and p are omitted in the next equations. Since the stochastic differential equation (51) is also linear one can apply the standard Ito formula and obtain the moment equations in closed form:

$$\frac{dm}{dt} = \mathcal{A}_0^w + \mathcal{A}^w m, \qquad m(t_0) = [x_0^T, 0^T]^T, \tag{52}$$

$$\frac{d\Gamma}{dt} = m\mathcal{A}_0^{wT} + m\mathcal{A}_0^w m^T + \Gamma\mathcal{A}^{wT} + \mathcal{A}^w\Gamma + \sum_{k=1}^{m}[\mathcal{H}_{k0}^w(\mathcal{H}_{k0}^w)^T + \mathcal{H}_k^w m(\mathcal{H}_{k0}^w)^T +$$

$$+\mathcal{H}_{k0}^w m^T(\mathcal{H}_k^w)^T + \mathcal{H}_k^w\Gamma(\mathcal{H}_k^w)^T], \quad \Gamma(t_o) = E[z(t_0)z^T(t_0)], \tag{53}$$

where

$$m = E[z], \quad \Gamma = E[zz^T], \quad z = [y^T, x^T]^T, \quad \mathcal{A}_0^w = [A_{0p}^T, A_o^T]^T,$$

$$\mathcal{A}^w = \begin{bmatrix} A & A_p \\ O & A \end{bmatrix}, \quad \mathcal{H}_{k0}^w = \begin{bmatrix} g_{kop} \\ g_{k0} \end{bmatrix}, \quad \mathcal{H}_k^w = \begin{bmatrix} g_k & g_{kp} \\ O & g_k \end{bmatrix},$$

$$A_{0p} = \frac{\partial A_0}{\partial p_j}, \quad A_p = \frac{\partial A}{\partial p_j}, \quad g_{k0p} = \frac{\partial g_{k0}}{\partial p_j}, \quad g_{kp} = \frac{\partial g_k}{\partial p_j}. \tag{54}$$

As an example we consider the linear oscillator under parametric and external white noise excitations described by equation (50) for $n = 2, \gamma = 3$: i.e.,

$$dx = Axdt + H_1xd\xi_1 + H_2xd\xi_2 + H_3d\xi_3, \quad x(t_0) = x_0, \tag{55}$$

where

$$x = \begin{bmatrix} x_1 \\ x_2 \end{bmatrix}, \quad A = \begin{bmatrix} 0 & 1 \\ a_{21} & a_{22} \end{bmatrix}, \quad H_1 = \begin{bmatrix} 0 & 0 \\ h_1 & 0 \end{bmatrix},$$

$$H_2 = \begin{bmatrix} 0 & 0 \\ 0 & h_2 \end{bmatrix}, \quad H_3 = \begin{bmatrix} 0 \\ \delta \end{bmatrix}, \quad x_o = \begin{bmatrix} x_{10} \\ x_{20} \end{bmatrix}. \tag{56}$$

$-a_{21}, a_{22}, h_1, h_2$ and δ are positive constants and the $\xi_i(t)$ are independent standard Wiener processes. Hence the output sensitivity processes with respect to, for instance, a parameter a_{22} and the state vector satisfy the stochastic differential equation

$$dz^w = A^w z^w dt + H_1^w z^w d\xi_1 + H_2^w z^w d\xi_2 + H_3^w d\xi_3, \quad z^w(t_0) = z_0^w, \tag{57}$$

where

$$A^w = \begin{bmatrix} 0 & 1 & 0 & 0 \\ a_{21} & a_{22} & 0 & 1 \\ 0 & 0 & 0 & 1 \\ 0 & 0 & a_{21} & a_{22} \end{bmatrix}, \quad H_1^w = \begin{bmatrix} 0 & 0 & 0 & 0 \\ h_1 & 0 & 0 & 0 \\ 0 & 0 & 0 & 0 \\ 0 & 0 & h_1 & 0 \end{bmatrix}, \quad H_2^w = \begin{bmatrix} 0 & 0 & 0 & 0 \\ 0 & h_2 & 0 & 0 \\ 0 & 0 & 0 & 0 \\ 0 & 0 & 0 & h_2 \end{bmatrix},$$

$$z^w = [y_1 \quad y_2 \quad x_1 \quad x_2]^T, \qquad z_0^w = [0 \quad 0_0 \quad x_{10} \quad x_{20}]^T, \qquad H_3^w = [0 \quad 0 \quad 0 \quad \delta]^T,$$

$$y_1 = \partial x_1/\partial a_{22}, \qquad y_2 = \partial x_2/\partial a_{22}. \tag{58}$$

The first two moment equations are hence derived as

$$\frac{dm}{dt} = A^w m, \qquad m(t_0) = z_0^w, \tag{59}$$

$$\frac{d\Gamma^w}{dt} = \Gamma A^{w^T} + A^w \Gamma + H_1^w \Gamma H_1^{w^T} + H_2^w \Gamma H_2^{w^T} + H_3^w (H_3^w)^T, \quad \Gamma^w(t_0) = \Gamma_0 \tag{60}$$

Here

$$m = E[z^w], \quad \Gamma^w = E[z^w (z^w)^T], \quad \Gamma_0 = z_0^w (z_0^w)^T. \tag{61}$$

3.3 Duffing Oscillator

As an example of the nonlinear oscillator we consider the Duffing oscillator with white noise coefficients and white noise excitations described by the Ito stochastic differential equation

$$dx = [Ax + b_1(x)]dt + H_1 x d\xi_1 + H_2 x d\xi_2 + H_3 d\xi_3, \quad x(t_0) = x_0, \tag{62}$$

where all quantities except $b_1(x)$ are the same as in the linear case and the vector $b_1(x)$ is defined by

$$b_1(x) = \begin{bmatrix} 0 \\ \varepsilon(x_1)^3 \end{bmatrix}, \quad (\varepsilon \text{ is a negative constant}). \tag{63}$$

Hence the output sensitivity process with respect to, for instance, a parameter a_{22} and the state vector satisfy the following stochastic differential equation

$$dz^w = (A^w z^w + B_1^w)dt + H_1^w z^w d\xi_1 + H_2^w z^w d\xi_2 + H_3^w d\xi_3, \quad z^w(t_0) = z_0^w, \tag{64}$$

where $z^w, A^w, H_1^w, H_2^w, H_3^w$ are as defined in equations (58) and

$$B_1^w = \begin{bmatrix} 0 \\ 3\varepsilon(x_1)^2 y_1 \\ 0 \\ \varepsilon(x_1)^3 \end{bmatrix}. \tag{65}$$

Hence one derives the first two moment equations in closed form by using the cumulative closure technique:

$$\frac{dm}{dt} = A^w m + C_1^w, \qquad m(t_0) = z_0^w, \tag{66}$$

$$\frac{d\Gamma^w}{dt} = \Gamma^w A^{wT} + A^w \Gamma^w + H_1^w \Gamma H_1^{wT} + H_2^w \Gamma H_2^{wT} + H_3^w (H_3^w)^T + D_1^w, \quad \Gamma^w(t_0) = \Gamma_0 \tag{67}$$

where $m, \Gamma^w, z_0^w, \Gamma_0$ are defined as in the linear case and the vector C_1^w and matrix D_1^w are defined as follows

$$C_1^w = E[B_1^w] = \begin{bmatrix} 0 \\ \varepsilon(m_1\Gamma_{33} + 2m_3\Gamma_{13} - 2(m_3)^2 m_1) \\ 0 \\ \varepsilon(3m_3\Gamma_{33} - 2(m_3)^3) \end{bmatrix} \tag{68}$$

$$D_1^w = \varepsilon E \left[\begin{bmatrix} 0 & 3(x_1)^2(y_1)^2 & 0 & (x_1)^3 y_1 \\ 3(x_1)^2(y_1)^2 & 6(x_1)^2 y_1 y_2 & 3(x_1)^3 y_1 & (x_1)^3 y_2 + 3(x_1)^2 x_2 y_2 \\ 0 & 3(x_1)^3 y_1 & 0 & (x_1)^4 \\ (x_1)^3 y_1 & (x_1)^3 y_2 + 3(x_1)^2) x_2 y_2 & (x_1)^4 & (x_1)^3 x_2 \end{bmatrix} \right] \tag{69}$$

Similar procedure can be applied to the systems with coloured noise coefficients and excitations when the coloured noise can be expressed as an output of linear filter excited by white noise. Such analysis was done in [12].

Comparisons between the output sensitivity measures in the case when the coefficients are deterministic, white noise and coloured noise for the linear oscillator are illustrated in Figure 1 and for the nonlinear Duffing oscillator in Figure 2. The integral measure in this example is defined as follows:

$$\rho_1^{a_{22}}(T) = \int_0^T E\left[\left(\frac{\partial x_1}{\partial a_{22}}(t)\right)^2\right] dt. \tag{70}$$

The linear filters are described by equation

$$d\eta_i = \alpha_i \eta_i dt + \beta_i d\xi_i, \quad \eta_i(t_0) = 0, \qquad i = 1, 2, \tag{71}$$

where η_i is modeled coloured noise,and ξ_i is standard Wiener process.

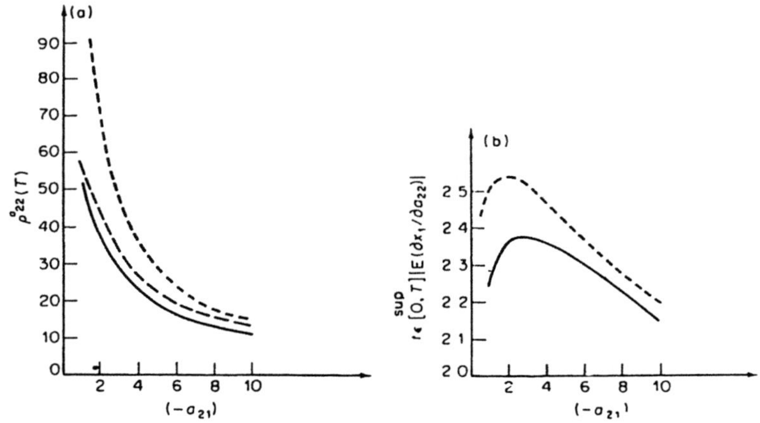

Figure 1. Output sensitivity measure of a linear oscillator, equation (55), under parametric noise excitations. (a) Integral measure; (b) supremum measure $\sup_{t\epsilon[0,T]} |E(\partial x_1/\partial a_{22})|$; $a_{22} = -2, h_1 = 0, h_2 = 2, \alpha = 0.5, \alpha_2 = 1, \beta_1 = 1, \beta_2 = 1, \delta = 0, t_0 = 0, T = 40, x_1(t_0) = 10, x_2(t_0) = 0$. — , Deterministic case; – –, white noise case; - - - -, coloured noise case.

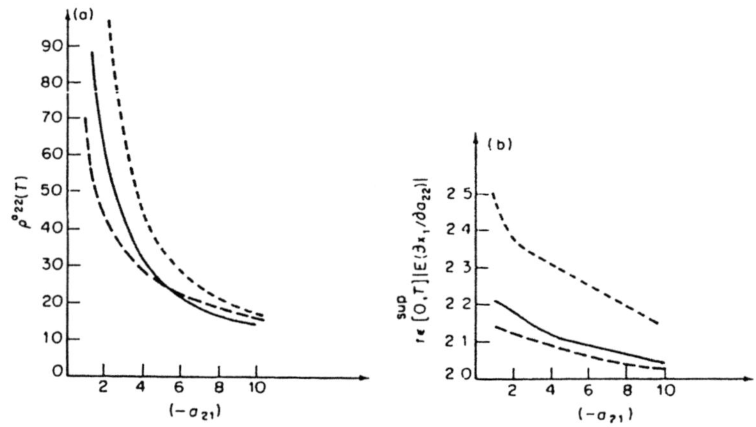

Figure 2. Output sensitivity measure of nonlinear Duffing oscillator, equation (62), under parametric excitations. (a) Integral measure; (b) supremum $\sup_{t\epsilon[0,t]} |E(\partial x_1/\partial a_{22})|$; $\alpha_{22} = -2, h_1 = 0, h_2 = 1, \alpha_1 = 0.5, \alpha_2 = 1, \beta_1 = 1, \beta_2 = 1, \delta = 0, \varepsilon = -0.05, t_0 = 0, T = 50, x_1(t_0) = 10$, $x_2(t_0) = 0$. — , Deterministic case; – –, white noise case; - - - -, coloured noise case.

The output sensitivity measures of linear oscillators for different types of disturbances have similar properties for a given set of parameters. For the measures $\sup_{t\in[0,T]} |E(\partial x_1/\partial a_{22})|$ for example, a maximum appears for the parameter $(-a_{21})$ in the interval [1,2]. The maximal values differ between themselves.

Since the operations of differentiating and averaging are linear one can prove that the moment sensitivity equations are the same as output sensitivity equations only for linear systems. A comparison of moment and sensitivity measures for linear and nonlinear stochastic dynamic systems was given in [17]. For illustration we show the moment equations (first and second order) and corresponding moment sensitivity equations.

Moment Equations:

$$\frac{dm}{dt} = A_0 + Am, \qquad m(t_0) = x_0, \tag{72}$$

$$\frac{d\Gamma}{dt} = mA_0^T + A_0 m^T + \Gamma A^T + A\Gamma + \sum_{k=1}^{M} [g_{k0}(g_{k0})^T + g_k m (g_{k0})^T +$$

$$+ g_{k0} m^T (g_k)^T + g_k \Gamma (g_k)^T], \quad \Gamma(t_0) = x_0 (x_0)^T, \tag{73}$$

where

$$m(t) = E[x(t)], \quad \Gamma(t) = E[x(t)x^T(t)]. \tag{74}$$

Moment Sensitivity Equations:

$$\frac{dm_p}{dt} = A_{0p} + A_p m + A m_p, \qquad m_p(t_0) = 0, \tag{75}$$

$$\frac{d\Gamma_p}{dt} = m_p A_o^T + m A_{op}^T + A_{op} m^T + A_o m_p^T + \Gamma_p A^T + \Gamma A_p^T + A_p \Gamma + A \Gamma_p +$$

$$+ \sum_{k=1}^{M} [g_{kop}(g_{ko})^T + g_{ko}(g_{kop})^T + g_{kp} m (g_{ko})^T + g_k m_p (g_{ko})^T +$$

$$+ g_k m (g_{kop})^T + g_{kop} m^T (g_k)^T + g_{ko} m_p^T (g_k)^T + g_{ko} m^T (g_{kp})^T +$$

$$+ g_{kp} \Gamma (g_k)^T + g_k \Gamma_p (g_k)^T + g_k \Gamma (g_{kp})^T], \quad \Gamma_p(t_o) = 0, \tag{76}$$

where subindex $(\ )_p$ denotes derivatives with respect to a parameter p_i.

Socha [17] also derived moment sensitivity equations for linear systems with coloured noise parameters and for Duffing oscillator with both white and coloured noise parameter and excitations. Unfortunately there is not done a comparison with simulations.

Although the quoted results were derived for parameter of the system it can be also done for initial conditions, i.e., instead of sensitivity functions $\frac{\partial x}{\partial p}$ or $\frac{\partial v}{\partial p}$ one can consider the case when p is an initial condition. An example of sensitivity analysis with respect to initial conditions for Duffing oscillator excited by random noise has been presented by Szopa [19].

We have discussed only the first and third class of sensitivity methods, i.e., the output and moment sensitivity. Unfortunately, the quantitative methods of the study of area sensitivity were not developed in the literature. One of the possibilities of study domain sensitivity measures is to apply the joint possibility density $\varphi(z,t) = \varphi(y,x,p,t)$ which satisfies the corresponding Fokker-Planck equation. However, in this case except trivial examples only the numerical calculations are necessary.

It should be stressed that the application of the sensitivity analysis in the time domain require of solutions of large number of differential equations and in the case of nonlinear system application of closure techniques.

4. Approximate Methods

Sues, Wen and Ang [20] used moment sensitivity variables to describe the uncertainty in the structural response to a given ground motion intensity. They considered the uncertainty associated with the structural ground motion parameters (e.g., stiffness, filter frequency, etc.), the uncertainty inherent in the mathematical idealizations, and the uncertainty associated with the randomness of the earthquake time history. The contribution of the various uncertainties to the overall variance of the structural response may be considered approximately by a first-order analysis as follows:

$$Var[X] \cong (E[\hat{X}])^2 \cdot Var[N] + (E[N])^2 \cdot Var[\hat{X}] \tag{77}$$

in which $E[\hat{X}]$ = the mean response obtained from the model using mean parameter values; $E[N]$ and $Var[N]$ represent the expected bias and variance of the error in the response due to the mathematical idealization of the structure; and $Var[\hat{X}]$ = the sum of the variances of the response associated with the parameter variables and the randomness of the loading history.

The variance of the response associated with the parameter variables may be obtained by a first-order approximation as

$$Var_P[\hat{X}] \cong \sum_i \sum_j \left(\frac{\partial \hat{X}}{\partial p_i}\right)_{\bar{P}} \left(\frac{\partial \hat{X}}{\partial p_j}\right)_{\bar{P}} \rho_{ij}\sigma_{p_i}\sigma_{p_j} \tag{78}$$

in which $\bar{P}$ = the set of mean model parameters; ρ_{ij} = the correlation coefficient of the ith and jth parameter; and σ_{p_i} = the standard deviation of the ith parameter.

To obtain the derivatives of the response statistics with respect to the various model parameters Sues, Wen and Ang [20] proposed first to apply equivalent linearization and then to find the differential equation for covariance matrix and next the moment sensitivity equations. This procedure has been applied to the analysis of a four-story steel frame building.

Another approach of approximation has been presented by Szopa [21]. He considered van der Pol equation in the form

$$\ddot{x} - \rho(1 - \gamma x^2)\dot{x} + c_1 x + c_3 x^3 = B\cos\xi t, \tag{79}$$

where $\rho \neq 0, \gamma, c_1, c_3, B,$ and ξ are constant with respect to the time t, and $x(0) = x_0$ and $\dot{x}(0) = x_1$ are initial conditions; $(\dot{})$ denotes differentiation with respect to time. The excitation is assumed to be a random process of the form

$$B \cos \xi t = B(\omega) \cos[\xi(\omega) t], \tag{80}$$

where $B(\omega)$ and $\xi(\omega)$ are random variables with uniform distribution in the intervals $[B_1, B_2]$ and $[N_1, N_2]$ respectively.

The stochastic sensitivity equations first and second order with respect to parameter p have the form

$$\ddot{y} - p(1 - \gamma x^2)\dot{y} + (2p\gamma x\dot{x} + c_1 + 3c_3 x^2)y = (1 - \gamma x^2)\dot{x}, \tag{81}$$

$$\ddot{\nu} - p(1 - \gamma x^2)\dot{\nu} + (2p\gamma x\dot{x} + c_1 + 3c_3 x^2)\nu = (1 - \gamma x^2)\dot{y} - 2\gamma x\dot{x}y. \tag{82}$$

where

$$y = \frac{\partial x}{\partial p}, \qquad \nu = \frac{\partial^2 x}{\partial p^2}, \tag{83}$$

If the solution of the stochastic differential equation (79) and corresponding sensitivity equations (81),(82) is known for $p = p_0$, then the solution for arbitrary p can be presented in the Taylor series form to a

$$x(t, p, \omega) = x(t, p_0, \omega) + \sigma(t, p_0, \omega)\Delta p + (1/2!)\nu(t, p_0, \omega)(\Delta p)^2 + \dots, \tag{84}$$

where $\Delta p = p - p_0$.

To verify the accuracy of this approximation Szopa [21] used the simulations and considered the following measures $E[R_i], Var[R_i] \quad i = 0, 1, 2$, where

$$R_0(t, p_0, p, \omega = x(t, p, \omega) - x(t, p_0, \omega),$$

$$R_1(t, p_0, p, \omega = x(t, p, \omega) - [x(t, p_0, \omega) + \sigma(t, p_0, \omega)\Delta p],$$

$$R_2(t, p_0, p, \omega = x(t, p, \omega) - [x(t, p_0, \omega) + \sigma(t, p_0, \omega)\Delta p + (1/2!)\nu(t, p_0, \omega)(\Delta p)^2]. \tag{85}$$

Some other applications were given by Szopa [22]. This approach has been extended and developed by Pawleta and Socha [23] for the two dimensional linear system with stationary coloured noise parameters. In general case if the solution of the equation (1) is known for nominal value of the parameter vector p_0 then an approximate solution can be presented in the form

$$x(t, p) \cong x(t, p_o) + \sum_{i=1}^{q} \left.\frac{\partial x(t, p)}{\partial p_i}\right|_{p=p_o} \Delta p_i + \frac{1}{2}\sum_{k=1}^{q}\sum_{i=1}^{q} \left.\frac{\partial^2 x(t, p)}{\partial p_i \partial p_k}\right|_{p=p_0} \Delta p_i \Delta p_k + \dots, \tag{86}$$

where the partial derivatives $\frac{\partial x}{\partial p_i}, \frac{\partial^2 x}{\partial p_i \partial p_k}$ one can find from sensitivity equations by applying one of the closure techniques.

Another approach of studying moment sensitivity has been proposed by Jumarie [24,25]. First he proved for special case of the system (34) ($n = 1, M = 1$)

$$dx = f(x,p,t)dt + g(x,p,t)d\xi, \tag{87}$$

that the k-order moments of solution of (87) $m_k(t) = E[x^k(t)]$ satisfy the relations

$$\dot{m}_1(t) = \int_{-\infty}^{+\infty} f(x,t)\phi(x,p,t)dx, \tag{88}$$

$$\dot{m}_k(t) = \int_{-\infty}^{+\infty} \left[kx^{k-1}f(x,t) + \frac{1}{2}k(k-1)x^{k-2}g^2(x,t)\right]\phi(x,p,t)dx \quad k \geq 2, \tag{89}$$

where $\phi(x,p,t)$ denote the probability density of the solution $x(t)$ and can be found from corresponding Fokker-Planck equation

$$-\frac{\partial\phi}{\partial t} = -\frac{\partial}{\partial x}(f\phi) + \frac{1}{2}\frac{\partial^2}{\partial x}(g^2\phi). \tag{90}$$

Next, Jumarie assumed that

$$m_k(t) + \mu_k(t) = E[x^k(t, p_o + \Delta p)] \tag{91}$$

and

$$f(x, p_o + \Delta p, t) \approx f(x, p_o, t) + \left.\frac{\partial f}{\partial p}\right|_{p=p_o} \Delta p, \tag{92}$$

where $\Delta p = p - p_o$ and p_o is a nominal value of parameter then using the equalities (88) and (89) he obtained the following approximate relations

$$\dot{\mu}_1(t) = \int_{-\infty}^{+\infty} \left.\frac{\partial\phi(x,p,t)}{\partial p}\right|_{p_o} f(x,p,t)\Delta p dx + \int_{-\infty}^{+\infty} \left.\frac{\partial f(x,p,t)}{\partial p}\right|_{p_o} \phi(x,p_o,t)\Delta p dx, \tag{93}$$

$$\dot{\mu}_k(t) = \int_{-\infty}^{+\infty} \left[kx^{k-1}f(x,p_o,t) + \frac{1}{2}k(k-1)x^{k-2}g^2(x,p_o,t)\right] \times \left.\frac{\partial\phi(x,p,t)}{\partial p}\right|_{p_o} \Delta p dx +$$
$$+ \int_{-\infty}^{+\infty} \left[kx^{m-1}\left.\frac{\partial f(x,p,t)}{\partial p}\right|_{p_o} + \frac{1}{2}k(k-1)x^{k-2} \times \left.\frac{\partial g^2(x,u,t)}{\partial p}\right|_{p_o}\right]\phi(x,p_o,t)\Delta p dx, \quad k \geq 2. \tag{94}$$

This result seems to be incorrect because in equality (92) the derivative $\frac{\partial f}{\partial x}$ was dropped, i.e., the correct form should be

$$f(x, p_o + \Delta, t) \approx f(x, p_o, t) + \frac{\partial f}{\partial p}\Delta p + \frac{\partial f}{\partial x}\frac{\partial x}{\partial p}\Delta p. \tag{95}$$

But in this case, the output sensitivity equation should be taken into account..

Unfortunately the analytical consideration of accuracy of approximation by sensitivity functions, except the book written by Evlanov and Konstantinov [16] have not been done in the literature. The problem how big can be Δp is still open. However, from numerical examples it follows that reasonable approximations one can obtain if $\frac{\Delta p}{p_o}$ is less than 10%, where p_o is a nominal value of parameter.

5. Quantitative Methods in Frequency Domain

The analysis of the dynamic systems with random parameters under white noise excitations by using sensitivity functions on the basis of the frequency domain was presented by Igusa and Kiureghian [26] and also by Chen and Soong [27,28]. Igusa and Kiureghian used the sensitivity approach to the study of the reliability of structural systems with time-invariant uncertain parameters subjected to stochastic excitations. They have found the sensitivity of reliability index β with respect to variation of some parameters. To compute the partial derivatives $\frac{\partial \beta}{\partial p_i}$ they have used the spectral moment sensitivity functions. The obtained results were applied to the sensitivity analysis of two degree of freedom systems and also for three degree of freedom shear-building system and a seven degree of freedom primary-secondary system. The application of spectral moment sensitivity functions has been later developed by Chen and Soong [27,28]. They considered both linear and nonlinear n-degree of freedom primary-secondary structural system.

In the linear case the equation of motion has the form

$$M(p)\ddot{x}(t) + C(p)\dot{x}(t) + K(p)x(t) = F(t), \tag{96}$$

where $x(t)$ is n-dimensional displacement vector, $F(t)$ is the input vector, $M(p), C(p)$ and $K(p)$ are respectively $n \times n$ mass, damping and stiffness matrices dependent upon q-dimensional vector of parameters $p \varepsilon Q \subset R^q$. The components of vector p represent system parameters with uncertainties. It is written in the form

$$p = \text{diag}\ \{1 + \varepsilon_i\} p_0, \tag{97}$$

where p_0 represents nominal system parameter values. It is assumed that the random uncertain vector $\varepsilon^T = [\varepsilon_1, \ldots, \varepsilon_q]$ has zero mean and covariance matrix $E[\varepsilon\varepsilon^T] = [\mu_{ij}]$. In steady state case the spectral density of the solution of (96) is

$$S_{XX}(\omega, p) = H^*(j\omega, p) S_{FF}(\omega, p) H^T(j\omega, p), \tag{98}$$

where $S_{FF}(\omega, p)$ and $S_{XX}(\omega, p)$ are the spectral density functions of the excitations and the structural responses respectively; $H(j\omega, p)$ is the frequency response and the superscript $(\cdot)^*$ denotes complex conjugate matrix. Chen and Soong [27,28] showed that

$$E[S_{XX}(\omega, p)] = S_{XX}(\omega, p_0) + S_1(\omega, p_0), \tag{99}$$

where

$$[S_1(\omega, p_0)]_{ij} = (vec I)^T [S_{FF}(\omega) \otimes \sum_{r=1}^{q} \sum_{s=1}^{q} \frac{p_{or} p_{os}}{2} R_{rs}(\omega, p_0) \mu_{rs}] vec I \tag{100}$$

with

$$R_{r,s}(\omega,p_0) = h_i^*(\omega,p_0)\frac{\partial^2 h^T(\omega,p_0)}{\partial p_{or}\partial p_{os}} + \frac{\partial^2 h_i^*(\omega,p_0)}{\partial p_{or}\partial p_{os}} H_j^T(\omega,p_0) + \frac{\partial h_i^*(\omega,p_0)}{\partial p_{or}}\frac{\partial H_j^T(\omega,p)}{\partial p_{os}}. \tag{101}$$

Here the symbol $\otimes$ denotes the Kronecker product of two matrices, $vecI = [1,\ldots 0,1,\ldots 0,\ldots 1,\ldots 0]^T$.

Taking into account the relation (99) Chen and Soong proposed the following sensitivity measures

$$[\Gamma_k]_{ij} = \frac{[\Delta\Lambda_k]_{ij}}{[\Lambda_{ok}]_{ij}}, \qquad k = 0,1,2,\ldots, \tag{102}$$

where

$$\Delta\Lambda_k = \int_0^\infty \omega^k S_1(\omega,p_0)d\omega, \qquad k = 0,1,2\ldots, \tag{103}$$

$$\Lambda_{ok} = \int_0^\infty \omega^k S_{XX}(\omega,p_0)d\omega \qquad k = 0,1,2\ldots, \tag{104}$$

in practice the maximum displacement responses of structures are more significant. Chen and Soong showed by suitable assumptions that

$$\gamma_0 = \frac{\int_0^\infty S_1(\omega,p_0d\omega}{\int_0^\infty S_{XX}(\omega,p_0)d\omega} = \frac{\sigma_1^2}{\sigma_{X0}^2} = \frac{S_{1,max}^2}{X_{0,max}^2} \tag{105}$$

or

$$\frac{E[X_{max}^2]}{X_{o,max}^2} = 1+\gamma_0. \tag{106}$$

In the case when a primary-secondary structure is subjected to ground acceleration $\ddot{X}_o(t)$, the input term in Eq. (96) becomes

$$F(t) = -Mr\ddot{X}_0(t), \tag{107}$$

where

$$r^T = [1,1\ldots 1]. \tag{108}$$

Then it is convenient to define

$$T(\omega,p) = (H(j\omega,p)Mr)^*(H(j\omega,p)Mr)^T. \tag{109}$$

The equations (98) and (100) then take the forms

$$S_{XX}(\omega,p) = T(\omega,p)S_{\ddot{x}_o\ddot{x}_o} \tag{110}$$

and

$$S_1(\omega,p_0) = \sum_{r=1}^{q}\sum_{s=1}^{q}\left[\frac{p_{or}p_{os}}{2}\frac{\partial^2 T(\omega,p)}{\partial p_{or}\partial p_{os}}\right]\mu_{rs}S_{\ddot{x}_o\ddot{x}_o}. \tag{111}$$

This approach has been modified by Soong and Chen [27]. They have showed that the expectation of spectral moments Λ_k can be approximated

$$E[\lambda_k(p)] = \lambda_{ok}(p_0) + \sum_{r=1}^{q}\sum_{s=1}^{q}\left[\frac{p_{or}p_{os}}{2}\frac{\partial^2\lambda_{ok}(p_0)}{\partial p_{or}\partial p_{os}}\right]\mu_{rs} + \ldots \tag{112}$$

In the above, the q-dimensional vector p represents uncertain structural parameters $p_0 = [p_{01}, \ldots p_{0q}]^T$. The ratios

$$\gamma_k = \frac{\Delta\lambda_k(p_0)}{\lambda_{ok}(p_0)} = \frac{1}{\lambda_{ok}(p_0)} \sum_{r=1}^{q} \sum_{s=1}^{q} \left[\frac{p_{or} p_{os}}{2} \frac{\partial^2 \lambda_{ok}(p_0)}{\partial p_{or} \partial p_{os}} \right] \mu_{rs}, \quad k = 0, 1, 2 \tag{113}$$

can be considered as sensitivity factors since the zeroth-order spectral moment is related to peak response characteristics.

Similarly Chen and Soong showed that

$$\frac{E[X_{max}]}{X_{o,max}} = 1 + \nu_0, \tag{114}$$

where

$$\nu_0 = \frac{1}{\sqrt{\lambda_{ok}(p_0)}} \sum_{r=1}^{q} \sum_{s=1}^{q} \left[\frac{p_{or} p_{os}}{2} \frac{\partial^2 \sqrt{\lambda_{ok}(p_0)}}{\partial p_{or} \partial p_{os}} \right] \mu_{rs}. \tag{115}$$

The difference between the maximum responses with and without parameter uncertainties is

$$\frac{E[(\Delta X(p))^2]}{X^2_{o\,max}} = \frac{E[X^2_{max}] - 2E[X_{max} X_{o,max}]}{X^2_{0,max}} + 1 = \gamma_0 - 2\nu_0. \tag{116}$$

The variance of the difference is

$$\frac{Var[\Delta X(p)]}{X^2_{o,max}} = \frac{\sigma^2_{\Delta X(p)}}{X^2_{o,max}} = \frac{E[(\Delta X(p))^2] - [E(\Delta X(p))]^2}{X^2_{o,max}} = \gamma_0 - 2\nu_0 - \nu_0^2. \tag{117}$$

In this procedure of determining the sensitivity factor, the effect of the interaction between primary and secondary systems and nonclassical damping have been accounted for. This is superior to a totally decoupled analysis.

An alternative approach in sensitivity analysis is to consider eigenvalue sensitivity. Soong and Chen [27] adopted this approach for systems with random parameters.

Let $\Omega_i(p)$ be the ith natural frequency of the structure. Expanding $\Omega(p)$ about p_0 one can obtain

$$\Omega_i(p) = \Omega_{i0}(p_0) + \sum_{r=1}^{q} \left[p_{or} \frac{\partial \Omega_i(p_0)}{\partial p_{or}} \right] \varepsilon_{or} + \sum_{r=1}^{q} \sum_{s=1}^{q} \left[\frac{p_{or} p_{os}}{2} \frac{\partial^2 \Omega_i(p_0)}{\partial p_{or} \partial p_{os}} \right] \varepsilon_r \varepsilon_s \ldots \tag{118}$$

Neglecting terms of orders higher than two, the rational factor for the expectation of difference can be calculated

$$\frac{E[\Omega_i(p) - \Omega_i(p_0)]}{\Omega_i(p_0)} = \frac{E[\Delta\Omega_i]}{\Omega_i(p_0)} = \frac{\Omega_{i1}(p_0)}{\Omega_i(p_0)} = \gamma_{\Omega_i} \tag{119}$$

where

$$\Omega_{i1}(p_0) = \sum_{r=1}^{q} \sum_{s=1}^{q} \left[\frac{p_{or} p_{os}}{2} \frac{\partial^2 \Omega_i(p_0)}{\partial p_{or} \partial p_{os}} \right] \mu_{rs}. \tag{120}$$

This can be used as the sensitivity factor for the average eigenvalue. Similarly the rational sensitivity factor for the variance of the difference one can calculate

$$\frac{Var[\Delta\Omega_i)]}{\Omega_i^2(p_0)} = \frac{\sigma^2_{\Delta\Omega_i}}{\Omega_i^2(p_0)} = \frac{1}{\Omega_i^2(p_0)}\sum_{r=1}^{q}\sum_{s=1}^{q} p_{or}p_{os}\frac{\partial\Omega_i(p_0)}{\partial p_{or}}\frac{\partial\Omega_i(p_0)}{\partial p_{os}}\mu_{rs} - \gamma^2_{\Omega_i}. \tag{121}$$

Another approach of sensitivity analysis for nonlinear dynamic systems has been proposed by Soong and Chen [27]. They proposed first to apply equivalent linearization. The linearized primary-secondary system can be represented by

$$M\ddot{X}(t) + C_E\dot{X}(t) + K_E X(t) = F(t), \tag{122}$$

where K_E and C_E consist of the primary system's equivalent linear stiffness and damping coefficients, K_e and C_e, and elastic secondary parameters K_s and C_s, respectively. Therefore, the transfer function $T_E(\omega, p)$ corresponding to the equation (122) defined by (109) can be treated as function of C_E and K_E. Then one can find the following partial derivatives

$$\frac{\partial T_E}{\partial Y},\quad \frac{\partial^2 T_E}{\partial Y^2}\quad \frac{\partial C_E}{\partial Y},\quad \frac{\partial K_E}{\partial Y},\quad \frac{\partial^2 C_E}{\partial Y^2}\quad \text{and}\quad \frac{\partial^2 K_E}{\partial Y^2}, \tag{123}$$

where Y can be either the yielding strength ratio R or ductility ratio μ. The yielding strength ratio is defined by

$$R = F_y/F_e \leq 1, \tag{124}$$

where F_y is the inelastic yielding force and F_e is the elastic force and the ductility ratio is

$$\mu = X_m/X_y, \tag{125}$$

where X_m is the maximum displacement and X_y is the yielding point displacement as shown in Fig. 3.

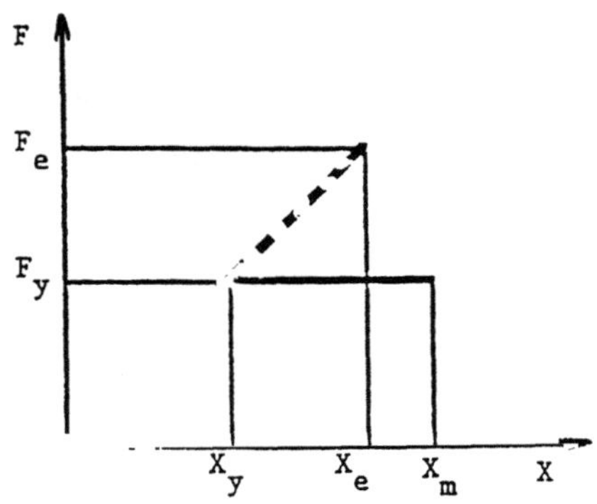

Figure 3. Force-Displacement Relationship

The above discussed method has been illustrated by a few numerical examples for single degree of freedom system and for 6-story single bay frame structure with rigid beams and negligible column and deformation (shear-beam type).

A different approach of the study of effect of parameters uncertainty on the response of linear vibratory systems to random excitation has been presented by Kotulski and Sobczyk [29] They have introduced several measures of these effects. However, they did not use the sensitivity functions. Several other methods were quoted in Evlanov and Konstantinov [16].

6. Qualitative Methods

As shown in the previous section, in order to derive the sensitivity measures it is necessary to solve the system of differential equations (moment equations) using one of the closure techniques. It should be stressed that even for such a simple example as an oscillator with two random coefficients (for instance, coloured noises) the number of sensitivity equation is sufficiently large. In the case of n-degree of freedom systems with k random coefficients the number of state and sensitivity equations is such that the number of mean values and second order moments will be greater or equal to

$$2(n+k)+2(2n+k)(2n+k+1).$$

It depends on the orders of the filters which are used to modulate the white noises (in the case of coloured noise coefficients). Therefore qualitative methods seem to be useful, particularly in multidimensional case. The concept of practical and exponential insensitivity for nonlinear stochastic dynamical systems was given by Socha [17] and Sočha and Pawleta [30] respectively. It is very close to the problem of stability of stochastic systems with respect to some of the variables. They have introduced the following definitions

Definition 2 (r-mean practical output insensitivity)

The system (1) is mean practically insensitive ($\gamma = 1$) or mean square practically insensitive ($\gamma = 2$) at the given estimations of values $(\delta, \varepsilon, t_f)$ it at any mining x_0 for which

$$|x_o| < \delta, |y_o| = 0 \Rightarrow \underset{t \epsilon [t_o, t_f]}{\forall} E|y(t)|^r < \varepsilon. \tag{126}$$

Definition 3 (mean square domain insensitivity)

An initial state x_o of the system (1) is said to be stochastically ε-insensitive in probability ρ in the normed square sense with respect to the set of initial conditions with the norm δ and to the specified target domain with the norm ε in the time interval $[t_0, t_f]$ if

$$P\{|y(t_f)\ ^2 > \varepsilon^2 | x(t_0) = x_o, y(t_0) = 0\} \leq 1 - \rho, \tag{127}$$

where $0 < \rho < 1$.

Definition 4 (r-th exponential moment insensitivity).

The null solution of a stochastic system is called r-moment exponential insensitive if there exist positive constants ε and α such that

$$|v^{[r]}| < \varepsilon|x(t_0)| \cdot exp(-\alpha(t - t_0)), \tag{128}$$

Definition 5 (r-mean exponential output insensitivity).

The null solution of a stochastic system is called mean exponential output insensitive ($r = 1$) or mean-square exponential output insensitive ($r = 2$) if there exist positive constants ε and α such that

$$E|y(t)|^r \le \varepsilon|x(t_0)|^r exp\{-\alpha(t - t_0)\}. \tag{129}$$

To derive the criteria of practical insensitivity Socha [17] applied Lyapunov approach. He obtained the following criteria for nonlinear systems described by equation (1).

Criterion 1

We assume that, there exists a scalar function $V(t, z)$ in the time interval $[t_0, t_f]$ with bounded continuous first and second derivatives with respect to every component of z and first derivative with respect to $t, t \neq t_f$ and there exist positive constants c_1, c_2, c_3 such that

$$c_1|y|^r \le V(t, z) \le c_2|z|^r, \tag{130}$$

$$L[V(t, z)] \le -c_3|z|^r, \tag{131}$$

where the operator $L(\cdot)$ is defined as follows

$$L(\cdot) = \frac{\partial}{\partial t}(\cdot) + \frac{\partial}{\partial x}(\cdot)^T F(t, z, p) + \frac{1}{2} tr \left[G^T(t, z, p) \frac{\partial}{\partial z} \left(\frac{\partial}{\partial z}\right)^T G(t, z, p) \right], \tag{132}$$

then the system (1) is practically r-mean insensitive at the given estimations of values $(\delta, \varepsilon, t_f)$ if

$$\frac{c_2}{c_1} \cdot exp\left\{-\frac{c_3}{c_2}(t_f - t_0)\right\} \le \left(\frac{\varepsilon}{\delta}\right)^r. \tag{133}$$

Criterion 2

The initial state x_o of the system (1) is stochastically ε-insensitive in probability ρ in the normed square sense with respect to the set of initial conditions with the norm δ and the terminal state within the time interval $[t_0, t_f]$ if the following conditions are satisfied:

(i) In the time interval $[t_0, t_f]$ a scalar function $V(t, z)$ is defined and has bounded continuous first and second derivatives with respect to every component of z and first derivative with respect to $t, t \neq t_f$.

(ii) $V(t, z)$ satisfies the terminal condition

$$V(t_f, z) \geq \frac{1}{\alpha} y^T(t_f) y(t_f), \tag{134}$$

where α is a positive number such that $\alpha << \varepsilon$.

(iii)

$$L[V(t, z)] \leq 0. \tag{135}$$

(iv) The given initial state $x(t_0) = x_0, y(t_0) = 0$ satisfies

$$V(t_0, z_0) \leq (1 - \rho)\frac{\varepsilon^2}{\alpha}, \tag{136}$$

where $z_0 = z(t_0), \qquad |z_0| < \delta$.

More specified criteria Socha [17] obtained for linear systems

$$dx(t) = Ax(t)dt + \sum_{k=1}^{M} R_k x(t) d\xi_k, \quad x(t_0) = x_0, \tag{137}$$

where A, R_k are $n \times n$ constant matrices.

The first criterion without assumption (133) gives us the sufficient condition of exponential output insensitivity. This fact has been stated by Socha and Pawleta [30], where they have derived criteria for moment insensitivity of linear triangular systems. It should be stressed that the criterion of exponential insensitivity (stability with respect to some variables) has been earlier obtained by Sharov [31]. In the proofs of these criteria the Lyapanov stochastic stability approach described, for instance by Khasminski [32] has been used. Some other results about practical stability with respect to some variables can be found in the paper written by Martynyuk [33].

7. Other Methods

In 1971 Fleming [34] proposed a numerical technique for small-noise stochastic control problems where white noise enter additively, i.e.,

$$dx_i = (a_{i1}x_1 + \ldots + a_{in}x_n)dt, \qquad i \leq j,$$

$$dx_i = (a_{i1}x_1 + \ldots + a_{in}x_n + b_{i1}u_1 + \ldots + b_{ik})dt + (2p)^{\frac{1}{2}} d\xi_i, \quad i > j, \tag{138}$$

where

$$0 \leq j \leq n \qquad k \leq n - j, \qquad x_i(0) = x_0 = \text{ constant}$$

The method uses the sensitivity functions with respect to initial conditions which are compatible from the optional solution to the corresponding deterministic control problem. Using this approach Holland [35] obtained numerical results for two dimensional linear regular problems with saturation and time optimal problem.

Another sensitivity method for indirect estimation of the accuracy of the statistical linearization was presented in 1978 by Artemév and Stepanov [36]. They introduced the sensitivity functions of the mean values and the second central moments of the state variables of the system with respect to the neglected quasimoments of the probability density.

8. Conclusions and Area for Further Research

As it was mentioned in the Introduction and also one can observe from review papers the sensitivity analysis of stochastic systems has not been developed as for deterministic systems. However, the quoted methods give one of the possibilities to simplify the analysis of the response of stochastic dynamic systems. Instead of considering the responses of a stochastic system for different values of the parameters and then comparing these as it has been done, for example by Deodatis [37] one can apply the stochastic system sensitivity analysis directly. From the engineering point of view it is interesting to find a set of parameters for which the deviation of the state variable with respect to a parameter will be the smallest in some stochastic sense. These information one can obtain directly from sensitivity analysis. This is the advantage of this approach both in time and frequency domain. Similar to the deterministic analysis the sensitivity approach in frequency domain is more useful for analysis of stationary solution of linear systems with time invariant coefficients subjected to stochastic excitation while the sensitivity approach in time domain is applicable for linear systems with stochastic coefficients and for nonlinear single and multi-degree of freedom systems subjected to stochastic excitations. However, as it was mentioned earlier the application of the sensitivity analysis in time domain require of solutions of a large number of differential and algebraic equations. In a few published numerical examples the accuracy of approximation obtained by first and second order sensitivity function was better for the linear system with stochastic coefficients than for nonlinear systems. The application of Taylor series to the determination approximate characteristics of the response of stochastic system has been earlier developed by Shinozuka and Astill [38] who applied rather perturbation than sensitivity approach to approximate random eigenvalues. This approach has been later developed by Nakagivi and his coauthors and also by Shinozuka and his coauthors who have studied stochastic finite element methods. It was described in survey paper by Vanmarcke, Shinozuka, Nakagivi, Schuëller and Grigoriu [39]. Similar to the deterministic theory one can expect that except for the development of the presented methods in this paper the following one will be created:

- λ sensitivity of stochastic systems that affect the order of mathematical model of the system. This analysis is connected with singular perturbation of the stochastic differential equations.

- Sensitivity of the overshoot in the time or frequency domain. This analysis in time domain is connected with the problem of peak distribution of the solution of sensitivity equation.
- Eigenvalue sensitivity in time domain. This approach is connected with mathematical analysis of random matrices.
- Performance index stochastic sensitivity. The necessity of the investigation of this approach will appear in the case of the study some optimization problem in structural dynamics. It will be connected, for example, with the problem of stochastic control.
- Sensitivity analysis of dynamic system described by stochastic partial differential equations.

Acknowledgement

This study is supported partially by Research Program RP.01.02 in Poland and by the National Center for Earthquake Engineering Research under Grant No. NCEER-89-2002A.

The manuscript was prepared during L. Socha visit to the Department of Civil Engineering in 1990. He would like to express his deepest appreciation to Professor T.T. Soong for scientific discussions and financial support.

References

1. Bode, H.W., Network Analysis and Feedback Amplifier Design, Van Nostrand-Reinhold, Princeton, New Jersey, 1945.
2. Kokotkovic, P.V. and Rutman, R.S., Sensitivity of Automatic Control Systems (Survey), *Avtomatika i Telemekhanika*, 26, (1965), 727-749.
3. Nuguyen Thuong Ngo, Sensitivity of Automatic Control Systems (Survey), *Autom. Remote Control (USSR)*, 32, (1971), 735-762.
4. Frank, P.M., Introduction to System Sensitivity Theory, New York, Academic Press, 1978.
5. Adelman, H.M. and Haftka, R.T., Sensitivity analysis of discrete structural systems, *AIAA Journal*, 25, (1986), 823-832.
6. Hsieh, C.C. and Arova, J.S., Design sensitivity analysis optimization of dynamic response, *Computer Methods in Applied Mechanics and Engineering*, 43, (1984), 195-219.
7. Zimoch, Z., Sensitivity analysis of vibrating systems, *Journal of Sound and Vibration*, 115, (1987), 447-458.
8. Nalecz, A.G. and Wicher, J., Design sensitivity analysis of mechanical systems in frequency domain, *Journal of Sound and Vibration*, 120, (1988), 517-526.

9. Sharp, R.S. and Brooks, P.C., Application of eigenvalue sensitivity theory to the improvement of the design of a linear dynamic system, *Journal of Sound and Vibration*, 114, (1987), 19-32.

10. Sharp, R.S. and Brooks, P.C., Sensitivities of frequency response functions of linear dynamic systems to variations in design parameter values, *Journal Sound and Vibration*, 126, (1988), 167-172.

11. Kuo, C.P. and Wada, B.K., Nonlinear sensitivity coefficients and corrections in system identification, *AIAA Journal*, 25, (1987), 1463-1468.

12. Socha, L., The sensitivity analysis of stochastic nonlinear dynamic systems, *J. of Sound and Vibration*, 110, (1986), 271-288.

13. Gihman, I.I. and Skorohod, A.V., Stochastic Differential Equations and Their Applications, Kiev: Naukova Dumca (in Russian), 1982.

14. Evlanov, L.G., Accuracy estimation of random parameter linear systems, *Automat. i Telemekhan*, 28, (1967), 382-390.

15. Evlanov, L.G., An approximate method of estimating the accuracy of nonlinear systems containing random parameters, *Automat. i. Telemekhan*, 28, (1967), 1968-1978.

16. Evlanov, L.G. and Konstantinov, V,M., Systems with random parameters, Moscou, Nauka, 1976.

17. Socha. L., Comparison of moment and output sensitivity measures for nonlinear stochastic dynamical systems, *Proceedings of the Conference MECO-85, Modelling and Simulation*, Lugano, (1985),328-331.

18. Socha, L., Practical insensitivity for nonlinear stochastic dynamic systems, in Applied Modelling and Simulation of Technological Systems, ed. P. Borne, North Holland, (1987), 329-335.

19. Szopa, J., Sensitivity of stochastic systems to initial conditions, *Journal of Sound and Vibration*, 97, (1984), 645-649.

20. Sues, R.H., Wen, Y.K. and Ang, A.H.S., Stochastic evaluation of seismic structural performance, *Journal of Structural Engineering*, 111, (1985), 1204-1218.

21. Szopa, J., The stochastic sensitivity of the Van der Pol equation, *Journal Sound and Vibration*, 100, (1985), 135-140.

22. Szopa, J., The methods of investigation and the sensitivity of the stochastic dynamical systems (in Polish), *Habilitation, Gliwice*, 1985.

23. Pawleta, M. and Socha, L., The application of sensitivity methods for determination of approximate characteristics of stochastic dynamic systems, *IMACS-88 XII World Congress on Scientific Computations*, July 18-22, (1988), 33-35.

24. Jumarie G., Simple general method to analyze the moment stability and sensitivity of non-linear stochastic systems with and without delay, *Int. J. System Sci.*, 19, (1988), 111-124.

25. Jumarie, G., Analysis of moment stability and sensitivity of non-linear stochastic distributed systems: a practical approach via a distributed Fokker-Planck equation, *Int. J. System Sci.*, 20, (1989), 2369-2385.

26. Igusa, T. and Der Kiureghian, A., Response of uncertain systems to stochastic excitations, *Journal of Engineering Mechanics*, 114, (1988), 812-832.

27. Soong, T.T. and Chen, Y.Q., Stochastic response sensitivity of secondary systems to primary structural uncertainties, *Structural Safety*, 6, (1989), 311-321.

28. Chen, Y.Q. and Soong, T.T., Seismic behavior and response sensitivity of secondary structural systems, *Technical Report NCEER-89-0030*, October, 1989.

29. Kotulski, Z. and Sobczyk, K., Effects of parameter uncertainty on the response of vibratory systems to random excitation, *Journal Sound and Vibration*, 119, (1987), 159-171.

30. Socha, L. and Pawleta, M., The exponential insensitivity of linear stochastic dynamic systems, *Advances in Modelling and Simulation*, 13, (1988), 45-55.

31. Sharov, V.F., Stability and stabilization of stochastic systems with respect to some of the variables, *Avtomatika i Telemekhanika*, 39, (1978), 1629-1636.

32. Khasminski, R.Z., Stochastic Stability of Differential Equations, Alphen aan den Rijn: Sijthoff and Noordhoff, 1980.

33. Martynyuk, A., Methods and problems of practical stability of motion theory, *Nonlinear Vibration Problems*, Warszawa, 22, (1984), 9-47.

34. Fleming, W., Stochastic control for small noise intensities, *SIAM J. Control*, 9, (1971), 473-517.

35. Holland, C.J., A numerical technique for small-noise stochastic control problems, *Journal Optimization Theory and Applications*, 13, (1974), 74-93.

36. Artemév, V.M. and Stepanov, V.L., Accuracy estimation of statistical linearization method, *Avtomat. i Telemekhan*, 37, (1978), 181-185.

37. Deodatis, G, Stochastic FEM sensitivity analysis of nonlinear dynamic problems, *Probabilistic Engineering Mechanics*, 4, (1989), 135-141.

38. Shinozuka, M. and Astill, C.J., Random eigenvalue problems in structural analysis, 10, (1971), 456-462.

39. Vanmarcke, E., Shinozuka, M., Nakagivi, S., Schuëller, G.I. and Grigoriu, M., Random fields and stochastic finite elements, *Structural Safety*, 3, (1986), 143-166.

Stochastic Analysis of Nonlinear Vehicle Systems using a Generalized Discrete Harmonic Linearization Technique

Hong SU, S. RAKHEJA and T. S. SANKAR

Department of Mechanical Engineering
Concordia University
1455 de Maisonneuve Blvd. W.
Montreal, Quebec, Canada H3G 1M8

Summary

The discrete harmonic linearization method which is applicable to stochastic systems could not provide an equivalent stiffness coefficient for a nonlinear restoring element. In this paper, this technique is generalized to obtain linear representations of both nonlinear restoring and damping elements, based on a principle of energy similarity of dynamic elements. A numerical iterative procedure is presented to computer the local linear coefficients of nonlinear dynamic elements. A nonlinear system is then represented by a set of its complex frequency response function matrices, as functions of the excitation frequency. Stochastic analysis of general multi-DOF nonlinear vehicle systems is established in terms of response PSD characteristics.

Introduction

Well established linear analytical tools provide a powerful and convenient approach to carry out stochastic analyses of either linear or linearized systems. Nonlinear mechanical systems are often expressed by their linear equivalents in order to carry out stochastic analyses by using the convenient linear analytical tools [1]. In equivalent linearization techniques, a nonlinear dynamic system is replaced by a related appropriate linear system, according to a specified criteria, such that the difference between the two systems is minimized as much as possible [2].

The deterministic linearization in the frequency domain was first developed by Thomson [3]. Various nonlinear damping elements in steady state harmonic vibratory systems are replaced by the equivalent viscous damping elements. The equivalent viscous damping coefficient of a nonlinear damping element is calculated by equating the energy dissipated per cycle of the nonlinear element to that of an equivalent viscous damper. This simple and practical technique has been widely used for deterministic analyses of nonlinear dynamic systems [4, 5].

Y. K. Lin · I. Elishakoff (Eds.)
Stochastic Structural Dynamics 1
New Theoretical Developments
2nd International Conference on Stochastic Structural Dynamics
May 9-11, 1990, Boca Raton, Florida

A discrete harmonic linearization (DHL) method, an extension of the deterministic frequency domain linearization, has been proposed to carry out stochastic analysis of nonlinearly damped mechanical systems [6]. The method replaces different nonlinear damping mechanisms by an array of equivalent viscous damping coefficients. A set of localized linear systems are functions of the local excitation frequency and amplitude. Various linearization methods in frequency domain and their accuracy, based on different equivalent criteria, have been discussed by Rakheja et al. [7]. The DHL method offers a simple approach to analyze stochastic response of multi-DOF systems with nonlinear damping. The DHL method, however, deals only with nonlinear damping elements with continuous force velocity characteristics. Mechanical systems with nonlinear restoring and sequential damping elements, however, cannot be analyzed by using the DHL technique.

In this paper, the DHL method is generalized to obtain linear representations of nonlinear restoring as well as sequential damping elements, based on a principle of energy similarity of dynamic elements. A processed energy function is proposed to characterize the energy properties of a dynamic element, whether it is dissipative or conservative. A dynamic system with nonlinear restoring and damping elements is, therefore, replaced by an equivalent linear system, using the generalized technique such that the difference in energy properties of the two systems is minimum. A numerical iterative procedure is formulated to compute the array of linear coefficients that can describe nonlinear dynamic elements over wide range of excitation frequencies. A nonlinear vehicle system is represented in terms of its complex frequency response function matrix, as function of the excitation frequency, excitation amplitude and nonlinearities. Effectiveness of the generalized DHL method is demonstrated by comparing the response characteristics of the linearized system with that of the nonlinear system for deterministic excitations. The stochastic response characteristics of a nonlinear vehicle system with gas-spring, orifice damping, and tunable sequential damping elements are presented and discussed in terms of response PSD.

Development of a Generalized Discrete Harmonic Linearization Technique

It has been established that an equivalent viscous damping coefficient of a nonlinear damping element can be obtained by equating the dissipated energy per cycle due to the nonlinear dissipative element to that of a equivalent viscous damper [3]. A conservative restoring element, however, does not dissipate energy, it rather stores and releases energy during a vibration cycle. The total energy due to the restoring element is thus zero.

Alternatively, the total amount of energy processed by a conservative element in one complete cycle can be measured by summing the absolute values of both the stored and released energy values. The energy properties of a conservative or dissipative element can be, in general, established as described below.

Consider a dynamic element subjected to a harmonic motion with period τ. Let $f(t)$ be the force function of the dynamic element, z(t) be the relative displacement across the element, and $\dot{z}(t)$ be the derivative of z with respect to time t. The energy processed by the dynamic element can be determined using following definitions.

Definition 1:

A function $f(t)$ is said to be in the same orientation in an interval $[t_{i-1}, t_i]$, if the sign of the product of $f(t)$ and $\dot{z}(t)$ remains the same for $t \in [t_{i-1}, t_i]$, where $|t_{i-1} - t_i| \le \tau$.

Definition 2:

A processed energy E_p associated with a function $f(t)$ is the sum of the absolute values of energy computed in all the same orientation intervals $[t_{i-1}, t_i]$ over the period τ, that is

$$E_p = \sum_{i=1}^{N} \left| \int_{t_{i-1}}^{t_i} f(t)\, \dot{z}\, dt \right| \tag{1}$$

where

$$\sum_{i=1}^{N} (t_{i-1} - t_i) = \tau \tag{2}$$

It is obvious that a dynamic element, whether dissipative or conservative, linear or nonlinear, can be characterized in terms of its processed energy. In order to obtain the equivalent linear coefficient of a nonlinear dynamic element based on its energy properties as compared with that of a related linear dynamic element, the following principle of energy similarity of dynamic elements is employed to generalize the discrete harmonic linearization technique:

Principle of Energy Similarity of Dynamic Elements:

> A dynamic element is considered to be equivalent to another dynamic element in a sense of energy similarity, if their force functions are of the same nature (dissipative or conservative) and their processed energies are identical.

A nonlinear mechanical system with nonlinear dissipative and/or conservative elements can be, therefore, replaced by an equivalent linear system, based on the above principle of energy similarity of dynamic elements such that the processed energy of elements within the linearized system does not deviate considerably from that of the elements within the nonlinear system.

Computation of Equivalent Linear Coeffcients

Consider a single DOF base excited vibration isolator, the equation of motion of the isolator is given by:

$$m\ddot{z}(t) + f(z,\dot{z},t) = -m\ddot{x}_i(t) \tag{3}$$

where **m** is the isolator mass, $\ddot{x}_i$ is the acceleration at the base, z is the relative displacement, $z=x-x_i$, x is displacement of mass, and $f(z,\dot{z},t)$ represents the nonlinear force generated by nonlinear elements of the isolator. Coefficients of equivalent linear elements are obtained by equating the processed energy of nonlinear dynamic elements to that of equivalent linear elements of the same nature (dissipative or conservative).

Assuming harmonic motion across a nonlinear dynamic element, the nonlinear force, $f(z,\dot{z},t)$, due to the nonlinear element can be approximated by local equivalent stiffness and damping coefficients. Here each local coefficient is valid in the vicinity of a selected discrete frequency and excitation amplitude. The nonlinear force can thus be approximated as:

$$f_j(z,\dot{z},t) = k_{eq}(\omega_j,Z)\, z(t) + c_{eq}(\omega_j,Z)\, \dot{z}(t), \qquad j = 1,\ldots N \tag{4}$$

where Z is the amplitude of displacement across the element at a discrete excitation frequency ω_j, given by

$$z(t) = Z \sin(\omega_j t) \tag{5}$$

N is the total number of discrete frequencies, $c_{eq}(\omega_j,z)$ and $k_{eq}(\omega_j,z)$

are the local equivalent damping and stiffness coefficients, respectively, corresponding to discrete frequency ω_j and relative displacement amplitude Z.

Damping Elements

For a dissipative damping element, it is obviously that the processed energy is the same as the energy dissipated per cycle. The processed energy can, therefore, be computed from the following expressions, correspondent to Coulomb, viscous and orifice damping elements, respectively [3]:

$$E_p = 4c_0Z; \qquad E_p = \pi c_1 Z^2\omega_j; \qquad E_p = 8\, c_2 Z^3\omega_j^2/\, 3 \tag{6}$$

where c_0, c_1 and c_2 are the coefficients of Coulomb, viscous and orifice damping elements, respectively. The equivalent viscous damping coefficient is then obtained by equating the processed energy to that of an equivalent viscous damper. The respective equivalent linear damping coefficients due to Coulomb and orifice damping elements are given by:

$$c_{eq}(\omega_j, Z) = 4c_0/\, \pi Z\omega_j; \qquad c_{eq}(\omega_j, Z) = 8c_2\omega_j\, Z/\, 3\pi \tag{7}$$

For a logical state control damper, such as a sequential variable damping element [3], the damping force is expressed as:

$$f_d = \begin{cases} c_2|\dot{z}|\dot{z} \;, & |f_d| < F_0 \\ F_0 \;, & \text{otherwise} \end{cases} \tag{8}$$

$$F_0 = \nu\, \alpha\, (p_{12})_0 \tag{9}$$

where $(p_{12})_0$ is the preset pressure limiting value, α is a constant and ν is the tuning factor. The time when the damping force approaches the limiting value F_0 at t_0 can be expressed as:

$$t_0 = \frac{1}{\omega_j} \sin^{-1}\left[\frac{1}{Z\omega_j} \sqrt{\frac{F_0}{c_2}}\,\right] \tag{10}$$

The processed energy can thus be expressed as:

$$E_p(\omega_j, Z) = 4 \left[\int_0^{t_o} c_2\, \dot{z}^3\, dt + \int_{t_o}^{\pi/2\omega_j} F_0\, \dot{z}\, dt \right] \tag{11}$$

The local equivalent damping coefficient is then obtained by equating the processed energy to that of a linear viscous damper:

$$c_{eq}(\omega_j, Z) =$$

$$\begin{cases} 8c_2\omega_j \; Z \,/\, 3\pi \,, & ; |f_d| < F_0 \\ \dfrac{4}{\pi\omega_j Z}\left[c_2 Z^3\omega_j^2(\frac{2}{3} - \cos\omega_j t_0 + \frac{1}{3}\cos^3\omega_j t_0) + F_0 Z\cos\omega_j t_0\right], & \text{otherwise} \end{cases} \tag{12}$$

<u>Spring Elements</u>

The force function f_k due to a linear spring with stiffness coefficient k is expressed by:

$$f_k(z,t) = k\; z(t) \tag{13}$$

It is easy to see that linear spring force f_k is an odd function of z and the sign of product of displacement z and velocity $\dot{z}$ changes four times during each harmonic cycle. There are, based on Definition 1, four same orientation intervals associated within a cycle $\tau=2\pi/\omega_j$, namely $[0, \pi/2\omega_j]$, $[\pi/2\omega_j, \pi/\omega_j]$, $[\pi/\omega_j, 3\pi/2\omega_j]$, and $[3\pi/2\omega_j, 2\pi/\omega_j]$. The processed energy of the linear spring, based on Definition 2, is then expressed by:

$$E_p(\omega_j, Z) = \left| \int_0^{\pi/2\omega_j} f_k(z,t)\,\dot{z}\,dt \right| + \left| \int_{\pi/2\omega_j}^{\pi/\omega_j} f_k(z,t)\,\dot{z}\,dt \right| +$$

$$\left| \int_{\pi/\omega_j}^{3\pi/2\omega_j} f_k(z,t)\,\dot{z}\,dt \right| + \left| \int_{3\pi/2\omega_j}^{2\pi/\omega_j} f_k(z,t)\,\dot{z}\,dt \right| \tag{14}$$

Integration of equation (14) yields

$$E_p = 2\;k\;Z^2 \tag{15}$$

Gas-spring is a component of hydraulic dampers and vehicle suspensions. The nonlinear restoring force due to a gas-spring is related to polytropic gas compression process and expressed as [9]:

$$f_a(z,\;t) = \left[\frac{(V_0 + A_r\;z)^{\gamma} - V_0^{\;\gamma}}{(V_0 + A_r\;z)^{\gamma}}\right] A_r\;p_0 \tag{16}$$

where γ is the polytropic exponent, $1 \le \gamma \le 1.4$, (γ=1 for isothermal process; γ=1.4 for adiabatic process). A_r, p_0 and V_0 are constants related to the damper geometry and static state. It is apparent, from equation (16), that the force f_a is a nonlinear asymmetric function of the relative displacement z across the gas-spring. A vibration cycle can be divided into four same orientation intervals as in the case of a linear spring. Considering two cases of polytropic exponent for $\gamma=1$ and $\gamma\neq1$, the processed energy of the gas-spring is computed, based on the Definition 2, and expressed by:

$$E_p = \begin{cases} 2p_0V_0\left[-\ln\left(1+\frac{A_rZ}{V_0}\right)-\ln\left(1-\frac{A_rZ}{V_0}\right)\right], & \text{for } \gamma=1 \\ \frac{2p_0V_0}{1-\gamma}\left[2-\left(1+\frac{A_rZ}{V_0}\right)^{1-\gamma}-\left(1-\frac{A_rZ}{V_0}\right)^{1-\gamma}\right], & \text{for } \gamma\neq1 \end{cases} \tag{17}$$

The local equivalent linear stiffness coefficient is then obtained by equating the processed energy of the gas-spring in (17) to that of a linear spring in equation (15):

$$k_{eq}(\omega_j, Z) = \begin{cases} \frac{p_0V_0}{Z^2}\left[-\ln\left(1+\frac{A_rZ}{V_0}\right)-\ln\left(1-\frac{A_rZ}{V_0}\right)\right], & \text{for } \gamma=1 \\ \frac{p_0V_0}{Z^2(1-\gamma)}\left[2-\left(1+\frac{A_rZ}{V_0}\right)^{1-\gamma}-\left(1-\frac{A_rZ}{V_0}\right)^{1-\gamma}\right], & \text{for } \gamma\neq1 \end{cases} \tag{18}$$

Random Response Analysis

The equations of motion of a multi-degree-of-freedom nonlinear vehicle model may be expressed in the following form:

$$[m]\{\ddot{x}\} + \{f(z,\dot{z},t)\} = \{f_t(x_i,\dot{x}_i,t)\} \tag{19}$$

where $[m]$ is the mass matrix, $\{x\}$ is the generalized coordinate vector, $\{f(z,\dot{z},t)\}$ is the suspension force vector, and $\{f_t(x_i,\dot{x}_i,t)\}$ is the excitation force vector due to tire-terrain interface. If $\{f_t\}$ is a random process vector, equation (19) is a nonlinear stochastic differential equation.

The nonlinear vehicle model in equation (19) can be represented by a set of linear models as complex frequency response function matrices, at each discrete frequency ω_j:

$$[H(\omega_j, X_i)]_{n\times m} = \Big([k_{eq}] - \omega_j^2[m] + i\omega_j[c_{eq}]\Big)^{-1}_{n\times n}\Big([k_t] + i\omega_j[c_t]\Big)_{n\times m} \tag{20}$$

$$(j = 1,2,\ldots,N)$$

where $i=\sqrt{-1}$, N is the total number of discrete frequencies, n is the number of degrees of freedom of the vehicle model, m the number of input variables, [H] the complex frequency response function matrix, $[k_{eq}]$ and $[c_{eq}]$ are equivalent stiffness and damping matrices, $[k_t]$ and $[c_t]$ are tire stiffness and damping matrices, respectively. The equivalent stiffness and damping matrices are evaluated by employing the generalized discrete harmonic linearization technique. The complex frequency response function matrix of the nonlinear vehicle model is thus established as a function of local excitation frequency, input amplitude and types of nonlinearities.

The stochastic response of the road vehicle is evaluated via power spectral density approach. The excitations and response are then characterized in terms of power spectral density (PSD) functions. The relation between excitation and response is expressed by:

$$[s_x(\omega_j)] = [H(\omega_j)]^* \, [s_i(\omega_j)] \, [H(\omega_j)]^T \qquad (21)$$

$$(j = 1,2,\ldots,N)$$

where $[H(\omega_j)]^*$ and $[H(\omega_j)]^T$ are the complex conjugate and transpose of $[H(\omega_j)]$, $[s_i(\omega_j)]$ and $[s_x(\omega_j)]$ are the excitation and response PSD matrices, respectively. The diagonal elements of $[s_x(\omega_j)]$ are the spectral densities of the generalized coordinates; while the off-diagonal elements are the cross spectral densities of the coordinates. The acceleration PSD of the corresponding coordinates can be determined by:

$$[s_{\ddot{x}}(\omega_j)] = (\omega_j)^4 [s_x(\omega_j)] \qquad (22)$$

Random Road Excitation

Road surfaces traversed by vehicles are random in nature. It has been established that most road surface irregularities are normally distributed and may be accurately described by a stationary random process [10]. The spectral density of road surface that closely approximates available experimental data can be described by [11]:

$$s(\mu) = \begin{cases} s(\mu_0)(\mu/\mu_0)^{-w_1}, & \mu \le \mu_0 \\ s(\mu_0)(\mu/\mu_0)^{-w_2}, & \mu > \mu_0 \end{cases} \qquad (23)$$

where $s(\mu)$ is the spatial spectral density of road roughness, μ is spatial frequency, $s(\mu_0)$ is roughness coefficient at reference spatial frequency μ_0, $\mu_0 = 1/2\pi$, w_1, w_2 are constants characterizing road waviness. Assuming a constant forward speed v, the temporal spectral density of road excitations, s(f), can be expressed as:

$$s(f) = s(\mu)/\, v \tag{24}$$

where $f = \mu v$ is the temporal frequency (Hz), and $f=\omega/2\pi$.

A multi-axles vehicle model is subjected to m vertical excitations at different tire-terrain interfaces. Assuming that the vehicle rear wheels follow the same profile as the front one, the cross-spectral density of between *p*th and *q*th wheels can be expressed in terms of a time delay function:

$$s_{pq}(f) = s(f)\ \exp(-i\ 2\pi\ f\ \tau_{pq}) \tag{25}$$

where $\tau_{pq} = \ell_{pq}/v$ is the time delay between the front and rear wheels, and ℓ_{pq} is the spacing between *p*th and *q*th wheels, and

$$s_{qp}(f) = s^{*}_{pq}(f) \tag{26}$$

The temporal displacement spectral density matrix, corresponding to discrete frequency ω_j, can be expressed as:

$$[s_{xi}(\omega_j)] = \frac{s(\mu)}{2\pi v} \begin{bmatrix} 1 & e^{-i\omega_j\tau_{12}} & \dots & e^{-i\omega_j\tau_{1m}} \\ e^{i\omega_j\tau_{12}} & 1 & & e^{-i\omega_j\tau_{2m}} \\ \vdots & \vdots & & \vdots \\ e^{i\omega_j\tau_{1m}} & e^{i\omega_j\tau_{2m}} & \dots & 1 \end{bmatrix} \tag{27}$$

A Numerical Iterative Algorith

In order to express the nonlinear vehicle model as a set of linear models as complex frequency response function matrices given in equation (20), a numerical iterative algorithm is developed to compute the local equivalent representation of a nonlinear vehicle model. The algorithm is summarized as follows:

Step 1 Assume initial values of the local equivalent coefficients at a selected excitation frequency, (c°_{eq} and k°_{eq}).

Step 2 Estimate an excitation amplitude vector from the input power spectral density PSD in the following manner [6]:

$$X_i(\omega_j) = \psi \left[\int_{\omega_j-\Delta\omega/2}^{\omega_j+\Delta\omega/2} s_i(\omega)\, d\omega \right]^{1/2} \tag{28}$$

where ψ is a constant.

Step 3 Synthesize a time history, $x_i(t)$ by using the sine series approximation, for generating an estimated PSD and improving the accuracy of the estimated amplitude vector.

$$x_i(t) = \sum_{j=1}^{N} (-1)^j X_i(\omega_j) \sin(\omega_j t) \tag{29}$$

Step 4 Compute the relative response quantities by solving the assumed linear system, using the estimated amplitude vector $X_i(\omega_j)$.

$$\{X(\omega_j)\} = [H(\omega_j, c^{\circ}_{eq}, k^{\circ}_{eq})] \{X_i(\omega_j)\} \tag{30}$$

$$\{Z(\omega_j)\} = [TZ] \{X(\omega_j)\} \tag{31}$$

where $\{Z(\omega_j)\}$ is relative response variable vector, [TZ] is the transform matrix from the generalized coordinate vector to relative variable vector.

Step 5 Determine local equivalent coefficients, c_{eq} and k_{eq}, corresponding to selected frequency ω_j, by employing the equations developed by using the principle of energy similarity of dynamic elements.

Step 6 Evaluate the error vector between the assumed and the computed values of local equivalent coefficients, for each nonlinear dynamic element

$$\bar{\varepsilon} = \begin{bmatrix} c^{\circ}_{eq}(\omega_j, Z) - c_{eq}(\omega_j, Z) \\ k^{\circ}_{eq}(\omega_j, Z) - k_{eq}(\omega_j, Z) \end{bmatrix} \tag{32}$$

Step 7 Update the assumed local equivalent coefficients and continue the iterative process at Step 4, until convergence is achieved.

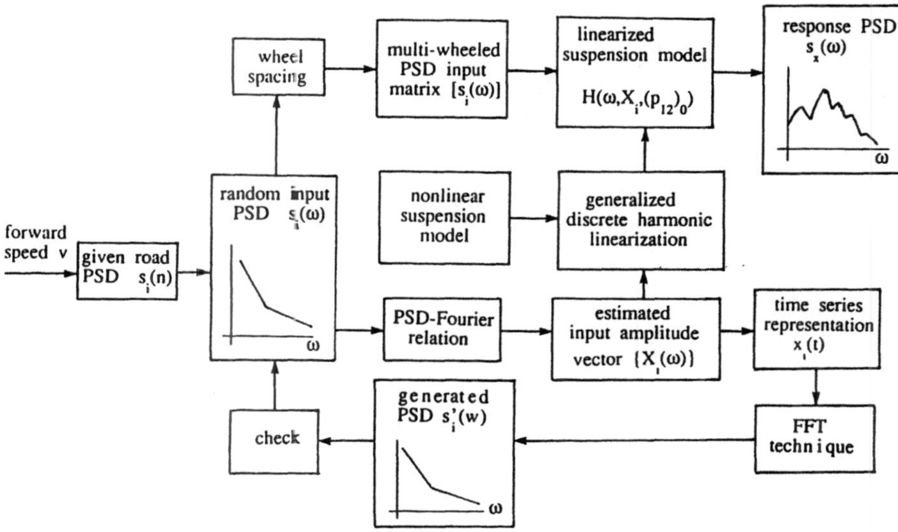

Fig. 1 A flow diagram of simulation procedure

Step 8 Repeat the iterative process for each discrete excitation frequency to generate a complete array of local equivalent linear systems (for each ω_j, $j = 1, \ldots N$).

When the linear representation of a nonlinear system is established in terms of a set of complex frequency response function matrix, the stochastic response can be evaluated based on equation (21). Fig. 1 illustrates a flow diagram of the simulation procedure.

Results and Discussion

One-DOF Vibration Isolator

The generalized discrete harmonic linearization technique is first applied to simulate a simple system given in equation (3), when it is subjected to a deterministic harmonic excitation. The results are compared to that obtained by using the fourth order Runge-Kutta method to illustrate the relative accuracy of the proposed technique. The one-DOF vibration isolation system employs a linear spring and a hydraulic damper. The force function $f(z,\dot{z},t)$ in (3) is given by:

$$f(z,\dot{z},t) = f_k(z,t) + f_a(z,t) + f_d(\dot{z},t) \tag{33}$$

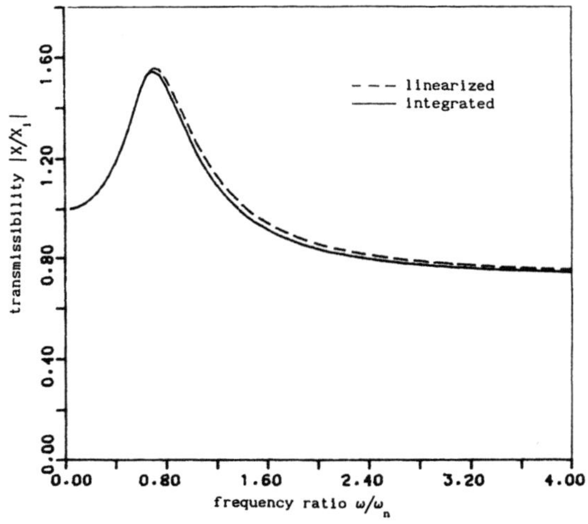

Fig.2 Displacement transmissibility of an isolator with gas-spring and fixed orifice damping

where the linear spring force $f_k(z,t)$ and gas-spring force $f_a(z,t)$ are given in (13) and (16), respectively, $f_d(\dot{z},t)$ is orifice damping force expressed by:

$$f_d(\dot{z},\ t) = c_2|\dot{z}|\dot{z} \tag{34}$$

$$c_2 = \frac{\rho}{2}\left[\frac{A_p}{n^2C_d^2}\left(\frac{A_p}{a_1}\right)^2 + \frac{A_r}{C_d^2}\left(\frac{A_r}{a_2}\right)^2\right] \tag{35}$$

The transmissibility characteristics of the vibration isolator are established by employing the linearization technique and the direct integration technique, with the following parameters: m=65 kg, k=4 kN/m, ρ=797 kg/m^3,, C_d=0.7, A_p=2×10^{-3} m^2, A_r=1.27×10^{-4} m^2, a_1=a_2=1.35×10^{-5} m^2, X_i=0.05 m and n=4. For the linearization technique the transmissibility of the isolator is the absolute value of the complex frequency response function. While for direct integration method the transmissibility is obtained by computing the amplitude ratio of the response and the excitation when the simulated response reaches to a steady state.

Figs. 2 and 3 show the mass displacement and relative displacement transmissibility characteristics of the hydraulic damper isolator

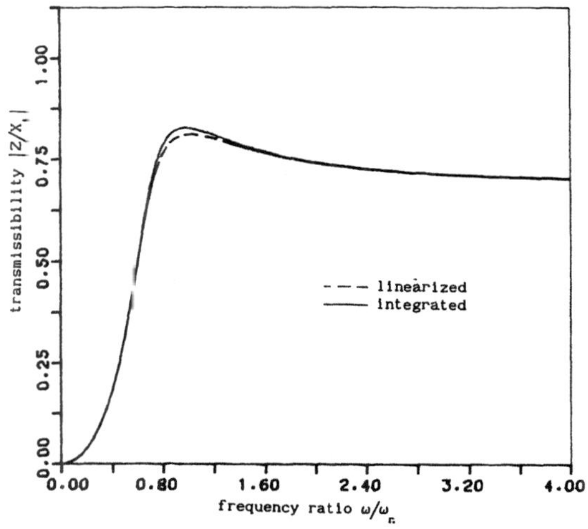

Fig.3 Relative displacement transmissibility of an isolator with gas-spring and fixed orifice damping

system, employing both generalized linearization technique and direct integration method, respectively. It is observed that the response values obtained by the generalized discrete harmonic linearization technique are very much close to that by the Runge-Kutta method. The generalized linearization method yields a slightly higher value of mass displacement transmissibility than that of Runge-Kutta method when excitation frequency is greater than the resonant frequency. The relative displacement response value by the linearization method is slightly smaller than that of the direct integration method around the resonant frequency. In many ranges, however, the two response results by the two different methods are identical. The linearization technique took maximum six iterations for a discrete frequency and about 4.5 s CPU time (VAX 11/780) to yield the simulation results; while the Runge-Kutta method produced the results by taking thirty five vibration cycle to reach a steady state and about 104 s CPU time.

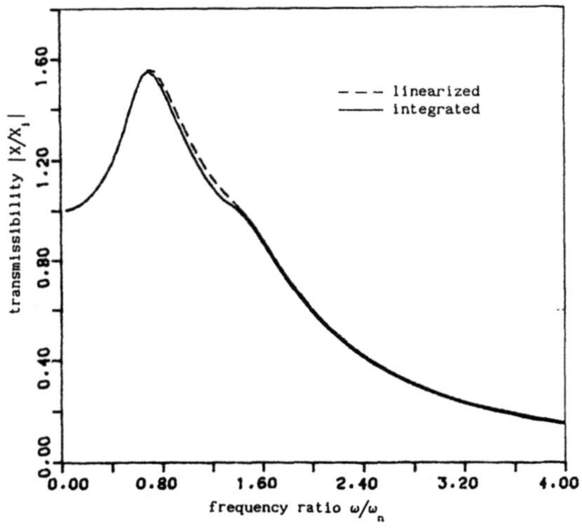

Fig. 4 Displacement transmissibility of an isolator with gas-spring and tunable sequential damping

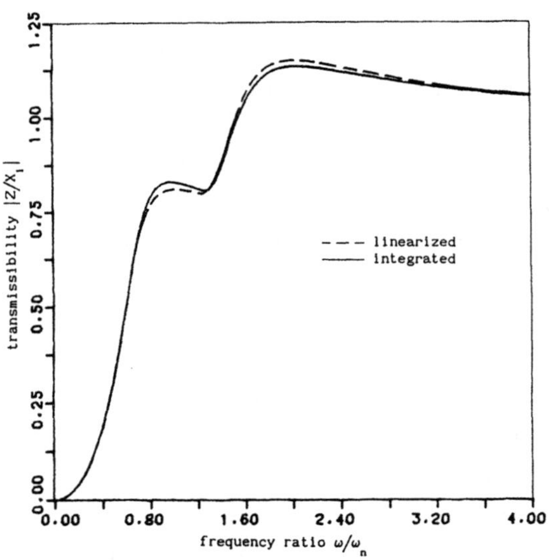

Fig. 5 Relative displacement transmissibility of an isolator with gas-spring and tunable sequential damping

A tunable hydraulic damper is also employed in the isolator system given in equations (3) and (33). Here the fixed damping force $f_d(\dot{z},t)$ is replaced by a sequential variable damping force given in equation (8). The mass and relative displacement transmissibility results are presented in Figs. 4 and 5, respectively. The results obtained by the generalized linearization technique are also shown to be very close to that by the Runge-Kutta method. The CPU time used by the generalized linearization technique and the direct integration method were about 5.5 and 230 s (VAX 11/780), respectively.

A Two-axle Vehicle System

A two-axle road vehicle, employing fixed orifice damping and tunable sequential damping shock absorbers, is modeled as an in-plane four-DOF dynamic system, as shown in Fig.6. The vehicle is subjected to random road excitations at tire-terrain interfaces. The equations of motion of the in-plane vehicle model, corresponding to equation (19), are expressed as follows:

$$m_s\ddot{x}_s(t) + k_{s1}z_1 + k_{s2}z_2 + f_{s1}(z_1,\dot{z}_1,t) + f_{s2}(z_2,\dot{z}_2,t) = 0$$

$$I_s\ddot{\theta}_p(t) - k_{s1}\,\ell_f\,z_1(t) + k_{s2}\,\ell_r\,z_2(t)$$
$$- f_{s1}(z_1,\dot{z}_1,t)\,\ell_f + f_{s2}(z_2,\dot{z}_2,t)\,\ell_r = 0$$

$$m_{u1}\ddot{x}_1(t) + c_{t1}\dot{x}_1(t) + k_{t1}x_1(t) - k_{s1}z_1(t) - f_{s1}(z_1,\dot{z}_1,t)$$
$$= c_{t1}\dot{x}_{i1}(t) + k_{t1}x_{i1}(t)$$

$$m_{u2}\ddot{x}_2(t) + c_{t2}\dot{x}_2(t) + k_{t2}x_2(t) - k_{s2}z_2(t) - f_{s2}(z_2,\dot{z}_2,t)$$
$$= c_{t2}\dot{x}_{i2}(t) + k_{t2}x_{i2}(t) \qquad (36)$$

The generalized coordinate, road displacement excitation, and relative displacement vectors of the vehicle model are, respectively, defined as:

$$\{x\} = [x_s\ \theta_p\ x_1\ x_2]^T$$
$$\{x_i\} = [x_{i1}\ x_{i2}]^T \text{ and } \{z\} = [z_1\ z_2]^T \qquad (37)$$

and

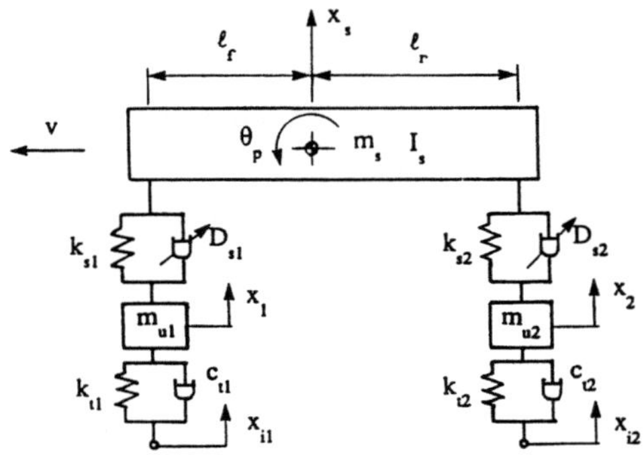

Fig.6 Schematic of an in-plane vehicle model

$$\{z\} = [TZ]\ \{x\}$$

$$[TZ] = \begin{bmatrix} 1 & -\ell_f & -1 & 0 \\ 1 & \ell_r & 0 & -1 \end{bmatrix} \tag{38}$$

In equation (36), f_{s1} and f_{s2} are dynamic forces generated by the front and rear shock absorbers and are comprised of a gas-spring and a damping components as:

$$f_{si} = f_{ai}(z_i, t) + f_{di}\ (\dot{z}_i, t) \qquad i=1,2 \tag{39}$$

where gas-spring force f_{ai} is given in equation (16), for fixed orifice damping shock absorber the damping force f_{di} is given in (34), while for tunable damping force f_{di} is expressed in (8).

The set of nonlinear differential equations (36), describing the ride dynamics of a two-axle vehicle employing fixed orifice and tunable hydraulic shock absorbers, are solved for stochastic excitations originating from the randomly distributed road roughness. Stochastic response of in-plane model of two-axle vehicle is evaluated using generalized harmonic linearization technique. The random road data are selected for poor road surfaces and given as $s(\mu_0) = 252.8\times10^{-6}$ m^3/cycle, $w_1=2.1$, $w_2=1.4$ and v=20 m/s. The given input PSD and a generated PSD by FFT based on the estimated input amplitude vector in equation (29) are illustrated in Fig. 7. It can be observed that the estimated PSD correlates quite well with the PSD of the road.

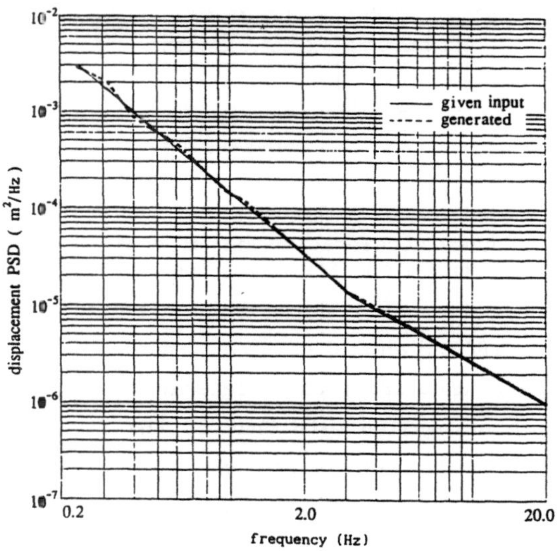

Fig.7 Displacement PSD of a random road input (v=20 m/s)

The dynamic ride performance of the tunable vehicle suspension is evaluated through PSD characteristics of the vehicle response. The bounce and pitch acceleration PSD response of the vehicle sprung mass is computed for the fixed orifice and tunable shock absorbers as shown in Figs. 8 and 9, respectively. Identical orifice sizes (n=2) have been selected for both tunable and fixed orifice shock absorbers to

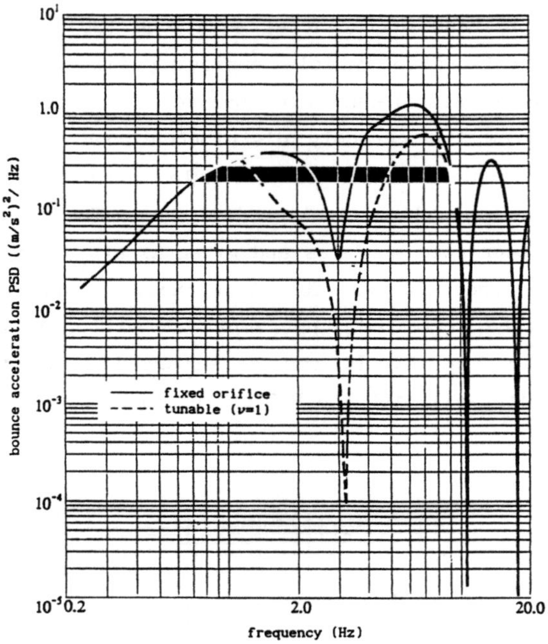

Fig.8 Bounce acceleration PSD of vehicle sprung mass with tunable and fixed orifice shock absorbers

demonstrate the effectiveness of the tunable damping modulation. A comparison of the bounce and pitch acceleration PSD of the vehicle model employing tunable suspension with that of the fixed orifice suspension reveals that the ride performance of the vehicle can be improved considerably via the tunable shock absorbers in the frequency range 1 to 10 Hz, that is known to be the frequency range to which human body is most fatigue sensitive [12]. The road excitation corresponding to frequencies above 10 Hz becomes insignificant as illustrated in Fig. 7. The fixed orifice and tunable shock absorbers, therefore, yield identical vehicle ride response at higher frequencies, as shown in Figs. 8 and 9.

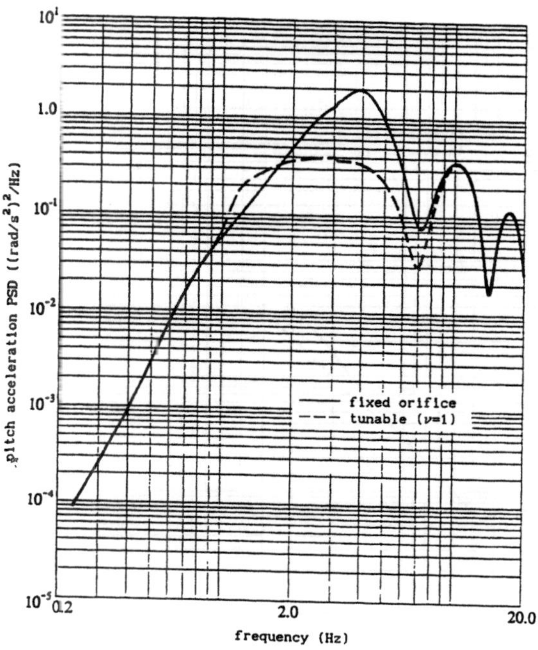

Fig. 9 Pitch acceleration PSD of vehicle sprung mass with tunable and fixed orifice shock absorbers

The vertical acceleration PSD response of the front and rear unsprung masses is presented in Figures 10 and 11, respectively. Tunable shock absorbers tend to reduce vertical acceleration response of sprung as well as unsprung masses at excitation frequencies above 1.1 Hz, where force limiting occurs. Figs. 10 and 11 clearly reveal that vertical acceleration response of unsprung masses with tunable shock absorbers is smaller than that with fixed orifice shock absorbers for frequency less than 4 Hz. The reduced damping due to tunable shock absorbers, however, can not suppress the unsprung mass resonance and thus yields larger acceleration PSD peak around 9.5 Hz.

Conclusions

Equivalent linear stiffness and damping coefficients of both nonlinear conservative and dissipative elements in mechanical dynamic systems can

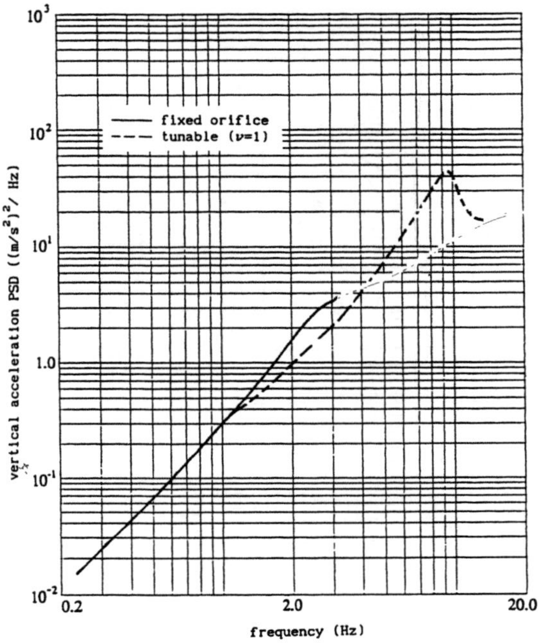

Fig. 10 Vertical acceleration PSD of front unsprung mass with tunable and fixed orifice shock absorbers

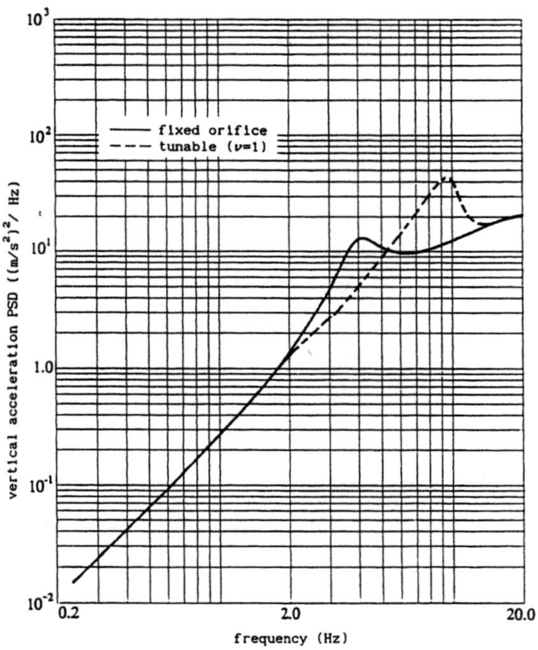

Fig. 11 Vertical acceleration PSD of rear unsprung mass with tunable and fixed orifice shock absorbers

be established by using the generalized discrete harmonic linearization technique presented in this paper. A numerical iterative procedure is presented to replace a nonlinear system by a set of its complex frequency response function matrices, as functions of the excitation frequency. The deterministic analysis results of a nonlinear system with orifice and sequential damping, and gas-spring elements, using the generalized discrete harmonic linearization technique are compared very well with that by employing the direct integration technique, while the linearization technique saves more than 20 times of computer simulation time. The stochastic analysis of a nonlinear vehicle system with fixed orifice and tunable dampers is presented in terms of response PSD by using the generalized linearization technique. This practical numerical technique is relatively simple and easy to apply to multi-DOF dynamic systems with nonlinear damping and spring elements, for both deterministic and stochastic analyses.

Acknowledgements

The financial support provided by the NSERC and the FCAR is gratefully acknowledged.

A version of this paper has been appeared in the Proceedings of the Second International Conference on Stochastic Structural Dynamics. The authors acknowledge the permission from Springer-Verlag, publisher of the proceedings, to publish this article in the present journal.

References

1. Roberts, J. B.,"Response of Nonlinear Mechanical Systems to Random Excitation — Part 2: Equivalent Linearization and Other Methods," *Shock and Vibration Digest*, Vol. 13, No.11, 1981, pp. 15-29.

2. Iwan, W. and Patula, E., "The Merit of Different Error Minimization Criteria In Approximation Analysis," *ASME Journal of Applied Mechanics*, Vol.39, 1972, pp. 257-262.

3. Thompson, W. T., *Theory of Vibration with Applications*, Prentice-Hall, Englewood Cliffs, N.J., 1965.

4. Bandstra, J P., "Comparison of Equivalent Viscous Damping and Nonlinear Damping in Discrete and Continuous Vibration Systems," *ASME Journal of Vibration, Acoustics, Stress and Reliability in Design*, Vol.105, 1983, pp. 382-392.

5. Ruzicka, J. E. and Derby, T. F.,"Vibration Isolation with Nonlinear Damping," *ASME Journal of Eng. Ind.*, 1971, pp. 627-635.

6. Rakheja, S., *Computer Aided Dynamic Analysis and Optimal Design of Suspension Systems for Off-Road Tractors*, Ph.D. Thesis, Mechanical Engineering Department, Concordia University, 1983.

7. Rakheja, S., Van Vliet, M. and Sankar, S., "A Discrete Harmonic Linearization Technique for Simulation Nonlinear Mechanical Systems," *Journal of Sound and Vibration*, 100(4), 1985, pp.511-526.

8. Su, H., Rakheja, S. and Sankar, T. S., "Vibration and Shock Isolation Performance of a Pressure Limited Hydraulic Damper," *Mechanical Systems and Signal Processing*, Vol.3, No.1, 1989, pp. 71-86.

9. Chu, Y. and Li, Z., "The Dynamic Response of Vehicles with Hydro-gas Suspension to Roadway Undulation," *ACTA Armamentari*, No.2, May 1984, pp. 30-42.

10. Roa, B. K. N., Jones, B. and Ashley, C., "Laboratory Simulation of Vibratory Road Surface Inputs," *Journal of Sound and Vibration*, Vol.31, No.2, 1973, pp. 175-183.

11. Robson, J. D. and Dodds, C. J., "Stochastic Road Inputs and Vehicle Response," *Vehicle System Dynamics*, Vol.5, 1975/76, pp. 1-13.

12. Stikeleather, L. F.,"Review of Ride Vibration Standards and Tolerance Criteria," *Trans. SAE*, Paper No. 760413, 1976, pp. 1460-1467.

Equivalent Nonlinearization of Nonlinear Systems to Random Excitations

C.W.S. To and D.M. Li

Department of Mechanical Engineering
The University of Western Ontario
London, Ontario, Canada, N6A 5B9

Abstract

The broadest class of solvable reduced Fokker-Planck equation is given and a new equivalent nonlinear method is presented to obtain an approximate probability density function for the response of a nonlinear oscillator to Gaussian white noise excitations. The method is based on the least mean-square criterion and Euler equation. It is shown that this method, which is simpler and more reasonable, generalizes Caughey's method and gives the same results as Cai and Lin under purely additive excitations. Examples are given to show the applications of the method. In one of the examples, this method leads to a better approximation than that obtained from the energy dissipation criterion.

1. Introduction

The response of nonlinear oscillators to stochastic excitations has been extensively studied in the last three decades. In general, no exact solutions can be found. When the excitations can be idealized as Gaussian white noise, in which case, the response of the system can be represented by a Markovian vector and the probability density function of the response is described by the Fokker-Planck equation, exact solution can be obtained. The solution of the Fokker-Planck equation has been reported in the literature [1-7]. The systematic method developed by Cai and Lin [7] gives a class of solvable reduced Fokker-Planck equations which contains all solvable equations previously obtained. However, it is difficult to find a real mechanical system corresponding to a solvable equation other than those already been reported. Therefore, it is necessary to apply approximation methods to deal with real mechanical systems. In this paper, we first follow the method

Y. K. Lin · I. Elishakoff (Eds.)
Stochastic Structural Dynamics 1
New Theoretical Developments
2nd International Conference on Stochastic Structural Dynamics
May 9–11, 1990, Boca Raton, Florida

presented by Cai and Lin, and apply the theory of elementary factor for first-order ordinary differential equations to obtain a class of solvable reduced Fokker-Planck equations which is broader than that given by Cai and Lin. Then, in terms of the least mean-square criterion and Euler equation, a new approximate technique of equivalent nonlinearization of nonlinear system to stochastic excitation is proposed to deal with an oscillator having nonlinear damping and spring force under additive, or multiplication excitation, or both. A simple formula for the approximate system is given. It must be noted that the least mean-square criterion has been used by Caughey [8] to provide solution for the approximate system that represents an oscillator with nonlinear damping and linear spring force under additive excitation. However, his method provides no simple formula and requires experience. Finally, some examples are included to illustrate the application of our method. Comparisons of results obtained by our method and those in the literature are made.

2. Solvability of the reduced Fokker-Planck equation

Consider the stochastic system

$$\ddot{x} + h(x, \dot{x}) = f_i(x, \dot{x})\, w_i(t) \qquad (i = 1,2,---n) \tag{1}$$

where $h(x_1, x_2)$ and $f_i(x_1,x_2)$ are generally nonlinear and $w_i(t)$ are Gaussian white noises with the delta-type correlation functions

$$E[w_i(t)\, w_j(t + \tau)] = 2\pi\, k_{ij}\delta(\tau) \tag{2}$$

The reduced Fokker-Planck equation is given by

$$x_2 \frac{\partial P_s}{\partial x_1} + \frac{\partial}{\partial x_2} ([-h(x_1,x_2) + \pi k_{ij} f_i(x_1,x_2) \frac{\partial f_j}{\partial x_2}] P_s)$$

$$- \pi k_{ij} \frac{\partial^2}{\partial x_2^2} [f_i(x_1,x_2)\, f_j(x_1,x_2)\, P_s] = 0 \tag{3}$$

where $x_1 = x$, $x_2 = \dot{x}$, and P_s is the stationary probability density of (1). According to the method described in reference [7] and using the same symbols, we get

$$A_1 = x_2$$

$$A_2 = -h(x_1,x_2) + \pi k_{ij} f_i(x_1,x_2)\frac{\partial}{\partial x_2} f_j(x_1,x_2)$$

$$B_{11} = B_{12} = B_{21} = 0. \quad B_{22} = 2\pi k_{ij} f_i(x_1,x_2) f_j(x_1,x_2)$$

Splitting the drift and diffusion coefficients into

$$A_1^{(1)} = 0 \qquad A_1^{(2)} = x_2$$

$$A_2^{(1)} = -h(x_1,x_2) + \pi k_{ij} f_i(x_1,x_2)\frac{\partial}{\partial x_2} f_j(x_1,x_2) - A_2^{(2)}$$

$$B_{22}^{(1)} = \tfrac{1}{2} B_{22} \quad , \qquad B_{12}^{(1)} = -B_{21}^{(2)}$$

Then equation (3) is solvable if the following equations are satisfied:

$$B_{12}^{(1)} \frac{\partial \phi}{\partial x_2} = \frac{\partial}{\partial x_2} B_{12}^{(1)} \tag{4}$$

$$B_{21}^{(2)} \frac{\partial \phi}{\partial x_2} + \pi k_{ij} f_i(x_1,x_2) f_j(x_1,x_2)\frac{\partial \phi}{\partial x_2} = \frac{\partial B_{21}^{(2)}}{\partial x_1} + h(x_1,x_2) + \pi k_{ij} f_i(x_1,x_2) \frac{\partial}{\partial x_2} f_j(x_1,x_2) + A_2^{(2)} \tag{5}$$

$$-x_2 \frac{\partial \phi}{\partial x_1} - A_2^{(2)} \frac{\partial \phi}{\partial x_2} + \frac{\partial A_2^{(2)}}{\partial x_2} = 0 \tag{6}$$

where $P_s(x_1,x_2) = C \exp(-\phi(x_1,x_2))$, (7)

C is a normalization constant.

By the characteristic function method, we get from equation (6)

$$dx_1/(-x_2) = dx_2/(-A_2^{(2)}) = d\phi/\left(-\frac{\partial A_2^{(2)}}{\partial x_2}\right) \tag{8}$$

From the first two equations of equation (8),

$$A_2^{(2)}\, dx_1 - x_2\, dx_2 = 0 \tag{9}$$

In order to obtain the exact solution of equation (9), we use the elementary factor method. Let $M = M(x_1,x_2)$ be the elementary factor of equation (9), then

$$\frac{\partial}{\partial x_1}[M\, x_2] = -\frac{\partial}{\partial x_2}[M\, A_2^{(2)}] \tag{10}$$

Therefore,

$$x_2 \frac{\partial M}{\partial x_1} + M \frac{\partial A_2^{(2)}}{\partial x_2} + A_2^{(2)} \frac{\partial M}{\partial x_2} = 0 \tag{11}$$

$$\frac{\partial A_2^{(2)}}{\partial x_2} = -\frac{1}{M}\frac{\partial M}{\partial x_2} A_2^{(2)} - \frac{x_2}{M}\frac{\partial M}{\partial x_1} \tag{12}$$

Equation (12) gives

$$A_2^{(2)} = \frac{1}{M}\left(C(x_1) - \int_0^{x_2} x_2 \frac{\partial M}{\partial x_1} dx_2\right) \tag{13}$$

where $C(x_1)$ and M are arbitrary functions.

Substituting (13) into (9), then

$$-\frac{1}{M}\left(C(x_1) - \int^{x_2} x_2 \frac{\partial M}{\partial x_1}\, dx_2\right) dx_1 + x_2 dx_2 = 0 \tag{14}$$

Integrating equation (14) leads to

$$r = \int^{(x_1,\, x_2)} \left\{\left[-C(x_1) + \int^{x_2} x_2 \frac{\partial M}{\partial x_1} dx_2\right] dx_1 + M\, x_2 dx_2\right\} = \text{constant} \tag{15}$$

Equation (15) is the implicit solution of the equation (9), there $A_2^{(2)}$ is defined by (13).

By the first and third equations of equation (8)

$$dx_1/(-x_2) = d\phi/\left(-\frac{\partial A_2^{(2)}}{\partial x_2}\right)$$

This gives

$$d\phi = \frac{1}{x_2}\frac{\partial A_2^{(2)}}{\partial x_2}\,dx_1 \tag{16}$$

Substituting (12) into (16), one has

$$d\phi = \frac{1}{x_2}\left(-\frac{x_2}{M}\frac{\partial M}{\partial x_1} - \frac{1}{M}\frac{\partial M}{\partial x_2}A_2^{(2)}\right)dx_1$$

$$= -\frac{1}{M}\frac{\partial M}{\partial x_1}\,dx_1 - \frac{1}{M}\frac{\partial M}{\partial x_2}\frac{\partial A_2^{(2)}}{\partial x_2}\,dx_1 \tag{17}$$

By (9)

$$A_2^{(2)}\,dx_1 = x_2 dx_2 \tag{18}$$

Then

$$d\phi = -\left[\frac{1}{M}\frac{\partial M}{\partial x_1}\,dx_1 + \frac{1}{M}\frac{\partial M}{\partial x_2}\,dx_2\right]$$

$$= -\frac{1}{M}\,dM$$

and

$$\phi = -\ell n\ M + \phi_0(r) \tag{19}$$

where $\phi_0(r)$ is an arbitrary function.

Substituting (19) into (5),

$$h(x_1,x_2) = B_{21}^{(2)}\frac{\partial\phi}{\partial x_1} + \pi k_{ij}f_if_j\frac{\partial\phi}{\partial x_2} - \frac{\partial B_{21}^{(2)}}{\partial x_1} - \pi k_{ij}f_i\frac{\partial f_j}{\partial x_2} - A_2^{(2)}$$

$$= B_{21}^{(2)}\frac{\partial\phi}{\partial x_1} - \frac{\partial B_{21}^{(2)}}{\partial x_1} + \pi k_{ij}f_if_j\left(-\frac{1}{M}\frac{\partial M}{\partial x_2} + \frac{d\phi_0}{dr}\frac{\partial r}{\partial x_2}\right)$$

$$-\pi k_{ij}\,f_i\frac{\partial f_j}{\partial x_2} - \frac{1}{M}\left(C(x_1) - \int^{x_2} x_2\frac{\partial M}{\partial x_1}dx_2\right) \tag{20}$$

By (4) one has

$$B_{21}^{(2)} = B_{12}^{(1)} = \frac{1}{M}\,C_1(x_1)\exp\left(\phi_0(r)\right) \tag{21}$$

where $C_1(x_1)$ is an arbitrary function.

Substituting (21) into (20) one can show that

$$B_{21}^{(2)}\frac{\partial\phi}{\partial x_1} - \frac{\partial B_{21}^{(2)}}{\partial x_1} = -B_{12}^{(1)}\frac{\partial\phi}{\partial x_1} + \frac{\partial B_{12}^{(1)}}{\partial x_1}$$

$$= \frac{1}{M}\,C_1'(x_1)\,e^{\phi_0}$$

Substituting the last equation into (20) gives

$$h(x_1, x_2) = \pi k_{ij} f_i f_j \left(-\frac{1}{M}\frac{\partial M}{\partial x_2} + \frac{d\phi_0}{dr}\frac{\partial r}{\partial x_2}\right)$$

$$- \pi k_{ij} f_i \frac{\partial}{\partial x_2} f_j - \frac{1}{M}\left(C(x_1) - \int^{x_2} x_2 \frac{\partial M}{\partial x_1} dx_2\right)$$

$$+ \frac{C_1'(x_1)}{M} e^{\phi_0} \tag{22}$$

If $M = \lambda_y(x\ , y)$ where $y = \frac{1}{2} x_2^2$, $x = x_1$ and $\lambda(x,y)$ is an arbitrary function, then by (15)

$$r = -\int^{x_1} C(x_1)dx_1 + \int^{(x_1,x_2)} \left[\left(\int^{x_2} x_2\lambda_{xy}dx_2\right) dx_1 + \lambda_y x_2 dx_2\right]$$

Setting $C(x_1) = 0$

$$r = \int^{(x_1,x_2)} \lambda x_1 dx_1 + \frac{\partial \lambda}{\partial x_2} dx_2 = \lambda(x,y) \tag{23}$$

By (19)

$$\phi = -\ell n\lambda_y + \phi_0(\lambda) \tag{24}$$

By equation (22)

$$h = \pi\, x_2 k_{ij} f_f f_j \left(\lambda_y \frac{d\phi_0}{d\lambda} - \frac{\lambda_{yy}}{\lambda_y}\right) - \pi\, k_{ij} f_i \frac{\partial f_j}{\partial x_2} \frac{\lambda_x}{\lambda_y} + \frac{D(x_1)}{\lambda_y} e^{\phi_0} \tag{25}$$

where $D(x_1) = C_1'(x_1)$ is an arbitrary function. Note that the results in [7] is just the special case of equations (23) through (25). The latter, up to now, is the broadest class of solvable equations.

3. Equivalent Nonlinear Method

In this section a new technique of equivalent nonlinearization method is developed. The basic idea is the governing equation replacing by a solvable equation given in Section 2. The criterion for this replacement is the least mean-square. For simplicity we divide the general theory into two parts.

3.1 Special Case

First consider the following stochastic system

$$\ddot{x} + H(x,\dot{x}) + x = w(t) \tag{26}$$

The approximate equation associated with (26) is

$$\ddot{x} + h(a)\,\dot{x} + x = w(t) \tag{27}$$

where

$$a = \tfrac{1}{2}(x^2 + \dot{x}^2) \quad \text{and } h(a) \text{ is to be determined.}$$

From Section 2, equation (27) is solvable and the probability density $P_e(x,\dot{x})$ is

$$P_e(x,\dot{x}) = C \exp\left(- \frac{1}{K} \int_0^a h(z)dz\right) \tag{28}$$

Now, suppose that $P(x,\dot{x})$ is the approximate stationary probability density of equation (26) and $p(x,\dot{x})$ depends only on a, i.e. $P(x,\dot{x}) = P(a)$. And let $D(x,\dot{x}) = h(a)\,\dot{x} - H(x,\dot{x})$

Then using the transformation

$$x = \sqrt{2a}\cos\theta \qquad \dot{x} = \sqrt{2a}\sin\theta$$

and we get

$$E(D^2(x,\dot{x})) = \iint_{-\infty}^{+\infty} P(x,\dot{x})\, D^2(x,\dot{x})\, dx\, d\dot{x}$$

$$= \int_0^{+\infty} P(a) \int_0^{2\pi} D^2(a,\theta)\, d\theta\, da$$

Now the Euler equation is used to get the minimum of the mean square, set

$$F(a, h(a)) = \frac{P(a)}{2\pi} \int_0^{2\pi} D^2(a, \theta)d\theta$$

Then $$\frac{\partial F}{\partial h} = \frac{P(a)}{2\pi} \int_0^{2\pi} \dot{x}\; D(a,\theta)d\theta = 0 \tag{29}$$

Hence

$$\frac{1}{2\pi} \int_0^{2\pi} \dot{x}\,(\, h(a)\dot{x} - H\,(x,\, \dot{x}\,))\, d\theta = 0$$

$$\frac{1}{2\pi}\int_0^{2\pi}\sqrt{2a}\,\sin\theta(h(a)\,\sqrt{2a}\sin\theta - H(a,\theta))\,d\theta = 0$$

$$\frac{\sqrt{2a}}{2}\,h(a) = \frac{1}{2\pi}\int_0^{2\pi} H(a,\theta)\,\sin\theta d\theta$$

$$h(a) = \frac{2}{\sqrt{2a}}\,\frac{1}{2\pi}\int_0^{2\pi} H(a,\theta)\sin\theta d\theta \tag{30}$$

3.2 General Case

Let the governing equation be

$$\ddot{x} + H(x,\dot{x}) = f_i(t)\,W_i(t) \tag{31}$$

The drift coefficients in the reduced Fokker-Planck equation corresponding to (31) are $A_1 = x_2$

$$A_2 = -\,H(x_1,x_2) + \pi k_{ij} f_i \frac{\partial f_j}{\partial x_2}$$

Without loss of generality, assuming

$$H(x_1,x_2) = g_o(x_1) + u(x_1,\,x_2)$$

$$\pi k_{ij}\,f_i\,\frac{\partial f_j}{\partial x_2} = g_1(x_1) + u_1(x_1,x_2) \tag{32}$$

Let $$g(x_1) = g_o(x_1) + g_1(x_1) \tag{33}$$

and $g(x_1)$ is called effective spring force [9]. We take the solvable approximate equation

$$\ddot{x} + h(x,\dot{x}) = f_i(t)\,w_i(t)$$

where $$h(x_1,x_2) = \pi k'_{ij}\,x_2 f_i f_j \phi'_o\,(\lambda) - \pi k_{ij} f_i \frac{\partial f_j}{\partial x_2} + g(x_1)$$

$$\lambda = \frac{1}{2}\,x_2^2 + \int^{x_1} g(x_1)\,dx_1 \tag{34}$$

Similarly suppose that $P(x,\dot{x})$ is the approximate probability density of equation (31) and $P(x,\dot{x})$ only depends on λ. Setting

$$D(x_1,x_2) = H(x_1,x_2) - h(x_1,x_2)$$

$$= u(x_1,x_2) + u_1(x_1x_2) - \pi k_{ij}\, x_2 f_i f_j\, \phi_o'(\lambda) \tag{35}$$

Let $x_1 = x_1,\quad x_2^{\pm} = \pm\,[2\lambda - 2\int g(x_1)dx_1]^{1/2}$ (36)

$$I(\phi_o') = E(D^2(x_1,x_2))$$

$$= \int\int_{-\infty}^{+\infty} P(x_1,x_2)\, D^2(x_1,x_2)\, dx_1 dx_2$$

Then using (36),

$$I(\phi_o') = \int_0^{+\infty} P(\lambda)d\lambda \int_{\mu\lambda_1}^{\mu\lambda_2} \left(\frac{D^2}{x_2}\Big|_{x_2^+} - \frac{D^2}{x_2}\Big|_{x_2^-}\right)dx_1$$

where $\lambda = \int_0^{\mu_j} g(x_1)dx_1 \qquad (j = (1,2)$

Let $\quad F(\lambda,\ \phi_o') = P(\lambda) \int_{\mu\lambda_1}^{\mu\lambda_2} \left(\frac{D^2}{x_2}\Big|_{x_2^+} - \frac{D^2}{x_2}\Big|_{x_2^-}\right)dx_1$

By Euler equation, $\quad \dfrac{\partial F}{\partial \phi_o'} = 0$

then $\quad P(\lambda) \int_{\mu\lambda_1}^{\mu\lambda_2} \left[\frac{2D}{x_2}\frac{\partial D}{\partial \phi_o'}\Big|_{x_2^-} - \frac{2D}{x_2}\frac{\partial D}{\partial \phi_o'}\Big|_{x_2^-}\right]dx_1 = 0$

Here $\quad \dfrac{\partial D}{\partial \phi_o'} = -\pi\, x_2 k_{ij} f_i f_j$

Hence,

$$\int_{\mu\lambda_1}^{\mu\lambda_2} \left[k_{ij} f_i f_j D(x_1x_2)\Big|_{x_2^+} - k_{ij} f_i f_j D(x_1x_2)\Big|_{x_2^-}\right]dx_1 = 0 \tag{37}$$

Substituting (35) into (37)

$$\int_{\mu\lambda_1}^{\mu\lambda_2} [k_{ij} f_i f_j [u + u_1 - \pi k_{ij} x_2 f_i f_j \phi_o'(\lambda)]\Big|_{x_2^+} - k_{ij} f_i f_j D(x_1x_2)\Big|_{x_2^-}]dx_1 = 0$$

$$\int_{\mu\lambda_1}^{\mu\lambda_2} [k_{ij}f_if_j(u(x_1,x_2) + u_1(x_1,x_2))|_{x_2^+} - k_{ij}f_if_j(u(x_1,x_2) + u_1(x_1,x_2))|_{x_1^-}]dx_1$$

$$= \pi \int_{\mu\lambda_1}^{\mu\lambda_2} [x_2(k_{ij}f_if_j)^2|_{x_2^+} - x_2(k_{ij}f_if_j)^2|_{x_2^-}]dx_1 \, \phi_0'(\lambda)$$

Therefore,

$$\phi_0'(\lambda) = \frac{\int_{\mu\lambda_1}^{\mu\lambda_2} [k_{ij}f_if_j(u+u_1)|_{x_2^+} - k_{ij}f_if_j(u+u_1)|_{x_2^-}]dx_1}{\pi \int_{\mu\lambda_1}^{\mu\lambda_2} [x_2(k_{ij}f_if_j)^2|_{x_2^+} - x_2(k_{ij}f_if_j)^2|_{x_2^-}]dx_1} \tag{38}$$

If the excitation is only additive, i.e. $f_i(x,\dot{x})$ in (1) is a constant, then (38) reduces to

$$\phi_0'(\lambda) = \frac{\int_{\mu\lambda_1}^{\mu\lambda_2} [(u+u_1)|_{x_2^+} - (u+u_1)|_{x_2^-}]dx_1}{\pi(k_{ij}f_if_j) \int_{\mu\lambda_1}^{\mu\lambda_2} (x_2|_{x_2^+} - x_2|_{x_2^-})dx_1}$$

This result is compatible with the one in [9] although our method and that presented in [9] are based on different approaches. When the spring force is linear, we have formula (30). Applying this formula, all the results of the examples in Caughey's paper [8] can be obtained.

It should be emphasized, however, that a simple formula (30) is obtained in this section whereas in [8] no such formula was given. Moreover, our method has a good mathematical foundation in contrast to that in [8] where experience is required.

4. Examples

In this section the developed approximate method presented above will now be applied to several examples. Before doing this, some notations are needed. $W_i(t)$ are correlated Gaussian white noise with spectral densities

k_{ij}. All the symbols used in [9] are designated with an asterisk.

Example 1. $\ddot{x} + H(x,\dot{x}) = f_i(x,\dot{x})W_i(t)$ (39)

where $H(x_1,x_2) = \pi k_{ij}x_2 f_i(x_1,x_2)f_j(x_1,x_2)\phi'(\Lambda) - \pi k_{ij}f_i \frac{\partial f_j}{\partial x_2} + G(x_1)$

$$\Lambda = \frac{1}{2}\, x_2^2 + \int^{x_1} G(x_1)dx_1, \quad x_1 = x, \quad x_2 = \dot{x}$$

Equation (39) is solvable from (25) with $D(x_1) = 0$. It will be proven in the following that the approximate equation for this solvable equation is itself, which is expected.

By (32), $H(x_1,x_2) = g_0(x_1) + u(x_1,x_2)$ (40)

$$\pi k_{ij} f_i \frac{\partial f_j}{\partial x_2} = g_1(x_1) + u_1(x_1,x_2) \tag{41}$$

then $g_0(x_1) = -\, g_1(x_1) + G(x_1)$

and $g(x_1) = g_0(x_1) + g_1(x_1) = G(x_1)$

Finally, the approximate equation is given by

$$h(x_1,\, x_2) = \pi k_{ij}x_2 f_i f_j \phi_0'(\lambda) - \pi k_{ij} f_i \frac{\partial f_j}{\partial x_2} + g(x_1)$$

$$= \pi k_{ij}x_2 f_i f_j \phi_0'(\lambda) - \pi k_{ij} f_i \frac{\partial f_j}{\partial x_2} + G(x_1)$$

$$\lambda = \frac{1}{2}\, x_2^2 + \int^{x_1} g(x_1)dx_1 = \Lambda$$

From (40) and (41), we obtain

$$u(x_1x_2) + u_1(x_1x_2) = H - \pi k_{ij} f_i \frac{\partial f_j}{\partial x_2} - G(x_1) = \pi x_2 k_{ij} f_i f_j \phi'(\Lambda) \tag{42}$$

Substituting (42) into (38), one obtains

$$\phi_o'(\lambda) = \phi'(\lambda)$$

Example 2. $\ddot{x} + (\alpha + \beta x^2)\dot{x} + x = xW_1(t) + \dot{x}\, W_2(t) + W_3(t)$ (43)

where α, β are real constants. In this case, $f_1 = x_1$, $f_2 = x_2$, $f_3 = 1$

$$H(x_1,x_2) = (\alpha + \beta x_1^2)\, x_2 + x_1 = u(x_1,x_2) + g_o(x_1)$$

$$\pi k_{ij} f_i \frac{\partial f_j}{\partial x_2} = \pi k_{22} x_2 + \pi(k_{12} x_1 + k_{23}) = u_1(x_1,x_2) + g_1(x_1)$$

then, from (33),

$$g(x_1) = g_o(x_1) + g_1(x_1)$$

$$= (1 + \pi\, k_{12})\, x_1 + \pi\, k_{23}$$

and $$\lambda = \frac{1}{2}\, x_2^2 + \frac{1}{2}\, (1 + \pi k_{12})\, (x_1 + x_{10})^2 \tag{44}$$

where $$x_{10} = \frac{\pi k_{23}}{1 + \pi k_{12}}$$

Setting $$x_1 + x_{10} = \sqrt{\frac{2\lambda}{1+\pi k_{12}}}\, \cos\theta$$

$$x_2 = \sqrt{2\lambda}\, \sin\theta \tag{45}$$

Similar to the special case in Section 3, then

$$\frac{1}{2\pi} \int_0^{2\pi} x_2 k_{ij} f_i f_j (\alpha x_2 + \beta x_1^2\, x_2 + \pi k_{22} x_2) d\theta$$

$$= \frac{1}{2\pi} \int_0^{2\pi} \pi\, (x_2 k_{ij} f_i f_j)^2 d\theta\, \phi_o'(\lambda) \tag{46}$$

Now we calculate the integrations in equation (46).

$$k_{ij}f_if_j = k_{11}f_1^2 + 2k_{12}f_1f_2 + k_{22}f_2^2 + k_{33}f_3^2 + 2k_{13}f_1f_3 + 2k_{23}f_2f_3$$

$$= k_{11}x_1^2 + k_{22}x_2^2 + k_{33} + 2k_{12}x_1x_2 + 2k_{13}x_1 + 2k_{23}x_2 \tag{47}$$

$$x_2k_{ij}f_if_j (\alpha x_2 + \beta x_1^2 x_2 + \pi k_{22}x_2)$$

$$= (\alpha + \pi k_{22})\, x_2^2\, k_{ij}f_if_j + \beta\, x_1^2\, x_2^2\, k_{ij}f_if_j \tag{48}$$

Using transformation (45) and writing the terms integrated on the interval $[0,2\pi]$ being zero as the abbreviation i.z.t. one has

$$x_2^2\, k_{ij}f_if_j = [\frac{k_{11}}{2(1+\pi k_{12})} + \frac{3}{2}\; k_{22}]\lambda^2 + (k_{11}x_{10}^2 - 2k_{13}x_{10} + k_{33})\lambda + \text{i.z.t.} \tag{49}$$

and

$$x_1^2 = x_{10}^2 + \frac{\lambda}{1+\pi k_{12}} - 2x_{10}\sqrt{\frac{2\lambda}{1+\pi k_{12}}}\cos\theta + \frac{\lambda}{1+\pi k_{12}}\; \cos 2\theta \tag{50}$$

The second term on the LHS of (48) is (disregarding β)

$$x_1^2\, x_2^2\, k_{ij}f_if_j = x_1^2x_2^2\, (k_{11}x_1^2 + k_{22}x_2^2 + k_{33} + 2k_{13}x_1) + 2k_{12}x_1^3\, x_2^3 + 2k_{23}x_1^2x_2^3$$

$$= x_1^2\, x_2^2[k_{33} - 2k_{13}x_{10} + k_{11}x_{10}^2 + (k_{22} + \frac{k_{11}}{1+\pi k_{12}})\lambda +$$

$$2\,(k_{13} - k_{11}x_{10})\;\sqrt{\frac{2\lambda}{1+\pi k_{12}}}\;\cos\theta +$$

$$\lambda\,(\frac{k_{11}}{1+\pi k_{12}} - k_{22}\,)\cos 2\theta] + \text{i.z.t.} \tag{51}$$

Also,

$$x_1^2x_2^2 = 2\lambda\sin^2\theta\, (x_{10}^2 + \frac{\lambda}{1+\pi k_{12}} - 2x_{10}\sqrt{\frac{2\lambda}{1+\pi k_{12}}}\cos\theta + \frac{\lambda}{1+\pi k_{12}}\cos 2\theta)$$

$$= \lambda\, [\, x_{10}^2 + \frac{\lambda}{2(1+\pi k_{12})} - x_{10}\sqrt{\frac{2\lambda}{1+\pi k_{12}}}\cos\theta - x_{10}^2\cos 2\theta +$$

$$x_{10}\sqrt{\frac{2\lambda}{1+\pi k_{12}}}\cos 3\theta \quad - \frac{\lambda}{2(1+\pi k_{12})}\cos 4\theta] \tag{52}$$

By (51) and (52), it can be shown that

$$x_1^2x_2^2\, k_{ij}f_if_j = \lambda\,\{\frac{1}{2(1+\pi k_{12})}(k_{22}+\frac{k_{11}}{1+\pi k_{12}})\lambda^2 + [\frac{k_{33}}{2(1+\pi k_{12})}$$

$$-\frac{3k_{13}}{1+\pi k_{12}}x_{10} + 3\,(\frac{k_{11}}{1+\pi k_{12}}+\frac{1}{2}k_{22})x_{10}^2]\lambda +$$

$$x_{10}^2\,(k_{33} - 2k_{13}x_{10} + k_{11}x_{10}^2)\} + \text{i.z.t.} \tag{53}$$

Hence, by (49) and (53), one has

$$x_2\, k_{ij}f_if_j[(\alpha+\beta x_1^2)\, x_2 + \pi k_{22}x_2]$$

$$= \lambda\,\{\frac{\beta}{2(1+\pi k_{12})}\,(k_{22}+\frac{k_{11}}{1+\pi k_{12}})\,\lambda^2 + [(\alpha+\pi k_{22})(\frac{k_{11}}{2(1+\pi k_{12})}+\frac{3}{2}k_{22})$$

$$+\beta\,(\frac{k_{33}}{2(1+\pi k_{12})} - \frac{3k_{13}}{1+\pi k_{12}}x_{10} + 3\,(\frac{k_{11}}{1+\pi k_{12}}+\frac{1}{2}k_{22})\,x_{10}^2]\,\lambda$$

$$+ (k_{33} - 2k_{13}x_{10} + k_{11}x_{10}^2)(\alpha + \pi k_{22} + \beta x_{10}^2)\} + \text{i.z.t.} \tag{54}$$

On the other hand, the integrand on the right hand side of the equation (46) is

$$x_2^2\,(k_{ij}f_if_j)^2 = x_2^2\,[k_{11}^2x_1^4 + k_{22}^2x_2^4 + k_{33}^2 + x_1^2x_2^2(4k_{12}^2 + 2k_{11}k_{22}) +$$

$$(4k_{13}^2 + 2k_{11}k_{33})\,x_1^2 + (4k_{23}^2 + 2k_{22}k_{33})\,x_2^2\,]$$

$$+ 8k_{12}k_{23}x_1x_2^4 + 4k_{11}k_{13}x_1^3x_2^2 + 4k_{13}k_{22}x_1x_2^4$$

$$+ 4k_{13}k_{33}x_1x_2^2 + \text{i.z.t.} \tag{55}$$

Moreover,

$$x_1\, x_2^4 = -\,(2\lambda)^2\, x_{10}\sin^4\theta + \text{i.z.t.} = -\,\frac{3}{2}\,\lambda^2 x_{10} + \text{i.z.t.}$$

$$x_1^3\, x_2^2 = -\,x_{10}^3\,\lambda - \frac{3}{4}\,x_{10}\,\frac{2\lambda^2}{1+\pi k_{12}} + \text{i.z.t.} = -\,x_{10}^3\lambda - \frac{3x_{10}}{2(1+\pi k_{12})}\,\lambda^2 + \text{i.z.t.}$$

$$x_1 x_2^2 = -\,x_{10}\,2\,\lambda\sin^2\theta + \text{i.z.t.} = -\,\lambda x_{10} + \text{i.z.t.} \tag{56}$$

Combining (55) and (56), then

$$\pi\,\frac{1}{2\pi}\int_0^{2\pi} (x_2 k_{ij} f_i f_j)^2 d\theta$$

$$= \pi\,\frac{1}{2\pi}\int_0^{2\pi} x_2^2\,[k_{11}^2 x_1^4 + k_{22}^2 x_2^4 + k_{33}^2 + x_1^2 x_2^2(4k_{12}^2 + 2k_{11}k_{22})$$

$$+(4k_{13}^2 + 2k_{11}k_{33})\,x_1^2 + (4k_{23}^2 + 2k_{22}k_{33})\,x_2^2]\,d\theta$$

$$+\,\pi\,[\,-\,4k_{11}k_{13}(x_{10}^2\lambda + \frac{3x_{10}}{2(1+\pi k_{12})}\,\lambda^2) - 6\,k_{22}k_{13}x_{10}\lambda^2$$

$$-\,2\,k_{33}2k_{13}x_{10}\lambda \;-\, 4k_{12}\,3k_{23}\,x_{10}\lambda^2] \tag{57}$$

And $\quad \frac{1}{2\pi}\int_0^{2\pi} x_2^2\,d\theta = \frac{1}{2\pi}\int_0^{2\pi} 2\lambda\sin^2\theta d\theta = \lambda$

$$\frac{1}{2\pi}\int_0^{2\pi} x_2^4\,d\theta = \frac{1}{2\pi}\int_0^{2\pi} 4\lambda^2\sin^4\theta d\theta = \frac{3}{2}\,\lambda^2$$

$$\frac{1}{2\pi}\int_0^{2\pi} x_2^6\,d\theta = \frac{1}{2\pi}\int_0^{2\pi} 8\lambda^3\sin^6\theta d\theta = \frac{5}{2}\,\lambda^3 \tag{58}$$

By this expression $x_1^2 x_2^2$ in (52), we have

$$\frac{1}{2\pi}\int_0^{2\pi} x_1^2 x_2^2 d\theta = \lambda(x_{10}^2 + \frac{\lambda}{2(1+\pi k_{12})})$$

$$\frac{1}{2\pi}\int_0^{2\pi} x_1^2 x_2^4 d\theta = \lambda^2\,(\frac{3}{2}x_{10}^2 + \frac{\lambda}{2(1+\pi k_{12})})$$

$$\frac{1}{2\pi}\int_0^{2\pi} x_1^4 x_2^2 d\theta = \lambda \left[x_{10}^4 + \frac{3x_{10}^2}{1+\pi k_{12}}\lambda + \frac{\lambda^2}{2(1+\pi k_{12})^2} \right] \tag{59}$$

By (57), (58), and (59), one obtains

$$\pi \frac{1}{2\pi}\int_0^{2\pi} x_2^2 (k_{ij}f_if_j)^2 d\theta$$

$$= \pi [k_{11}^2\lambda (x_{10}^4 + \frac{3x_{10}^2}{1+\pi k_{12}}\lambda + \frac{\lambda^2}{2(1+\pi k_{12})^2}) + \frac{5}{2} k_2^2 \lambda_{22}^3 + k_{33}^2 \lambda$$

$$+ (4k_{12}^2 + 2k_{11}k_{22})(\frac{3}{2} x_{10}^2 + \frac{\lambda}{2(1+\pi k_{12})}) \lambda^2$$

$$+ (4k_{13}^2 + 2k_{11}k_{33})(x_{10}^2 + \frac{\lambda}{2(1+\pi k_{12})})\lambda + \frac{3}{2}\lambda^2 (4k_{23}^2 + 2k_{22}k_{33})]$$

$$+ \pi [-4k_{11}k_{13}(x_{10}^2\lambda + \frac{3x_{10}}{2(1+\pi k_{12})}\lambda^2) - 6k_{22}k_{13}x_{10}\lambda^2$$

$$- 4k_{13}k_{33}x_{10}\lambda - 12k_{12}k_{23}x_{10}\lambda^2] \tag{60}$$

Finally by (46), (54), and (60), and some algebraic manipulation one has

$$\lambda\{\bar{A}_1\lambda^2 + \bar{B}_1\lambda + \bar{C}_1\} = \pi\lambda \{\bar{A}_2\lambda^2 + \bar{B}_2\lambda + \bar{C}_2\} \phi_o'(\lambda)$$

Hence,

$$\phi_o'(\lambda) = \frac{\bar{A}_1\lambda^2+\bar{B}_1\lambda+\bar{C}_1}{\pi(\bar{A}_2\lambda^2 +\bar{B}_2\lambda+\bar{C}_2)} \tag{61}$$

where $\bar{A}_1 = \dfrac{\beta k_{22}}{2(1+\pi k_{12})} + \dfrac{\beta k_{11}}{2(1+\pi k_{12})^2}$

$$\bar{B}_1 = \frac{\beta k_{33}+k_{11}(\alpha+\pi k_{22})}{2(1+\pi k_{12})} + \frac{3}{2} k_{22}(\alpha+\pi k_{22}) - \frac{3k_{13}\beta}{1+\pi k_{12}} x_{10}$$

$$+ 3\beta(\frac{k_{11}}{1+\pi k_{12}} + \frac{1}{2} k_{22}) x_{10}^2$$

$$\bar{C}_1 = (k_{33} - 2k_{13}x_{10} + k_{11}x_{10}^2)(\alpha + \pi k_{22} + \beta x_{10}^2)$$

$$\bar{A}_2 = \frac{k_{11}^2}{2(1+\pi k_{12})^2} + \frac{k_{11}k_{22}+2k_{12}^2}{1+\pi k_{12}} + \frac{5}{2}\ k_{22}^2$$

$$\bar{B}_2 = \frac{2k_{13}^2+k_{11}k_{33}}{1+\pi k_{12}} + 3(2k_{23} + k_{22}k_{33}) - 6\ (\frac{k_{11}k_{13}}{1+\pi k_{12}} + k_{13}k_{22} + 2k_{12}k_{23})x_{10}$$

$$+ 3\ x_{10}^2\ (2k_{12} + k_{11}k_{22} - \frac{k_{11}^2}{1+\pi k_{12}})$$

$$\bar{C}_2 = k_{11}^2x_{10}^4 + (4k_{13}^4 - 4k_{11}k_{13} + 2k_{11}k_{33})x_{10}^2 - 4k_{13}k_{33}x_{10} + k_{33}^2$$

Applying (61), one can obtain

$$\phi_0(\lambda) = \frac{\bar{A}_1}{\pi\bar{A}_2}\{\lambda + \frac{1}{2}\ (\frac{\bar{B}_1}{\bar{A}_1} - \frac{\bar{B}_2}{\bar{A}_2})\ell n\ (\lambda^2 + \frac{\bar{B}_2}{\bar{A}_2}\lambda + \frac{\bar{C}_2}{\bar{A}_2})$$

$$+ [\frac{\bar{C}_1}{\bar{A}_1} - \frac{\bar{C}_2}{\bar{A}_2} - \frac{\bar{B}_2}{2\bar{A}_2}\ (\frac{\bar{B}_1}{\bar{A}_1} - \frac{\bar{B}_2}{\bar{A}_2}\ )\ \int \frac{1}{\lambda^2 + \frac{\bar{B}_2}{\bar{A}_2}\lambda + \frac{\bar{C}_2}{\bar{A}_2}}\,d\lambda\} \qquad (62)$$

The integral in (62) can be directly found or obtained from the integral table.

Now, two special cases are considered.

(i) $k_{33} \neq 0$ and $k_{ij} = 0$ then

$$\bar{A}_1 = 0 \qquad \bar{B}_1 = \frac{1}{2}\beta k_{33} \qquad \bar{C}_1 = \alpha k_{33} \qquad (63)$$

$$\bar{A}_2 = 0 \qquad \bar{B}_2 = 0 \qquad \bar{C}_2 = k_{33}^2$$

Substituting (63) into (61)

$$\phi_0'(\lambda) = \frac{2\alpha+\beta\lambda}{2\pi k_{33}} \tag{64}$$

Equation (64) can also be obtained by the energy dissipation method [9].

(ii) $k_{11} \neq 0$, and $k_{ij} = 0$, then

$$\bar{A}_1 = \frac{1}{2}\beta k_{11} \qquad \bar{B}_1 = \frac{1}{2}\alpha k_{11} \qquad \bar{C}_1 = 0$$

$$\bar{A}_2 = \frac{1}{2}k_{11}^2 \qquad \bar{B}_2 = 0 \qquad \bar{C}_2 = 0$$

And

$$\phi_0'(\lambda) = \frac{\frac{1}{2}\beta k_{11}\lambda^2 + \frac{1}{2}\alpha k_{11}\lambda}{\frac{\pi}{2}k_{11}^2\lambda^2} = \frac{\beta}{\pi k_{11}} + \frac{\alpha}{\pi k_{11}}\frac{1}{\lambda}$$

But from the method in [9]

$$\phi_0^{*\prime}(\lambda^*) = \frac{\beta}{\pi k_{11}} + \frac{2\alpha}{\pi k_{11}}\frac{1}{\lambda^*}$$

where $\lambda^* = \lambda$. Thus, the last term on the right hand side has a factor of 2 difference.

Example 3. $\ddot{x} + \alpha\,\dot{x} + b\,\dot{x}^2 + c\,\dot{x}^3 + x = \dot{x}\,W_1(t) + W_2(t)$ (65)

where a,b,c are real constants. Then by the method in [9], one has

$$f_1^* = x_2 \qquad f_2^* = 1 \qquad H^*(x_1,x_2) = a\,x_2 + b\,x_2^2 + c\,x_2^3 + x_1$$

$$g^*(x_1,x_2) = x_1 - \pi k_{12} \qquad \lambda^* = \frac{1}{2}x_2^2 + \frac{1}{2}(x_1 - \pi k_{12})^2$$

and $h^*(x_1,x_2) = \pi\,x_2(k_{11}x_2^2 + 2k_{12}x_2 + k_{22})\phi_0^{*\prime}(\lambda^*) - \pi(k_{11}x_2 + k_{12}) + x_1 - \pi k_{12}$

Taking the polar coordinate transformation

$$x_1 = \sqrt{2\lambda^*}\cos\theta + \pi k_{12} \qquad x_2 = \sqrt{2\lambda^*}\sin\theta$$

Then

$$\phi_0^{*\prime}(\lambda^*) = \frac{\frac{1}{2\pi}\int_0^{2\pi} \sin\theta\,[ax_2+bx_2^2+cx_2^3+\pi k_{11}x_2+2\pi k_{12}]d\theta}{\frac{1}{2}\int_0^{2\pi} \pi x_2 \sin\theta\, k_{ij}f_if_j d\theta}$$

$$= \frac{2a + 2\pi k_{11} + 3c\lambda^*}{2\pi k_{22} + 3\pi k_{11}\lambda^*}$$

If $a = \pi k_{22} - \pi k_{11}$, $\quad b = 2k_{12}\pi$, $\quad c = \pi k_{11}$ (66)

then

$$\phi_0^{*\prime}(\lambda^*) = 1$$

And the approximate equation from this method becomes

$$\ddot{x} + \pi x_2(k_{11}x_2^2 + 2k_{12}x_2 + k_{22}) - \pi k_{11}x_2 + x_1 - 2\pi k_{12} = 0 \tag{67}$$

which is different from equation (65).

On the other hand, substituting (66) into (65), and comparing (65) with (34), equation (65) is solvable with $\phi_0'(\lambda) = 1$, $g = x_1 + \pi k_{12}$ and $\lambda = \frac{1}{2}x_2^2 + \frac{1}{2}(x_1+\pi k_{12})^2$.

Using the result in example 1 above for the present example, the approximate equation is itself, which shows that our method leads to a better approximation than that given by the energy dissipation method.

Example 4. $\ddot{x} + \dot{x}(a_1+a_2x^2) - x + x^3 = xW(t)$ (68)

where $a_i(i=1,2)$ are real constants. And now

$$g_0(x_1) = -x_1+x_1^3 \qquad u(x_1,x_2) = x_2(a_1 + a_2x_1^2)$$

$$g_1(x_1) = 0 \qquad u_1(x_1,x_2) = 0$$

$$g(x_1) = g_0(x_1) = -x_1 + x_1^3$$

$$\lambda = \frac{1}{2}x_2^2 - \frac{1}{2}x_1^2 + \frac{1}{4}x_1^4 \qquad \text{and}$$

$$h(x_1,x_2) = \pi k \quad x_2 x_1^2 \quad \phi_0'(\lambda) - x_1 + x_1^3$$

$$-\frac{1}{2}\mu_\lambda^2 + \frac{1}{4}\mu_\lambda^4 - \lambda = 0 \qquad \text{and}$$

$$\mu_\lambda^2 = \frac{\frac{1}{2} \pm \sqrt{\frac{1}{4}+\lambda}}{\frac{1}{4}\times 2} = 1 \pm \sqrt{1+4\lambda}$$

Neglecting the negative root, we have

$$\mu\lambda_i^2 = 1 + \sqrt{1+4\lambda} \qquad (i = 1,2)$$

Substituting all the above equations into (38), then

$$\phi_0'(\lambda) = \frac{k\int_{\mu\lambda_1}^{\mu\lambda_2}(x_1^2 u|_{x_2^+} - x_1^2 u|_{x_2^-})dx_1}{\pi\int_{\mu\lambda_1}^{\mu\lambda_2}[x_2k^2x_1^4|_{x_2^+} - x_2k^2x_1^4|_{x_2^-}]dx_1}$$

$$= \frac{b_1}{\pi k} + \frac{a_1\int_0^{\sqrt{1+\sqrt{1+4\lambda}}} x_1^2\sqrt{2\lambda+x_1^2-\frac{1}{2}x_1^4}\,dx_1}{\pi k\int_0^{\sqrt{1+\sqrt{1+4\lambda}}} x_1^4\sqrt{2\lambda+x_1^2-\frac{1}{2}x_1^4}\,dx_1} \tag{69}$$

The approximate probability density $p(x_1,x_2)$ of equation (68) is given by

$$p(x_1,x_2) = C\exp(-\int_0^\lambda \phi_0'(\lambda)d\lambda) \tag{70}$$

where C is normalization constant.

5. Conclusions

In this paper, a broadest class of solvable reduced Fokker-Planck equation is given and a new technique of equivalent nonlinearization for systems subjected to stochastic excitations is proposed.

It is shown that the proposed method, which is simpler and more reasonable, generalizes Caughey's method and gives the same results in [9]. Exact solution for the approximated systems can be obtained if the effective spring is linear. For nonlinear spring, however, a numerical computation is needed. As shown in example 3, the proposed method leads to a better approximation than that obtained from the energy dissipation method [9].

ACKNOWLEDGEMENTS

The research reported in this paper was supported by the Natural Sciences and Engineering Research Council of Canada under Grant No. A5074 to the first author. Partial support from the National Science Foundation of China made to the second author is gratefully acknowledged.

REFERENCES

1. Andronov, A., Pontryagin, L., Witt, A. On the Statistical investigation of dynamical systems. Zh. Eksprim, i Theor. Fiz. 3, (1933), 165-180 (in Russian).

2. Caughey, T.K. On the response of a class of nonlinear oscillators to stochastic excitation. Proceedings Collog. Int. du Centre National de la Recherche Scientifique, No. 148, Marseille, Sept., (1964) 392-402.

3. Dimentberg, M.F. An exact solution to a certain nonlinear random vibration problem. Int. J. Nonlinear Mech. 17, (1982) 231-236.

4. Caughey, T.K. and Ma, F. The exact steady-state solution of a class of nonlinear stochastic systems. Int. J. Nonlinear Mech. 17, (1982) 137-142

5. Caughey, T.K. and Ma, F. The steady-state response of a class of dynamical systems to stochastic excitation. J. Applied Mech. 49, (1982) 629-632.

6. Yong, Y. and Lin, Y.K. Exact stationary reponse solution for second order nonlinear systems under parametric and external white noise excitations. J. Applied Mech. 54, (1987) 414-418.

7. Cai, G.Q. and Lin, Y.K. On exact stationary solutions of equivalent nonlinear stochastic systems. J. Nonlinear Mech. 23, No. 4, (1988) 315-325.

8. Caughey, T.K. On the response of nonlinear oscillators to stochastic excitation. Probabilistic Engineering Mech. 1, No. 1, (1986) 2-4.

9. Cai, G.Q. and Lin, Y.K. A new approximate solution technique for randomly excited nonlinear oscillators. J. Nonlinear Mech. 23, No. 5/6, (1988) 409-420.

Chaotic and Stochastic Dynamics for Inelastic Systems with Hysteresis and Degradation

C. Y. Yang and A. H-D. Cheng
DEPARTMENT OF CIVIL ENGINEERING, UNIVERSITY OF DELAWARE, NEWARK, DELAWARE, 19716
R. V. Roy
DEPARTMENT OF MECHANICAL ENGINEERING, UNIVERSITY OF DELAWARE NEWARK, DELAWARE 19716

Abstract

In this paper we attempt to introduce some new and fascinating discoveries in chaotic dynamics with particular emphasis on their relationship with stochastic dynamics. Our investigation centers around the deterministic structural system. We begin with a nonlinear hard or soft spring system governed by the well known Duffing's equation. We then add the important engineering material behavior of hysteresis and degradation. Under a set of unique initial conditions and a harmonic steady state excitation, the response for a certain deterministic structural system is found to be stochastic.

INTRODUCTION

Chaotic dynamics has emerged as a new exciting and sometimes confusing field of engineering research and education in the last decade. The most enthusiastic advocates in the field have used such strong words as, *"Chaos is a new revolution in science,"* and *"engineering in the 1990's"*.[1–4] Others may feel that *"chaos is nothing new but a subset of stochastic nonlinear dynamics."* In this paper we attempt to introduce some new and fascinating discoveries in chaotic dynamics with particular emphasis on their relationship with stochastic dynamics. Our investigation centers around the deterministic structural system governed by an extended version of the well known Duffing's nonlinear equation. In addition to strong nonlinear soft or hard spring, the system has the important engineering material behavior of hysteresis and degradation.[5–10] Under a unique set of initial conditions and a *deterministic* harmonic steady state excitation, the responses for a certain *deterministic* structural systems are found to be *stochastic*. Hence the term *chaotic and stochastic* dynamics is used here to describe the new phenomenon which in our opinion falls in between the typical deterministic and the typical stochastic. In this sense the new science of chaos, as it is known now, may well serve as a link to bring the two traditionally somewhat separate fields, dynamics versus stochastic dynamics, more closer together for mutual reinforcement.

The new science of chaos begins at the discovery of the 'butterfly effect' of Lorenz[11] in 1963 in his study of long time prediction of weather. Using a simple but nonlinear system of three first order differential equations of three variables for fluid velocity amplitude and temperature distributions, Lorenz found that long time weather prediction is impossible because a negligible disturbance in the initial conditions (butterfly effect) will cause drastic and practically unpredictable changes in the output for the state of weather. The phenomenon is now known as chaos due to the sensitive dependence of the nonlinear response to initial conditions. The most fascinating discovery from the Lorenz's weather system, besides the butterfly effect which seems to be negative and discouraging to the profession, is the predictability of the distribution of the state of the weather in a geometric form of the so-called Lorenz strange attractor. Lorenz's attractor is a three dimensional trajectory plot of the state of the weather with time as the

[1]This paper was presented in the Session dedicated to the memory of Professor Frank Kozin

Y. K. Lin · I. Elishakoff (Eds.)
Stochastic Structural Dynamics 1
New Theoretical Developments
2nd International Conference on Stochastic Structural Dynamics
May 9–11, 1990, Boca Raton, Florida

implicit independent variable. With negligible difference of the initial conditions, the trajectories of two different weather predictions are drastically different, but the long time trajectories in the three dimensional state space are attracted to a geometric form, the chaotic or strange attractor, which is found to be unique, very orderly and accurately reproducible. Hence a probabilistic response is generated by a completely deterministic system because of the instability of the response in the system with strong nonlinearity.

Since 1963, numerous discoveries of chaos have been reported in physics,[12] mathematics,[13,14] physiology,[15] biology[16] and engineering.[2,3,17] This wide range of discipline involved is not surprising because of the many unsolved significant nonlinear problems which had been left aside until the recent computer revolution and the awareness of chaos. Focusing our attention now to the specific area of structural dynamics, chaotic vibrations are reported recently by Poddar, Moon and Mukherjee,[18] Holmes,[19] Dowell and Pezeshki,[20] Reinhall, Caughey, and Storti,[21] Kapitaniak,[22] and Tongue.[23] A nonlinear structural system governed by the well known Duffing's equation with hard or soft spring is central to most of the above investigations. This equation has been used also by Ueda[24-26] to investigate a nonlinear electric circuit. He found the first group of well defined chaotic attractors in the phase plane and the broad banded power spectral densities of the response chaotic motion.

We begin our investigation with the numerical solution of the Duffing-Ueda[24] problem and generate chaotic vibrations under a unique set of initial conditions and harmonic excitations. The chaotic characteristics of the steady state response are demonstrated by the dramatic divergence of two displacement time series corresponding to two identical problems, with the exception of a negligible difference in the initial conditions. We then show the clearly defined chaotic attractors in the phase plane, by plotting the Poincaré maps from the trajectory solution of a single problem. Finally a group of several hundred identical problems with a uniformly distributed initial conditions of negligible difference in magnitude have been solved numerically to demonstrate the ergodicity of the chaotic process.

Our primary investigation lies in the chaotic and stochastic dynamics of the structural systems exhibiting inelastic, hysteretic and degrading material properties in addition to the nonlinear elastic spring in the Duffing-Ueda system described above. A structural model originated by Barber and Wen[5,6,27] and used recently by Sues, Mao and Wen[7] is used here. The nonlinear and inelastic restoring force in this hysteretic and degrading model is expressed in an analytic form of a nonlinear differential equation. Together with the equation of motion of a single degree of freedom system with harmonic excitation, the problem takes the mathematical form of the well known Lorenz[11] system of three first order nonlinear differential equations. For this nonlinear system, new chaotic vibrations are found through the examination of the sensitive dependence of system response to initial conditions, the existence of well defined and reproducible chaotic attractors and the probability density functions.

DUFFING'S NONLINEAR ELASTIC SYSTEM

A very thorough investigation has been carried out by Ueda[24,25] on the regular and chaotic vibrations of a class of nonlinear elastic systems governed by the well known Duffing's equation with a linear damping, a cubic nonlinear elastic restoring force and a harmonic forcing function of unit circular frequency and a varying amplitude. For our interest here the selected simple equation is

$$\ddot{x} + 0.05\dot{x} + x^3 = G\cos t \tag{1}$$

The above equation governs the displacement x, velocity $\dot{x}$, and acceleration $\ddot{x}$ of a single degree of freedom structural system subject to a harmonic acceleration excitation of constant amplitude G and unit circular frequency. In the range of small amplitude vibration, the system is essentially linear. The response consists of a transient part depending on the initial conditions and a steady

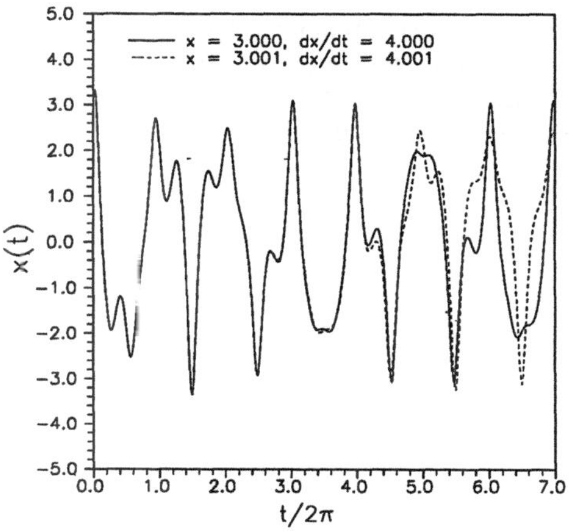

Figure 1: Sensitivity to initial conditions of Duffing's equation $\ddot{x} + 0.05\dot{x} + x^3 = 7.5 \cos t$.

state harmonic part with the forcing unit circular frequency. The interesting part of course is the strong nonlinear vibration when the amplitude G of the excitation acceleration is large. When G is increased, the steady-state response of the system departs from the linear harmonic vibration but remains to be regularly periodic with period equal to twice, three times and higher multiples of the forcing period. The process of changing periods is known as 'period doubling' which is one of the indications of the onset of chaos. See the recent work of Tongue[23] for a large number of such response solutions. Indeed for equation (1), chaotic vibration occurs when the acceleration amplitude is increased to 7.5 (see p. 102 Thompson and Stewart[2]).

The first indication of chaos is the sensitive dependence of the response to initial conditions. This is demonstrated by solving (1) using a fourth order Runge-Kutta numerical integrator. The result is illustrated in Figure 1 in which the displacement time series (x versus t) for two almost identical problems of the forced nonlinear vibration with $G = 7.5$ are shown. The two problems differ only in the initial conditions as $(x(0), \dot{x}(0)) = (3.000, 4.000)$ for problem 1 and $(x(0), \dot{x}(0)) = (3.001, 4.001)$ for problem 2. The numerical solution clearly indicates the drastic divergence of the two time series after about five forcing periods.

Next we show the chaotic phase portrait from the trajectory of the first problem in Figure 2. To eliminate the initial transient effect, the trajectory corresponds to the 31 to 50 cycles of the forcing period. This chaotic phase trajectory, which is not a closed loop, shows a distinct difference from the well known elliptic phase portrait of a linear steady-state harmonic vibration and also from the non-elliptic, simple or multiple limit cycles of nonlinear vibrations with subharmonics and ultraharmonics.[2,25,29]

For the chaotic response as shown in the above, a stroboscopic sampling of the phase portrait is known to produce the chaotic attractor. It is the mapping of coordinates of x and $\dot{x}$ in the phase plane at discrete multiples of the forcing period $t = 2n\pi$, n = integer, the so-called Poincaré map. For the chaotic vibration problem of equation (1), with $G = 7.5$, the Poincaré map shown in Figure 3 clearly has a fractal geometry. The attractor is called chaotic or strange as versus the single or multiple-point attractor of the linear harmonic steady state vibration, or the harmonic and subharmonic vibration of a nonlinear system.

Figures 4 through 8 plot the Poincaré map at various phase angle, namely, the samples are taken at $t = 2n\pi + \theta$, where θ is the phase shift. From these discrete sections, we observe a 'twisting' of the strange attractor. If we consider that these maps are arranged into a torroidal phase space (p. 92 Thompson and Stewart[2]), the 'center of mass' of the attractor appears to move on a wavy orbit. Since Duffing's equation is anti-symmetric ($x(t+\pi) = -x(t)$ and $\dot{x}(t+\pi) = -\dot{x}(t)$), the phase portrait at $\theta = \pi$ is the inverted image of that at $\theta = 0$. It is hence sufficient to show phase portraits from $0 \leq \theta \leq \pi$.

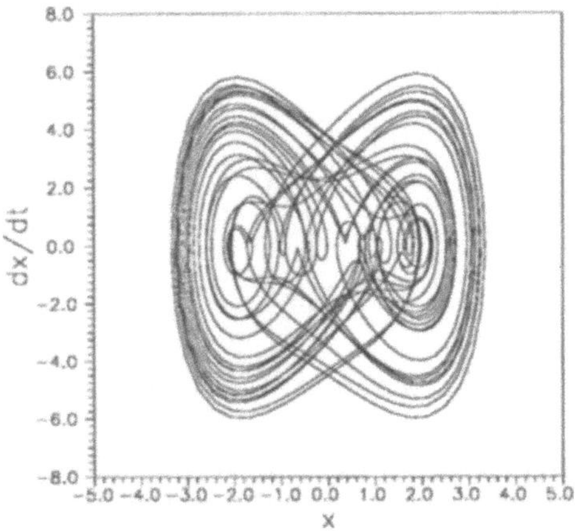

Figure 2: Phase trajectory of Duffing's equation (20 cycles).

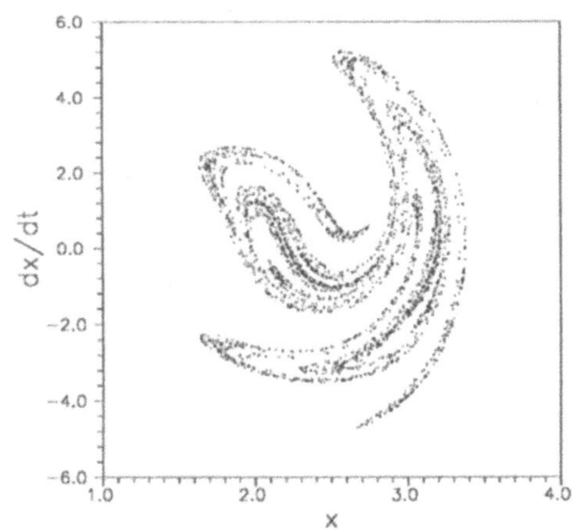

Figure 3: Poincaré map for Duffing's equation.

The exciting part of the chaotic attractors of Figures 3 to 8 is that the distribution of the coordinates $(x, \dot{x})$ which make up this attractor is unique, orderly and reproducible. It is unique, in the sense that two almost identical problems with negligible differences in initial conditions as investigated earlier, produce drastically divergent trajectories as shown in Figure 1, but settle down to the same chaotic attractor with a unique probability distribution! The divergence of solution is also observed if one uses different step size in the numerical integrator, different digital precision in the computer (double or single precision), or different numerical algorithm. But the final distribution of the strange attractor is still the same.

The joint probability distribution of the time history of x and $\dot{x}$ is approximately obtained as follows. Duffing's equations is solved for 500 loading cycles with 200 pairs of $(x, \dot{x})$ data collected per cycle. The first 20 cycles of data, which are believed to be transient, are discarded. The joint PDF is determined by overlaying a grid system on the $(x, \dot{x})$ plane and counting the number of occurrences of data falling in each grid, and then normalizing the distribution. The final result is presented as Figure 9 in a three-dimensional plot. It is of interest to observe the symmetry of the PDF with respect to both the x and $\dot{x}$ axes. The PDF displays a double peak feature. Similar characteristics has been reported earlier from experimentally obtained PDF of a buckled beam (p. 152 Moon[3]).

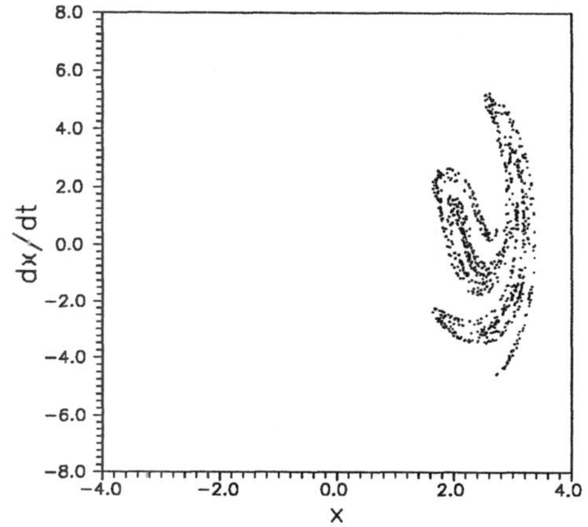

Figure 4: Poincaré map at phase angle 0.

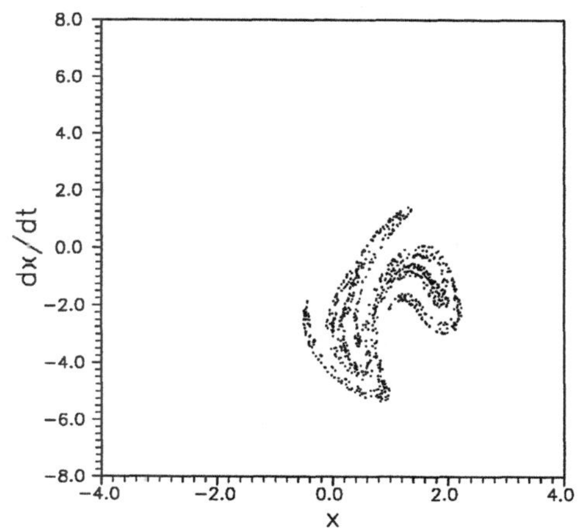

Figure 5: Poincaré map at phase angle $\pi/4$.

One of the most interesting result to us in the chaotic process is the 'ergodicity'. Notice that the above PDF was obtained from a time series, which is justified only if the process is ergodic. To test that assumption, an 'ensemble experiment' is conducted. We run the same Duffing's equation for roughly 3,000 cases with slightly different initial conditions. We select a small square in the phase plane, $2.999 \leq x \leq 3.001$ and $3.999 \leq \dot{x} \leq 4.001$. Within that square, 0.002 × 0.002 in size, 3,000 initial conditions are uniformly distributed. Each initial condition may be considered as a particle. We then examine the trace of the group of particles as we take snap shots at time equal to multiples of the forcing period. It should be reminded that under non-chaotic conditions, we expect the group of particles to stick together throughout the simulation. Figure 10 shows the initial group of particles, which under the current graphic resolution, appear as a single dot. At the end of the first and second loading cycles ($t = 2\pi$ and 4π), the particles have drifted in the phase plane but they stick together. For the ensuing loading cycles ($t = 6\pi$ to 10π), the particles, initially in a square formation, are now stretched to look like a curve. For $t = 12\pi$ to 16π (Figures 11 to 13), the drifting has ceased but the stretching continues and in the same time it is folded. The process of stretching and folding is known as Smale horseshoe effect[28] (see Thompson and Stewart, p. 245). After 15 loading cycles ($t = 30\pi$), Figure 14, the particles have settled into essentially the same strange attractor as Figure 3.

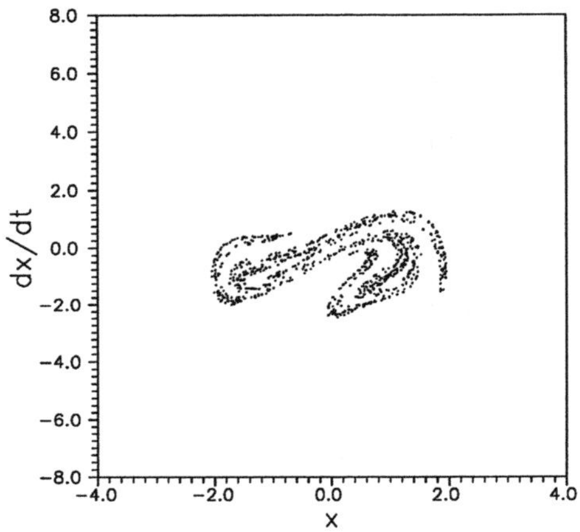

Figure 6: Poincaré map at phase angle $\pi/2$.

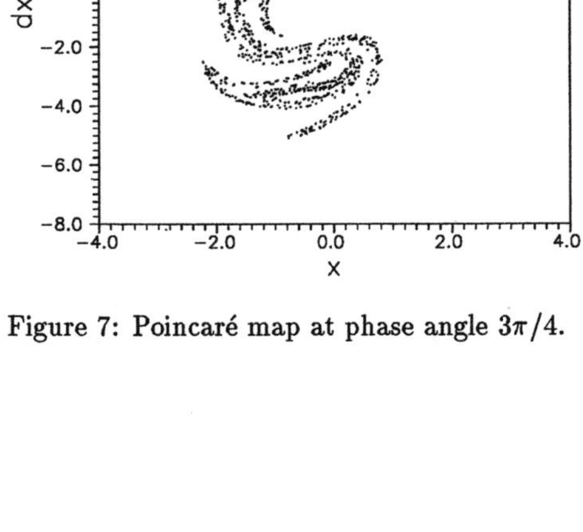

Figure 7: Poincaré map at phase angle $3\pi/4$.

Observing from the stroboscopically obtained still pictures, we conclude the following:

- Macroscopically, after an initial transient period, the particles quickly settle into a definite shape as in Figure 14. The stationarity condition is achieved.
- Microscopically, however, the folding continues indefinitely to give the fractal geometry.
- From the point of view of obtaining the joint PDF as in Figure 9, the chaotic process, after the initial transient period, is ergodic.

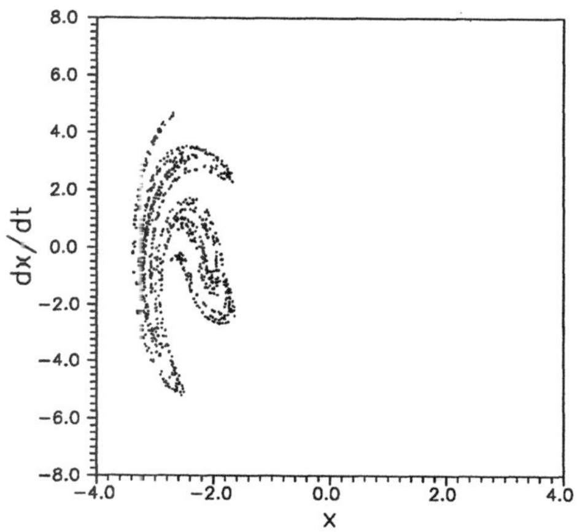

Figure 8: Poincaré map at phase angle π.

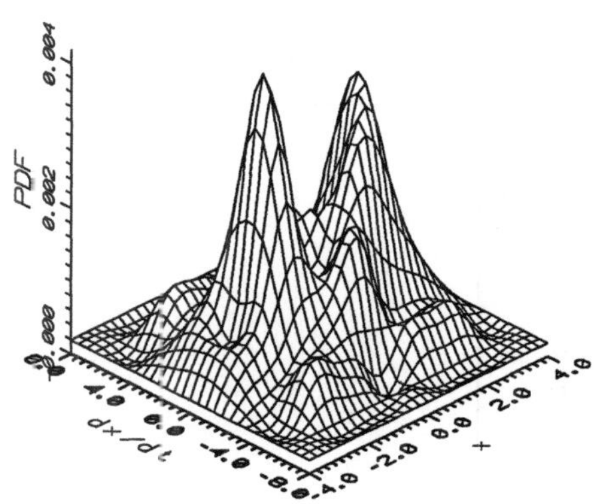

Figure 9: Joint probability density function of Duffing's equation.

INELASTIC, HYSTERETIC AND DEGRADING SYSTEM

It is well known that for the most important structural materials of reinforced concrete and steel, the mechanical behavior under cyclic loading at high amplitudes is inelastic. The restoring force versus displacement diagram shows in general hysteretic loops with decreasing slopes,[10,27] representing energy dissipation and material damage. For such material behavior, several well known models are available.[27] A model originated by Barber and Wen[6] is selected here because of its versatility and suitability for analytical work. With this model, the equation of motion for the single degree of freedom system with mass m, damping coefficient c, a nonlinear and inelastic restoring force q, and a harmonic force excitation of amplitude f, and frequency ω is given by

$$m\ddot{x} + c\dot{x} + q = f\cos\omega t \tag{2}$$

where $\dot{x}$ is the velocity, and $\ddot{x}$ is the acceleration. The restoring force q is defined as

$$q = \alpha k x + (1-\alpha) k z \tag{3}$$

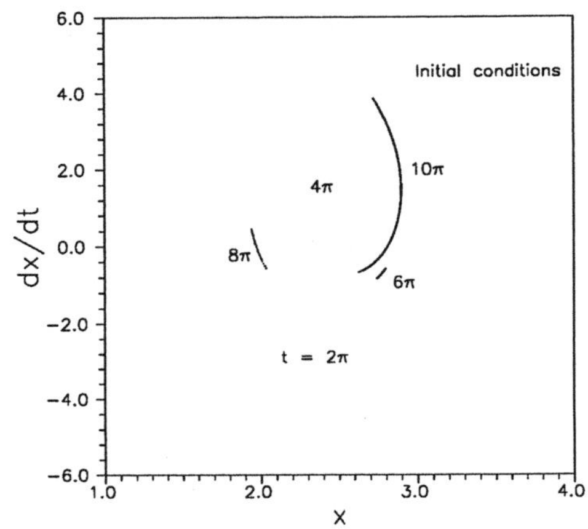

Figure 10: Snap shot of ensemble simulation, t = 0 to 10 π.

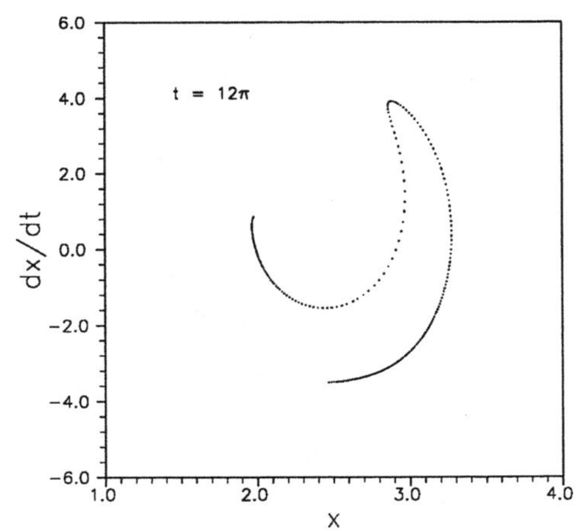

Figure 11: Snap shot of ensemble simulation, t = 12 π.

where k is the spring constant, x the displacement, and z the equivalent displacement. Clearly α is a nondimensional constant which partitions the restoring force q into a linear elastic and a nonlinear inelastic part defined by the equivalent displacement z. Its time derivative is given by

$$\dot{z} = (1 - \delta\epsilon)\left[\dot{x} - \beta|\dot{x}|z x_y^{-1} - \gamma\dot{x}|z|x_y^{-1}\right] \tag{4}$$

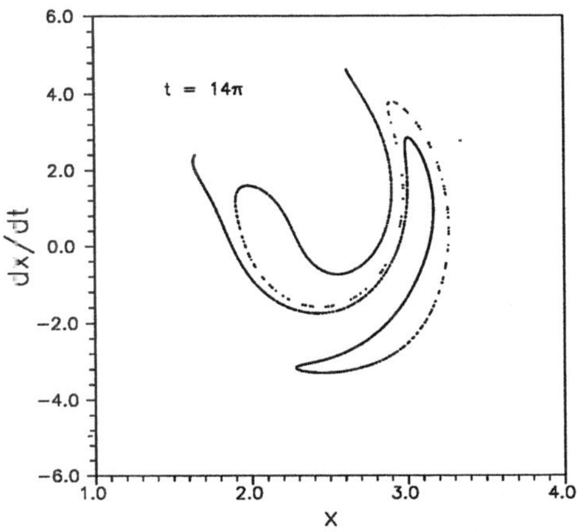

Figure 12: Snap shot of ensemble simulation, $t = 14\ \pi$.

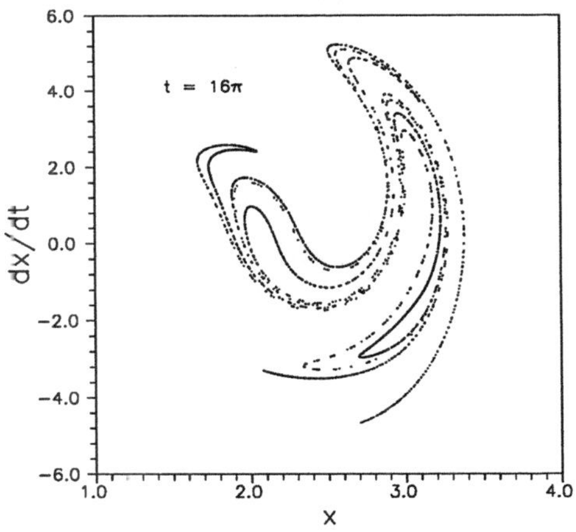

Figure 13: Snap shot of ensemble simulation, $t = 16\ \pi$.

Equation (4) is the key nonlinear differential equation characterizing a range of the inelastic, hysteretic and degrading material behavior by the three non-dimensional parameters δ, β and γ. The dimensional constant x_y is the displacement at yield. The nondimensional energy dissipation ϵ is defined as a function of time of the loading process by

$$\begin{aligned} \epsilon &= k^{-1}x_y^{-2}\int_0^x (1-\alpha)kz\,dx \\ &= (1-\alpha)x_y^{-2}\int_0^t z\dot{x}\,dt \end{aligned} \tag{5}$$

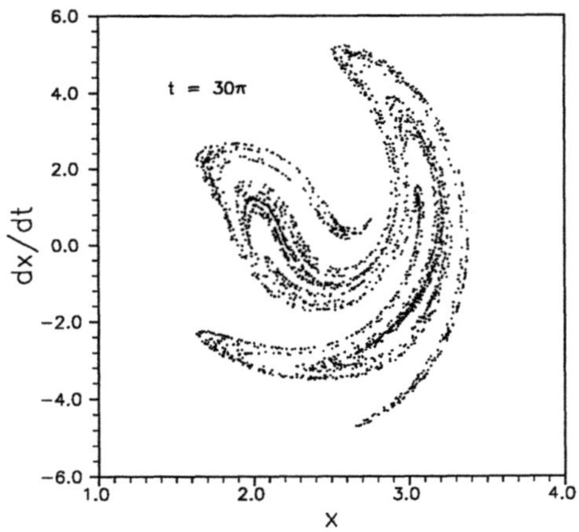

Figure 14: Snap shot of ensemble simulation, $t = 30\ \pi$.

where $(1-\alpha)kz$ is the inelastic part of the restoring force q given by (3). Constant parameters β and γ govern the size and shape of the hysteretic loop of the model. The degrading parameter δ governs the damage behavior of the material, and equals to zero for non-degrading material.

For our investigation of chaotic forced vibrations for the inelastic, hysteretic and degrading structural system of equations, (2) through (5), the equations and variables are normalized as follows.

Normalized equations:

$$\begin{aligned} \dot{X} &= V && (6)\\ \dot{V} &= F\Omega^{-2}\cos\tau - 2\xi\Omega^{-1}V - \alpha\Omega^{-2}X \\ &\quad -(1-\alpha)\Omega^{-2}Z && (7)\\ \dot{Z} &= (1-\delta\epsilon)\left[V - \beta|V|Z - \gamma V|Z|\right] && (8)\\ \dot{\epsilon} &= (1-\alpha)ZV && (9) \end{aligned}$$

Normalized variables:

$$\begin{aligned} X &= \frac{x}{x_y}\\ V &= \frac{\dot{x}}{\omega x_y}\\ Z &= \frac{z}{x_y}\\ \tau &= \omega t\\ \dot{X} &= \frac{\dot{x}}{\omega x_y}\\ \dot{V} &= \frac{\ddot{x}}{\omega^2 x_y}\\ \dot{Z} &= \frac{\dot{z}}{\omega x_y} && (10) \end{aligned}$$

Normalized parameters:

$$F = \frac{f}{kx_y}$$
$$\xi = \frac{c}{c_{cr}}$$
$$\Omega = \frac{\omega}{\omega_n} \tag{11}$$

Normalized parameters ξ and Ω are the linear damping ratio and the forcing frequency ratio respectively, and the dimensional constants c_{cr} and ω_n are the critical linear damping and the linear natural frequency respectively. For our numerical investigation, the set of constants selected are as follows:

$$\alpha = 0.04$$
$$\beta = 0.1$$
$$\gamma = -0.5$$
$$\xi = 0.025$$
$$\Omega = 1.0$$

The selection is based on simplicity, practicality and the likelihood of generating chaotic behavior for various forcing amplitudes F. We notice that the current condition $0 > \beta+\gamma > \gamma-\beta$ suggests that this is a strain hardening spring.[7]

For simplicity of analysis, we begin with the special case of non-degrading material, $\delta = 0$. As a consequence, equation (9) drops out and we have a three-variable system, X, V and Z, as defined in (6) through (8), which is similar to the Lorenz system.[11] To find the chaotic range, a scanning is performed using various F values. Starting from $F = 0$, we find a number of chaotic bands. The first burst of chaos is roughly located between $72 < F < 81$. As F increases, the occurrence of chaos become more frequent and the band wider. To demonstrate the drastic difference between chaotic versus non-chaotic behaviors, two problems, $F = 70$ (non-chaotic) and $F = 75$ (chaotic), are tested. To compare the sensitivity of solution to initial conditions, we run the chaotic problem for two sets of initial conditions $(X(0), V(0), Z(0)) = (0.000, 0.000, 0.000)$ and $(0.000, 0.001, 0.000)$. To show the correlation of the results, we sample and plot the X versus X data pairs for the two cases, taken at the same time, in Figure 15. Roughly 2,000 samples are taken within 20 loading cycles ($\tau < 40\pi$). For fully correlated results, we expect data points to fall on a 45° straight line. It is clear from Figure 15 that the results become uncorrelated within 20 cycles. For the non-chaotic case, the same scattergram is presented for 200 loading cycles in Figure 16. The correlation of solution is still excellent after 200 cycles.

Another way to visualize chaos is to examine the trajectory in the phase plane. The non-chaotic case is found to have an extremely long transient. After about 300 loading cycles, a stationary phase is established. In Figure 17 we plot the final limiting cycle in the X versus V phase plane. The limiting cycle, though looks quite complicated, is indeed a closed loop! The general motion is observed to be a period-four sub-harmonic, but also contains higher harmonics such that a number of looping take place within one loading cycle. In Figure 18 we present the liming cycle in the X–Z plane for the non-chaotic case. We notice the steepening of the curve at the upper and lower tips which characterizes the strain hardening behavior. For the chaotic case, limiting cycle is never observed. Figure 19 shows the X–V trajectory for loading cycles 500 to 520. The behavior is apparently chaotic.

For the chaotic behavior, order is normally found in the Poincaré map. However, there does not seem to be a standard technique for displaying such a map for a forced, three-variable system. The well known Lorenz system is autonomous, where no explicit time t appears in the formulation. The Poincaré map is created by using a selected plane cutting through the trajectories in the three-dimensional space. For a forced system, it is more natural to sample data at the multiples of forcing period. These data points are scattered in the three-dimensional space. They can be projected onto the phase plane, X–V or X–Z, for visualization. We shall use both techniques to search for possible chaotic structure.

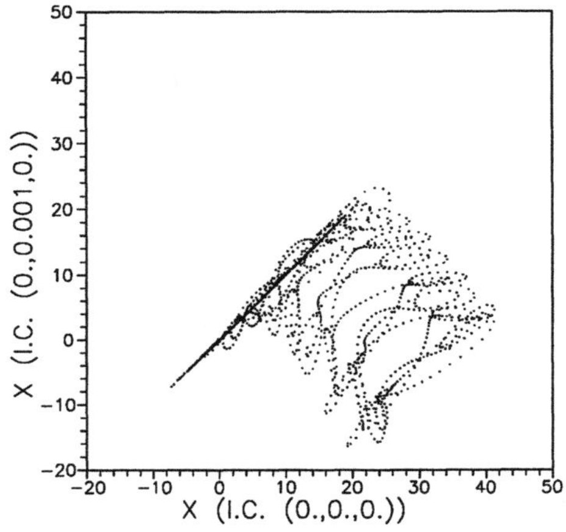

Figure 15: Scattergram for chaotic case (F = 75), 20 cycles.

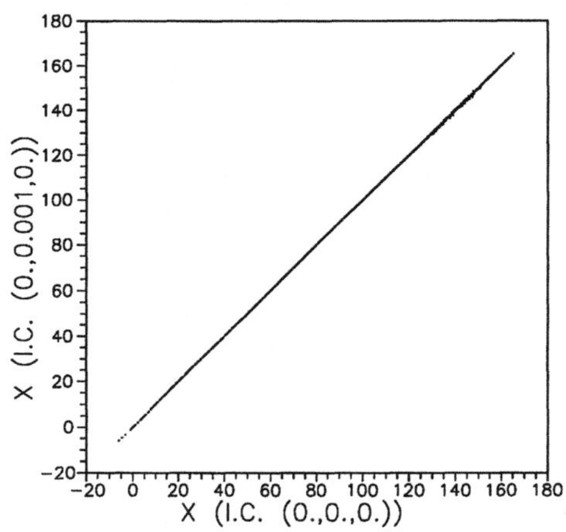

Figure 16: Scattergram for non-chaotic case (F = 70), 200 cycles.

Figure 20 presents the (X, V, Z) data sampled at multiples of forcing period ($t = 2n\pi$), and then projected onto the X–V plane. The result for the non-chaotic case emerges as four dots, which are enlarged in the diagram for visual effect. This appearance indicates a period-four motion. The chaotic result, shown as small dots, is scattered. There is no apparent pattern, or the fractal-like strange attractor, but they are apparently bounded. For a second way of showing Poincaré map, we record the points whenever the three-dimensional trajectory pierces through the X–V plane, as Figure 21. Again, the non-chaotic case gives four points (in big dots), while the chaotic case appears to be random but bounded.

For stochastic engineering applications, it is of interest to know the joint probability density function of displacement X versus velocity V. Similar to the technique used for producing Figure 9, about 10^5 samples of X–V pairs are taken from the time history. Through simple counting, the final PDF for the chaotic case is displayed as Figure 22.

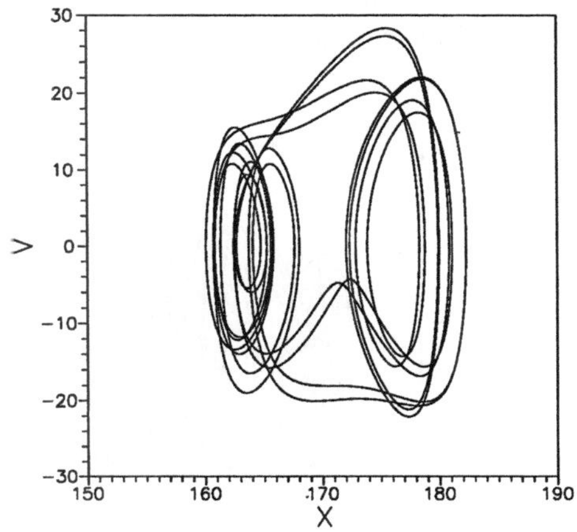

Figure 17: Limiting cycle in X–V phase plane for non-chaotic case.

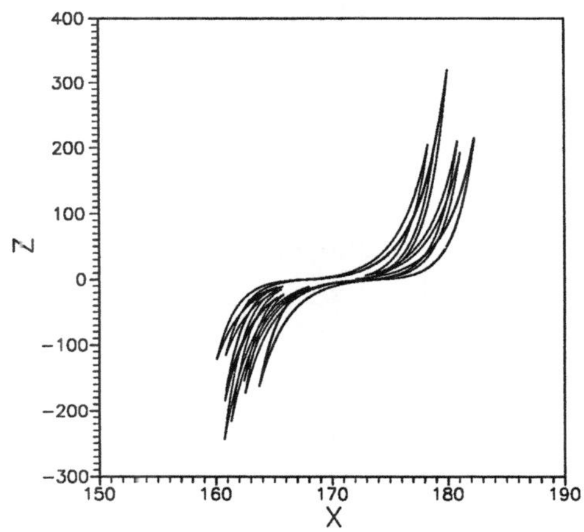

Figure 18: Limiting cycle in X–Z phase plane for non-chaotic case.

Finally, we examine the degrading, four-variable system, given by (6) through (9). The parameters α, β, γ, ξ and Ω are the same as before, except that the degrading parameter δ is now chosen as 0.01, instead of zero. Once again we scan for chaos for various F values based on the sensitivity of solution to initial conditions. The first occurrence of chaos is found to be around $F = 4$. A closer examination of the time series reveals that the chaos is transient. In Figure 23 we show the time series X versus τ for two problems with slightly different initial conditions, $(X, V, Z, \epsilon) = (0.000, 0.000, 0.000, 0.000)$ and $(0.000, 0.001, 0.000, 0.000)$. The two solutions clearly diverge after about 13 loading cycles, but then start to converge after 25 cycles. 'Transient chaos' has been reported before,[2] but its implication has not been fully explored. To see the diverging/converging behavior more clearly, we plot the 'absolute error' of the solutions in Figure 24. The error is defined as the difference between the two solutions, taken as absolute value, and plotted in a time series. We clearly see the error initially grows in an exponential fashion (note that the vertical scale is logarithmic), and then diminishes. After 100 cycles, the magnitude of error is about 0.1, while the maximum amplitude of solution is roughly ± 7. In the same diagram, we present the error for a non-chaotic case comparison. We use the same pair of initial conditions but with a forcing magnitude $F = 3$. The error remains small throughout the 100 cycles of simulation.

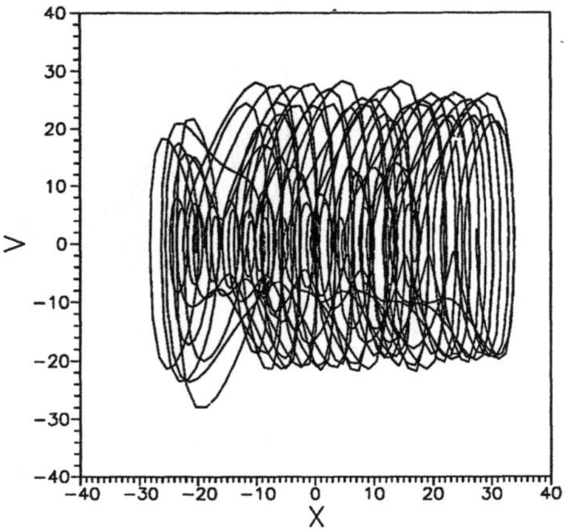

Figure 19: X–V phase diagram for chaotic case.

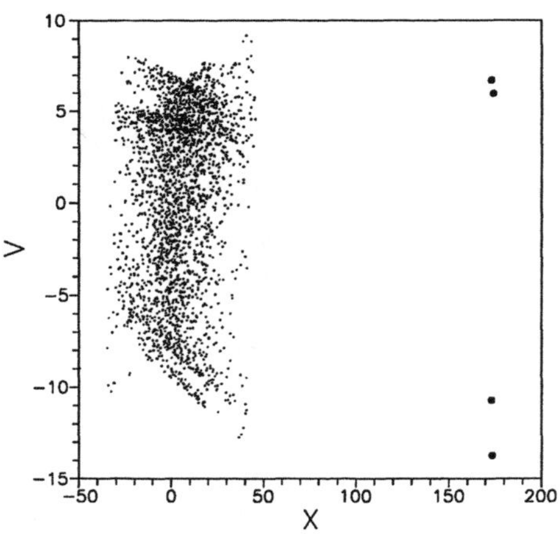

Figure 20: Poincaré map sampled at multiples of forcing period, projected onto X–V plane.

SUMMARY AND CONCLUSIONS

We draw the following conclusions from the above study:

1. The most important result demonstrated here is that for a completely deterministic structural system with a deterministic forcing function, the response can be random-like, or so-called chaotic. In the deterministic sense the individual motion of a chaotic response is unpredictable due to its sensitivity to initial condition and disturbance. In the stochastic sense, however, the collective (ensemble) motion is predictable, unique, and reproducible in the form of a joint probability density function. This hybrid of random response induced by deterministic excitation is a new phenomenon uncovered by the study of chaos.

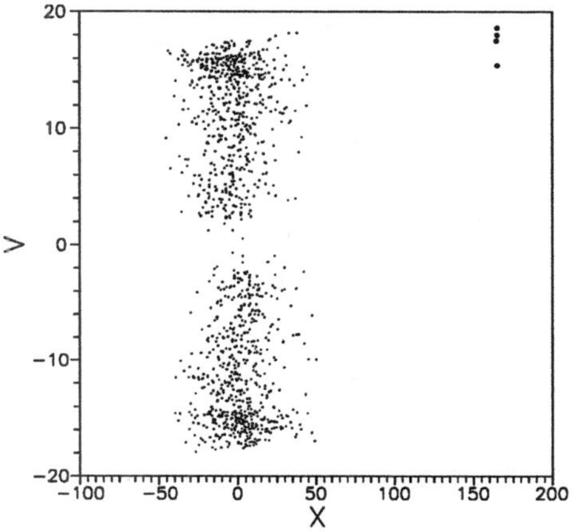

Figure 21: Poincaré map sampled at X–V plane.

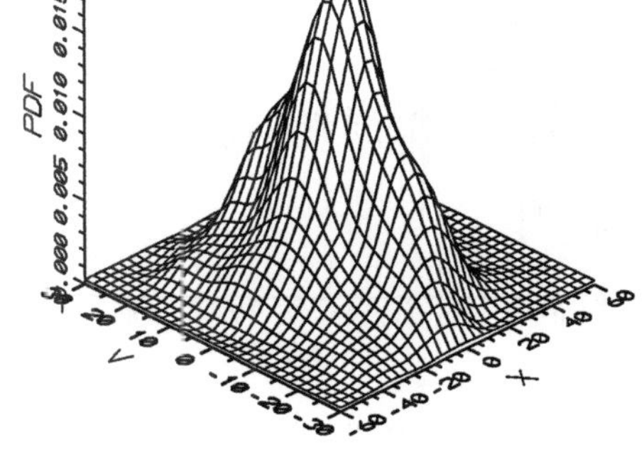

Figure 22: Joint probability density function for hysteretic structural system.

2. The chaotic vibration associated with the nonlinear elastic system governed by Duffing's equation has been thoroughly studied by Ueda[24] and others. The ergodicity of the system is often implied which allows the probability density functions to be obtained from a single time series. By performing an ensemble experiment, we have explicitly demonstrate the validity of the ergodicity assumption.

3. For nonlinear, inelastic, hysteretic and degrading structural systems, occurrence of the chaotic vibration is demonstrated here by the sensitive dependence of the system response to initial conditions. The chaotic attractor and the joint probability density function for the response velocity and displacement are presented here for the first time.

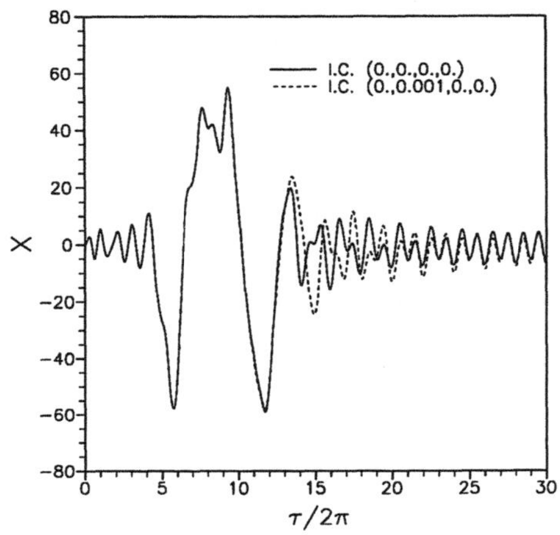

Figure 23: Time history for degrading, chaotic case ($F = 4.0$).

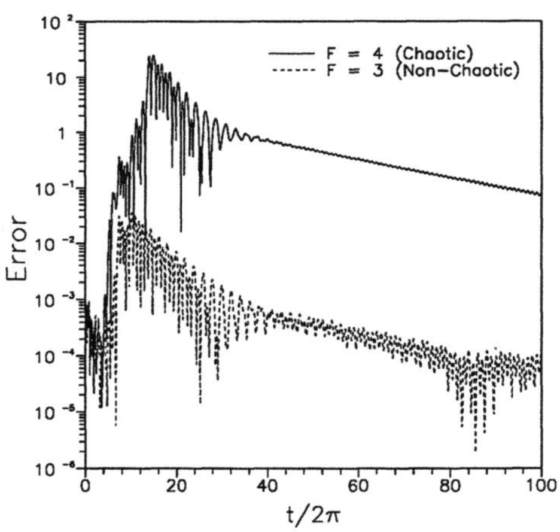

Figure 24: Growth of error from an initial disturbance for $F = 4.0$ (chaotic) and $F = 3$ (non-chaotic).

4. For chaotic vibrations, a power spectral density function normally shows a broad frequency band.[29-31] For the inelastic, hysteretic and degrading system, we have been working on cases to be presented later.

5. The engineering significance of chaos lies in the identification of its occurrence in term of the system parameters as in the case of the determination of the buckling load of columns and the Reynolds number in fluids. Once these parameters are known, and the chaotic motion is predictable stochastically, then an engineering system can be designed to either avoid the chaotic motion or to safely operate under such predicted chaotic conditions.

6. In the new science of chaos, the important areas of topology in mathematics, the determination of instability by Lyapunov's exponents[32,33] and the concept of fractal dimensions[34,35] have attracted a great deal of attention among scientists and engineers. These important areas have not been studied here.

It is hopeful that progress in these areas will help the engineers to deal with nonlinear dynamics more efficiently in the near future.

References

[1] Gleick, J., *Chaos*, Penguin Books, 1987.

[2] Thompson, J.M.T. and Stewart, H.B., *Nonlinear Dynamics and Chaos*, Wiley, 1986.

[3] Moon, F.C. *Chaotic Vibrations*, Wiley Interscience, 1987.

[4] Goldstein, G. and Moon, F., "Coming to terms with chaos", Mechanical Engineering Magazine, ASME, January, 1990.

[5] Wen, Y.K., "Methods for random vibration of hysteretic systems," J. Eng. Mech. Div., ASCE, **102**, 249-263, 1976.

[6] Barber, T.T. and Wen, Y.K., "Random vibration of hysteretic degrading systems," J. Eng. Mech. Div., ASCE, **107**, 1069-1087, 1981.

[7] Sues, R.H., Mau, S.T. and Wen, Y.K., "System identification of degrading hysteretic restoring forces," J. Eng. Mech. Div., ASCE, **114**, pp. 883-846, 1988.

[8] Iwan, W.D. and Lutes, L.D., "The response of the bilinear hysteretic system to stationary random excitation," J. Acoust. Soc. Am., **43**, 1968.

[9] Lutz, L.D., "Approximate technique for treating random vibration of hysteretic systems," J. Acoust. Soc. Am., **48**, 299-306, 1970

[10] Sozen, M.A., "Hysteresis in structural elements," Proc. Conf. Appl. Mech. Earthquake Eng., ASME Ann. Meeting, AMD, **8**, 63-98, 1974.

[11] Lorenz, E.N., "Deterministic nonperiodic flow," J. Atmospheric Sci., **20**, 130-141, 1963.

[12] Hao, B.L., *Chaos*, World Scientific Pub., Singapore, 1984.

[13] Holmes, P.J., *New Approaches to Nonlinear Problems in Dynamics*, (ed.), SIAM, 1980.

[14] Feigenbaum, M.J., "Quantitative universality for a class of nonlinear transformations," J. Statistical Phys., **19**, 25-52, 1978.

[15] Goldberger, A.L., Bhargava, V. and West, B.J., "Nonlinear dynamics of the heartbeat," Physica, **17D**, 207-214, 1985.

[16] May, R.M., "Simple mathematical models with very complicated dynamics," Nature, **261**, 459-467, 1976.

[17] Berges, P., Pomeau, Y. and Vidal, C., *Order and Chaos, Toward a Deterministic Approach to Turbulence*, Wiley Interscience, 1987.

[18] Poddar, B., Moon, F.C. and Mukherjee, S., "Chaotic motion of an elastic- plastic beam," J. Appl. Mech., ASME, **55**, 184-189, 1988.

[19] Holmes, P.A., "Nonlinear oscillator with a strange attractor," Phil. Trans., Roy. Soc. London, **292A**, 419-448, 1979.

[20] Dowell, E.H. and Pezeshki, C., "On the understanding of chaos in Duffing's equation including a comparison with experiments," J. Appl. Mech., ASME, **53**, 6-10, 1986.

[21] Reinhall, P.G., Caughey, T.K. and Storti, D.W., "Order and chaos in a discrete Duffing oscillator: Implications on numerical integration," J. Appl. Mech., ASME, **56**, 162-167, 1989.

[22] Kapitaniak, T., *Chaos in Systems with Noise*, World Scientific Pub., Singapore, 1988.

[23] Tongue, B.H., "Characteristics of numerical simulations of chaotic systems," J. Appl. Mech., ASME, **54**, 695-699, 1987.

[24] Ueda, Y., "Random phenomena resulting from nonlinearity in the system described by Duffing's equation," Int. J. Non-linear Mech., **20**, 481- 491, 1985.

[25] Ueda, Y., "Steady motions exhibited by Duffing's equation: A picture book of regular and chaotic motions," in *New Approaches to Nonlinear Problems in Dynamics*, ed. P. Holmes, SIAM, 1980.

[26] Ueda, Y., "Randomly transitional phenomenon in the system governed by Duffing's equation," J. of Statistical Phys., **20**, 181-196, 1979.

[27] Wen, Y.K., "Methods of random vibration for inelastic structures," Appl. Mech. Rev., **42**, 39-52, 1989.

[28] Smale, S., "Differentiable dynamical systems," Bull. Am. Math. Soc., **73**, 747-817, 1967.

[29] Abraham, R.H. and Shaw, C.D., *Dynamics–The Geometry of Behavior, Part 2: Chaotic Behavior*, Aerial Press, 1984.

[30] Crutchefied, J.P., Donnelly, J.D., Farmer, G., Jones, N. Packard, and Shaw, R., "Power Spectra Analysis of a Dynamical System," Phys. Letters, **76A**, 1-4.

[31] Huberman, B.A. and Zisook, A.B., "Power spectra of strange attractors," Phys. Rev. Letters, **46**, 626-628, 1981.

[32] Tongue, B.H. and Smith, D., "Determining Lyapunov exponents by means of interpolated mapping," J. Applied Mech., ASME, **56**, 691-696, 1989.

[33] Wolf, A., Swift, J.B., Swinney, H.L. and Vastano, J.A., "Determining Lyapunov exponents from a time series," Physica, **16D**, 285-317, 1985.

[34] Mandelbrot, B.B., *The Fractal Geometry of Nature*, W.H. Freeman, San Francisco, 1983.

[35] Barnsley, M.F. and Demko, S.G., *Chaotic Dynamics and Fractals, Notes and Reports in Mathematics in Science and Engineering, Vol. 2*, (ed.), Academic Press, 1986.

Stochastic Earthquake Modeling with Discretized Line Source

Ruichong Zhang, Yan Yong, and Y. K. Lin

Center for Applied Stochastics Research
College of Engineering
Florida Atlantic University
Boca Raton, FL 33431, USA

Summary

An earthquake model is developed, in which the earth is idealized as horizontally stratified layers and the seismic source as propagating dislocation along a line. The line source is further discretized into point sources at equal intervals, and the times at which seismic signals are emitted from the source points are assumed to be Poisson events. The strengths of individual sources are assumed to be independent and identically distributed random variables. A generalized version of the random-pulse-train theory is then used to compute the mean and covariance functions of the ground motion at one site, and the cross-covariance function at two sites. The covariance and cross-covariance functions are converted to the evolutionary spectral density and cross-evolutionary density of the ground motion, which are useful in the computation of the statistics of structural response to earthquake excitations. Numerical examples are given for illustration.

Introduction

As a first approximation to the highly inhomogeneous media in which seismic waves are transmitted, the earth may be idealized as being composed of horizontally stratified layers, with uniform physical properties in each layer, as shown schematically in Fig. 1. By using a combined Fourier-Hankel transform, the transformed differential equations for each layer represent a depth-invariant system, whose solution represents a steady-state motion. Continuity of solutions from layer to layer may be expressed in terms of transfer matrices (e.g. Pestel and Leckie [21]). This procedure was first employed by Haskell [11]. By properly combining the unknown variables [12], two sets of transfer relationships can be found; one set corresponds to the coupled P and SV waves, the other to the SH waves. However, the use of the transfer-matrix technique to

Y. K. Lin · I. Elishakoff (Eds.)
Stochastic Structural Dynamics 1
New Theoretical Developments
2nd International Conference on Stochastic Structural Dynamics
May 9–11, 1990, Boca Raton, Florida

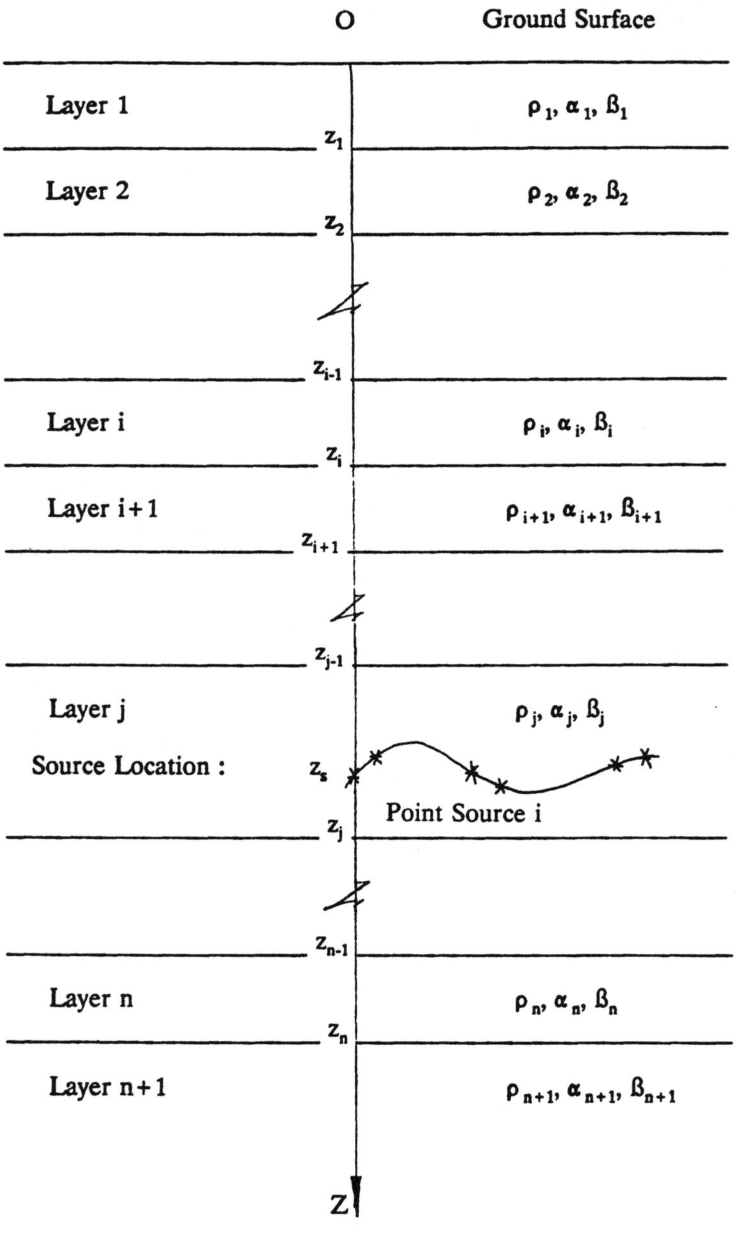

Fig. 1 A horizontally stratified earth medium

compute state variables of displacements and forces can become numerically unstable in some cases, due to the fact that the computation involves both exponentially increasing and exponentially decreasing terms. To circumvent this difficulty, Kennett [15] suggested a transformation from the state variables to wave variables, and the use of scattering matrices which characterize the wave reflection and transmission properties in the sequential elastic media. A similar technique was also employed by Yong and Lin [27] in a study of dynamic response of a structural network consisting of piece-wise periodic units.

Once a complete solution is obtained in the transformed Fourier-Hankel domain, the ground motion at a given site can be evaluated through inverse transformation. Several approximation techniques [3,15,25] have been devised to evaluate such inverse transforms. Yet, another type of numerical difficulties often arises in carrying out the inversion, due to branch points and singularities in the wave number domain, and highly oscillatory Bessel functions at large arguments. This latter type of difficulties can be circumvented by substituting the usual Hankel transform by a finite Hankel transform as in the Refs. 2, 4, 5, 16, and 20, in which case the corresponding inverse transform becomes a summation.

Records of many past major earthquakes, e.g., the Guatemala earthquake [13] show that they may consist of a series of subevents which are the results of seismic dislocation propagating along a long fault. An interesting semi-empirical approach was proposed by Kanamori [14], in which a line-source generated earthquake was considered. The line source was discretized into several point sources with random seismic moments, and the actual records of small real earthquakes were used as the Green's functions for individual pulses. This semi-empirical theory is physically enlightening, although it cannot be used readily to obtain statistical properties of the ground motion.

In the present paper, a brief account is given of a new stochastic earthquake model proposed recently by Zhang, Yong, and Lin [28,29] in which use has been made of Kanamori's concept of discretized line source. In this model, the discretization points are evenly spaced, and the epoch times for the discretized sources to emit signals are assumed to be independent events. Each seismic pulse arriving at a given site on the ground surface is computed using the finite Hankel transform approach. Stochastic earthquake models of this type belong to the general class of random pulse train, and they have been proposed by Lin [17], Cornell [8], Lin and Yong [19], Yong [26], Deodatis, Shinozuka, and Papageorgiou [9,10], although the geophysical nature of the pulses was not incorporated in the earlier works (Lin [17], Cornell [8]). As shown by Lin [18] such a random pulse train model has an evolutionary spectrum.

A Generalized Version of Random Pulse-Train Theory

Let $U(\underline{v})$ be a continuous random process, and $\underline{v}$ be a vector of continuous parameters. The characteristic functional of $U(\underline{v})$ is defined as

$$M_{(U)}[\theta(\underline{v})]=E\left\{\exp\left[i\int_V \theta(\underline{v})\, U(\underline{v})\, d\underline{v}\right]\right\} \tag{1}$$

where E indicates an ensemble average, V is the range of $\underline{v}$, and the function $\theta(\underline{v})$ belongs to a family for which the stochastical integral $\int_V \theta(\underline{v}) U(\underline{v}) d\underline{v}$ is meaningful.

A useful class of random processes may have a general form as follows

$$U(\underline{v})=\sum_{k=1}^{N(t)} Y_k h(\underline{v};\tau_k)\ , \qquad \tau_k \le T \tag{2}$$

where $h(\underline{v};\tau_k)$ is a deterministic function describing the shape of a pulse which begins at a random time τ_k, N(t) is a random counting process and may be replaced by N(T) in equation (2) because of the physical causality of the pulses, T is the range of τ_k, and Y_k are identically distributed random variables which are mutually independent and independent of the distribution of τ_k. In particular, if τ_k are independent, then N(T) is a Poisson process, and it will be so assumed in what follows.

Substitution of equation (2) into equation (1) yields

$$\begin{aligned} M_{\{U\}}[\theta(\underline{v})] &= E\left\{\exp[i\int_V \theta(\underline{v}) \sum_{k=1}^{N(T)} Y_k h(\underline{v};\tau_k) d\underline{v}]\right\} \\ &= E\left\{E\{\exp[i\int_V \theta(\underline{v}) \sum_{k=1}^{N(T)} Y_k h(\underline{v};\tau_k) d\underline{v}] \,|\, N(T)\}\right\} \\ &= \sum_{n=0}^{\infty} P_{\{N\}}(n,T) E\{\exp[i\int_V \theta(\underline{v}) \sum_{k=1}^{n} Y_k h(\underline{v};\tau_k) d\underline{v}]\} \end{aligned} \tag{3}$$

where $P_{\{N\}}(n,T)$ is the probability function of the Poisson process given as

$$P_{\{N\}}(n,T) = \frac{1}{n!}\left[\int_0^T \lambda(\tau) d\tau\right]^n \exp[-\int_0^T \lambda(\tau) d\tau] \tag{4}$$

in which λ is the expected arrival rate. Making use of the assumptions about Y_k and τ_k, we obtain for each n,

$$\begin{aligned} & E\left\{\exp[i\int_V \theta(\underline{v}) \sum_{k=1}^{n} Y_k h(\underline{v};\tau_k) d\underline{v}]\right\} \\ &= E\left\{\prod_{k=1}^{n} \exp[i\int_V \theta(\underline{v}) Y_k h(\underline{v};\tau_k) d\underline{v}]\right\} \\ &= \prod_{k=1}^{n} E\left\{\exp[i\int_V \theta(\underline{v})\, Y_k h(\underline{v};\tau_k) d\underline{v}]\right\} \\ &= \prod_{k=1}^{n} \left[1 + E\{\sum_{m=1}^{\infty} \frac{i^m}{m!} [\int_V \theta(\underline{v}) Y_k h(\underline{v};\tau_k) d\underline{v}]^m\} \right] \\ &= (1+\alpha)^n \end{aligned} \tag{5}$$

where

$$\begin{aligned} \alpha &= E\left\{\sum_{m=1}^{\infty} \frac{i^m}{m!} [\int_V \theta(\underline{v}) Y_k h(\underline{v};\tau_k) d\underline{v}]^m\right\} \\ &= \sum_{m=1}^{\infty} \frac{i^m}{m!} E[Y^m] E\left\{\int_V d\underline{v}_1 \cdots \int_V d\underline{v}_m h(\underline{v}_1;\tau_k) \cdots h(\underline{v}_m;\tau_k) \theta(\underline{v}_1) \cdots \theta(\underline{v}_m)\right\} \\ &= \sum_{m=1}^{\infty} \frac{i^m}{m!} E[Y^m] \int_V d\underline{v}_1 \cdots \int_V d\underline{v}_m E[h(\underline{v}_1;\tau_k) \cdots h(\underline{v}_m;\tau_k)] \theta(\underline{v}_1) \cdots \theta(\underline{v}_m) \\ &= \sum_{m=1}^{\infty} \frac{i^m}{m!} E(Y^m) \int_V d\underline{v}_1 \cdots \int_V d\underline{v}_m \theta(\underline{v}_1) \cdots \theta(\underline{v}_m) \frac{\int_0^T h(\underline{v}_1;\tau) \cdots h(\underline{v}_m;\tau) \lambda(\tau) d\tau}{\int_0^T \lambda(\tau) d\tau} \end{aligned} \tag{6}$$

It follows from combining equation (3) through (6),

$$M_{\{U\}}[\theta(\underline{v})]=\sum_{n=0}^{\infty}\frac{1}{n!}\left[\int_0^T\lambda(\tau)d\tau\right]^n\exp[-\int_0^T\lambda(\tau)d\tau](1+\alpha)^n$$
$$=\exp[\alpha\int_0^T\lambda(\tau)d\tau] \tag{7}$$

Taking natural logarithm of equation (7), we obtain

$$\ln\{M_{\{U\}}[\theta(\underline{v})]\}=\alpha\int_0^T\lambda(\tau)d\tau$$
$$=\sum_{m=1}^{\infty}\frac{i^m}{m!}E(Y^m)\int_V d\underline{v}_1\cdots\int_V d\underline{v}_m\theta(\underline{v}_1)\cdots\theta(\underline{v}_m)\int_0^T h(\underline{v}_1;\tau)\cdots h(\underline{v}_m;\tau)\lambda(\tau)d\tau \tag{8}$$

It is known (e.g. Stratonovich [24]) that the log-characteristic functional has a series representation

$$\ln\{M_{\{U\}}[\theta(\underline{v})]\}=i\int_V K_1[U(\underline{v})]\theta(\underline{v})d\underline{v}+\frac{i^2}{2!}\int_V d\underline{v}_1\int_V d\underline{v}_2\theta(\underline{v}_1)\,\theta(\underline{v}_2)K_2[U(\underline{v}_1)U(\underline{v}_2)]+\cdots \tag{9}$$

where $K_j[U(\underline{v})]$ denotes the j-th cumulant function of $U(\underline{v})$. Comparing equation (8) and (9), we obtain

$$K_m[U(\underline{v}_1)\cdots U(\underline{v}_m)]=E(Y^m)\int_0^T h(\underline{v}_1;\tau)\cdots h(\underline{v}_m;\tau)\lambda(\tau)d\tau \tag{10}$$

The first and the second cumulant functions coincide, respectively, with the mean and the covariance functions, i.e.,

$$\mu_U(\underline{v})=\mu_Y\int_0^T h(\underline{v};\tau)\,\lambda(\tau)d\tau \tag{11}$$

$$K_{UU}(\underline{v}_1;\underline{v}_2)]=E(Y^2)\int_0^T h(\underline{v}_1;\tau)\,h(\underline{v}_2;\tau)\lambda(\tau)d\tau \tag{12}$$

Usually, the parameter vector $\underline{v}$ is composed of the spatial coordinates (x,y,z) and time t. When carrying out actual calculations the causality condition $h(x,y,z,t;\tau)=0$ for $\tau>t$ must be imposed on equation (11) and (12).

A Stochastic Earthquake Model with Discretized Line Source

Most seismic faults are narrow, i.e., the widths are much smaller than the lengths. For mathematical analysis, it is reasonable to use a line source approximation, shown in Fig. 1, and assume that the dislocation propagates along the fault line, starting at point S and ending at point E. The line source may be further discretized into a series of point sources at equal distances. Since resistance to dislocation varies from point to point, the time intervals required to propagate from one source point to the next may be treated as being random. We shall assume that these time intervals are independent; namely, the time instants at which the point sources are activated are Poisson events. Obviously, the strengths of the activated seismic moments are also random, and we shall assume that they are independent and identically distributed random variables. These assumptions can now be incorporated in the following expression

$$\begin{aligned} U_d(x,y,z=0,t) &= \sum_{i=1}^{N} Y_i h_d(x,y,z=0,t;x_i,y_i,z_i,\tau_i) \\ &= \sum_{i=1}^{N} Y_i h_d(x-x_i,y-y_i,z=0,t;z_i,\tau_i)\,, \qquad d=x\,,y\,,or\,z \end{aligned} \tag{13}$$

where $U_d(x,y,z=0,t)$ is the ground motion in the d-th direction at the ground site $(x,y,z=0)$ and at time t due to N point sources along line SE; Y_i is the i-th seismic moment which is the product of the rigidity of the fault region, the contributory fault area, and the average slip of the discretized dislocation; $h_d(x,y,z=0,t;x_i,y_i,z_i,\tau_i)$ is the ground motion in the d-th direction at the ground site $(x,y,z=0)$ at time t due to a point source with a unit seismic moment activated at time τ_i at location (x_i,y_i,z_i). In accordance with the assumed horizontal homogeneity, this h_d function must be a function of $x-x_i$ and $y-y_i$, instead of x, x_i, y, y_i, separately, and it can be expressed as follows[1]

$$\begin{aligned} h_d(x-x_i,y-y_i,z=0,t;z_s,\tau) &= m(t,\tau)*u_d(x-x_i,y-y_i,z=0,t) \\ &= \int_0^t m(t',\tau)\,u_d(x-x_i,y-y_i,z=0,t-t')\,dt' \\ &= \int_{-\infty}^{\infty} m(t',\tau)\,u_d(x-x_i,y-y_i,z=0,t-t')\,dt' \\ &= \frac{1}{2\pi}\int_{-\infty}^{\infty} m(\omega,\tau)\,u_d(x-x_i,y-y_i,z=0,\omega)\,e^{i\omega t}d\omega \end{aligned} \tag{14}$$

where * denotes a convolution operation, $u_d(x-x_i,y-y_i,z=0,t)$ is the Green's function which describes the ground motion in the

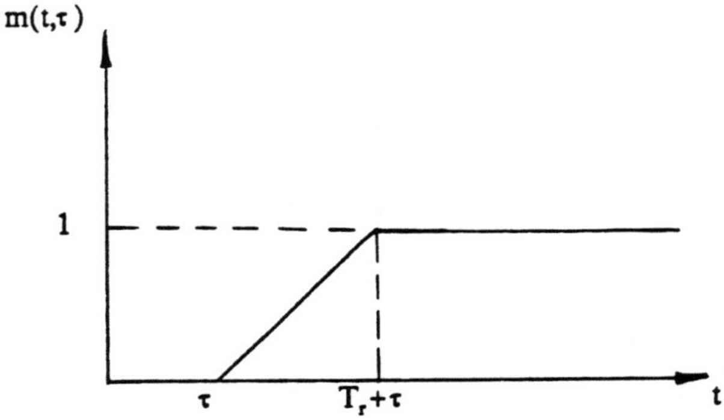

Fig. 2 Time variation of an idealized seismic moment

d-th direction due to a dislocation pulse occured at the point source location, m(t,τ) describes the time-variation of the seismic moment, which is assumed to be a ramp function shown in Fig 2, and where use is made of the fact that $u_d(x,y,z,t-t')=0$ for $t'<0$ and $t'>t$. The form of the Green's function will be discussed in the next section. Equation (14) may be substituted into equations (11) and (12) to determine the mean and covariance of a simulated earthquake motion.

When modeling an earthquake process of short duration, we may let $\lambda(\tau)=0$ for $\tau<0$ and $\tau>T$. Equation (12) can then be rewitten as follows

$$K_{UU}(x_1,y_1,z_1,t_1;x_2,y_2,z_2,t_2)=E(Y^2)\int_{-\infty}^{\infty}h(x_1,y_1,z_1,t_1;\tau)h(x_2,y_2,z_2,t_2;\tau)\lambda(\tau)d\tau \tag{15}$$

Let

$$b(x,y,z,t;\omega)e^{i\omega t}=\int_{-\infty}^{\infty}h(x,y,z,t;\tau)\sqrt{\lambda(\tau)}e^{i\omega\tau}d\tau \tag{16}$$

Then its inverse should be

$$h(x,y,z,t;\tau)\sqrt{\lambda(\tau)}=\frac{1}{2\pi}\int_{-\infty}^{\infty}b(x,y,z,t;\omega_1)e^{i\omega_1 t}e^{-i\omega_1\tau}d\omega_1 \tag{17}$$

Since the left-hand side of equation (17) is a real-value function, the right-hand side may be replaced by its complex conjugate form. Substituting equation (17) and its complex conjugate form into equation (15), we arrive at

$$\begin{aligned}&\mathbf{K}_{UU}(x_1,y_1,z_1,t_1;x_2,y_2,z_2,t_2)\\&=\frac{1}{2\pi}E(Y^2)\int_{-\infty}^{\infty}b(x_1,y_1,z_1,t_1;\omega)\,b^*(x_2,y_2,z_2,t_2;\omega)\,e^{i\omega(t_1-t_2)}d\omega\end{aligned} \tag{18}$$

The evolutionary cross-spectral density of $U(x,y,z,t)$ at sites (x_1,y_1,z_1) and (x_2,y_2,z_2) is then clearly [18,29]

$$\Phi_{UU}(x_1,y_1,z_1,t;x_2,y_2,z_2,t;\omega)=\frac{1}{2\pi}E(Y^2)b(x_1,y_1,z_1,t;\omega)\,b^*(x_2,y_2,z_2,t;\omega) \tag{19}$$

The evolutionary spectral density may be obtained by letting $x_1=x_2=x$, $y_1=y_2=y$, and $z_1=z_2=z$.

<u>Green's Function for the Dislocation Pulse</u>

In this paper, the earth is modeled as a horizontally stratified medium, each layer being homogeneous, isotropic and linearly elastic for which the governing equations of motion and the constitutive relations between displacements and stresses are known (e.g. Kennett [15]). The equations of motion may be expressed in terms of two sets of state variables in the cylindrical coordinates [12]. One set corresponds to coupled compressional wave and vertical shear wave (called the P and SV wave). The other set corresponds to a horizontal shear wave (called the SH wave). To remove the partial differentiations with respect to r, φ and t from the two sets of governing equations, the following Fourier finite-Hankel transform [23] is performed

$$\overline{\psi}(k_n,m,z,\omega)=\int_{-\infty}^{\infty}dt\,e^{-i\omega t}\int_0^R dr\,rJ_m(k_nr)\frac{1}{2\pi}\int_0^{2\pi}d\varphi\,e^{-im\varphi}\psi(r,\varphi,z,t) \tag{20}$$

where $\psi(r,\varphi,z,t)$ is any one of the original state variables and $\overline{\psi}$ denotes its transformed counterpart, and where k_n is the wave number, m is the azimuthal order, and J_m is the Bessel function of the first kind of the m-th order. The corresponding inverse transform is obtained as

$$\psi(r,\varphi,z,t)=\frac{1}{2\pi}\int_{-\infty}^{\infty}d\omega\, e^{i\omega t}\sum_{m=-\infty}^{\infty}e^{im\varphi}\sum_{n=1}^{\infty}\frac{2J_m(k_n r)}{R^2[J_m'(k_n R)]^2}\overline{\psi}(k_n,m,z,\omega)\ ,\quad r\leq R \tag{21}$$

where, for a given m, k_n is a positive root of the transcendental equation

$$J_m(k_n R)=0 \tag{22}$$

It is of interest to note that when R goes to infinity, any real value is a root of equation (22). In other words, k_n becomes continuous, which is of course the case with the usual Hankel transform. From equations (21) and (22) it is seen that

$$\psi(R,\varphi,z,t)=0 \tag{23}$$

Since ψ represents a physical variable at (R,φ,z,t), the validity of equation (23) should be carefully justified. Clearly, equation (23) is exact if $t\leq T_w$ where T_w is the time lapsed for the fastest seismic wave to propagate from the source to a distance R. In practice, equation (23) always can be rendered approximately valid, if R is chosen large enough such that any seismic signal beyond R becomes negligible.

When carrying out the inverse Hankel transform numerically according to equation (21), the summation over n must be truncated. The number of terms to be kept in the truncation is equal to the number of roots of equation (22) up to $k_N R$ where k_N is a suitable cut-off wave number. Thus a larger R value implies more terms in the truncated series.

Although the inverse of a conventional Hankel transform must be computed numerically as an approximate sum, and the terms associated with large wave numbers must also be truncated, the errors involved in the two approaches are very different. The only error associated with the finite-Hankel transform is the truncated part for large wave numbers. In contrast, the error associated with the conventional Hankel transform also includes that of changing from an integral to summation. In view of the presence of singularities and highly oscillatory nature of the integrand, the finite-Hankel transform procedure is expected to be numerically more stable and generally more accurate.

After performing equation (20) on the two sets of partial differential equations, two sets of ordinary differential equations are obtained which may be cast in the following form,

$$\frac{d}{dz}\begin{Bmatrix} w \\ f \end{Bmatrix} = [A]\begin{Bmatrix} w \\ f \end{Bmatrix} - \begin{Bmatrix} 0 \\ F \end{Bmatrix} \tag{24}$$

where for the P-SV waves

$$w=[U \;\; V]^T \;, \qquad f=[P \;\; S]^T \;, \qquad F=[F_z \;\; F_V]^T \tag{25}$$

$$[A]=\begin{bmatrix} 0 & k_n(1-\frac{2\beta^2}{\alpha^2}) & \frac{\omega}{\rho\alpha^2} & 0 \\ -k_n & 0 & 0 & \frac{\omega}{\rho\beta^2} \\ -\rho\omega & 0 & 0 & k_n \\ 0 & \rho\omega(\frac{\gamma k_n^2}{\omega^2}-1) & -k_n(1-\frac{2\beta^2}{\omega^2}) & 0 \end{bmatrix} \tag{26}$$

and for the SH waves

$$w=W \;, \qquad f=T \;, \qquad F=F_H \tag{27}$$

$$[A]=\begin{bmatrix} 0 & \frac{\omega}{\rho\beta^2} \\ \rho\omega(\frac{\beta^2 k_n^2}{\omega^2}-1) & 0 \end{bmatrix}$$

In the above equations, $\alpha=(\lambda+2\mu/\rho)^{1/2}$ is the P wave speed, $\beta=(\mu/\rho)^{1/2}$ is the S wave speed, $\gamma=4\beta^2(1-\beta^2/\alpha^2)$. To account for damping, the wave speeds α and β may be replaced by a pair of complex values, i.e., $\alpha(1+i\gamma_\alpha \text{sgn}\ \omega)$ and $\beta(1+i\gamma_\beta \text{sgn}\ \omega)$, where sgn ω denotes the sign of frequency ω. Upon solving equation (24), the ground motion is obtained in the ω-k_n-m domain, which can then be inverted to yield

$$u_d(x,y,z=0,t)=\frac{1}{2\pi}\int_{-\infty}^{\infty}d\omega\, e^{i\omega t}\sum_{m=-\infty}^{\infty}e^{im\varphi}\sum_{n=1}^{\infty}\frac{2}{R^2J_{m+1}^2(k_nR)}b_d \tag{29}$$

where the subscript d represents the direction in which the ground motion is evaluated. In the directions of x, y, and z, b_d are given, respectively, as

$$b_x=[a_1\cos\varphi-a_2\sin\varphi]V+[a_2\cos\varphi+a_1\sin\varphi]W \tag{30}$$

$$b_y=[a_1\sin\varphi+a_2\cos\varphi]V+[a_2\sin\varphi-a_1\cos\varphi]W \tag{31}$$

$$b_z=J_m(k_nr)U \tag{32}$$

in which

$$a_1=\frac{m}{k_nr}J_m(k_nr)-J_{m+1}(k_nr)\ , \qquad a_2=\frac{im}{k_nr}J_m(k_nr) \tag{33}$$

$$r=(x^2+y^2)^{\frac{1}{2}}\ , \qquad \varphi=\tan^{-1}(\frac{y}{x})\ , \qquad 0\le r\le R \tag{34}$$

<u>Wave Motion in a Multi-Layer Medium</u>

For a multi-layer medium, equation (24) may be solved for each layer. Consider the i-th layer which is governed by the differential equation

$$\frac{d}{dz}\begin{Bmatrix}w_{i^*}\\ f_{i^*}\end{Bmatrix}=[A_i]\begin{Bmatrix}w_{i^*}\\ f_{i^*}\end{Bmatrix}\ , \qquad i-1^+\le i^*\le i^- \tag{35}$$

where A_i is defined in equations 26 and 28. The subscript i^* indicates the location z_i^* at which the state vector $\{w_i,f_i\}^T$ is evaluated, and $i-1^+$ and i^- mark the boundaries of the i-th layer, shown in Fig. 1. Introduce the transformation

$$\begin{Bmatrix} w_{i^*} \\ f_{i^*} \end{Bmatrix} = [D_i] \begin{Bmatrix} \mu_u(i^*) \\ \mu_d(i^*) \end{Bmatrix} , \qquad i-1^+ \le i^* \le i^- \tag{36}$$

where the columns in D_i are the eigenvectors of A_i. Each D matrix can be partitioned into

$$[D] = \begin{bmatrix} M_u & M_d \\ N_u & N_d \end{bmatrix} \tag{37}$$

where for the P-SV waves

$$M_{u,d} = \begin{bmatrix} \mp i q_\alpha \varepsilon_\alpha & \frac{k_n}{\omega}\varepsilon_\beta \\ \frac{k_n}{\omega}\varepsilon_\alpha & \mp i \frac{k_n}{\omega}\varepsilon_\beta \end{bmatrix} \tag{38}$$

$$N_{u,d} = \begin{bmatrix} \rho(2\beta^2 \frac{k_n^2}{\omega^2} - 1)\varepsilon_\alpha & \mp 2i\rho\beta^2 \frac{k_n}{\omega} q_\beta \varepsilon_\beta \\ \mp 2i\rho\beta^2 \frac{k_n}{\omega} q_\alpha \varepsilon_\alpha & \rho(2\beta^2 \frac{k_n^2}{\omega^2} - 1)\varepsilon_\beta \end{bmatrix} \tag{39}$$

and for the SH waves

$$M_{u,d} = \frac{\varepsilon_\beta}{\beta} , \qquad N_{u,d} = \mp i\rho\beta q_\beta \varepsilon_\beta \tag{40}$$

In equation (38) through (40),

$$\varepsilon_\alpha = (2\rho q_\alpha)^{-\frac{1}{2}} , \qquad \varepsilon_\beta = (2\rho q_\beta)^{-\frac{1}{2}} \tag{41}$$

where

$$q_\alpha = (\alpha^{-2} - \frac{k_n^2}{\omega^2})^{\frac{1}{2}} , \qquad q_\beta = (\beta^{-2} - \frac{k_n^2}{\omega^2})^{\frac{1}{2}} \tag{42}$$

The branch cuts for the radicals in the expressions for q_α and q_β are taken to be $\mathrm{Im}(\omega q_\alpha) \ge 0$ and $\mathrm{Im}(\omega q_\beta) \ge 0$.

Equation (36) transforms the state vector to a wave vector $\{\mu_u, \mu_d\}^T$ where the subscripts u and d represent the upward and downward propagation directions, respectively, as shown schematically in Figs. 3 and 4. In Fig 3, $\mu_d(i-1^+)$ and $\mu_u(i^-)$ may be treated as incoming wave vector with respect to layer i, while $\mu_u(i-1^+)$ and $\mu_d(i^-)$ treated as outgoing wave vectors. The wave vector shown in Fig. 4 also may be treated similarly as that in Fig. 3. Substituting equation (36) into equation (35) and rearranging the wave vector in an input-output relationship, we obtain

$$\begin{Bmatrix} \mu_u(i-1^+) \\ \mu_d(i^-) \end{Bmatrix} = \begin{bmatrix} R(i-1^+,i^-) & T(i^-,i-1^+) \\ T(i-1^+,i^-) & R(i^-,i-1^+) \end{bmatrix} \begin{Bmatrix} \mu_d(i-1^+) \\ \mu_u(i^-) \end{Bmatrix} \tag{43}$$

where

$$T(i-1^+,i^-)=T(i^-,i-1^+)=\begin{cases} diag[e^{i\omega q_\alpha \Delta z_i} \;\; e^{i\omega q_\beta \Delta z_i}] \,, & \text{for P-SV waves} \\ e^{i\omega q_\beta \Delta z_i} \,, & \text{for SH waves} \end{cases} \tag{44}$$

$$R(i-1^+,i^-)=R(i^-,i-1^+)=0 \tag{45}$$

and where diag indicates a diagonal matrix and $\Delta z_i = |z_i - z_{i+1}|$.

On the two sides of the interface between layers i and i+1, the two state vectors should be the same. Then, transformation of these two state vectors into the corresponding two wave vectors by using equation (36) and rearrangement of the wave vectors in an input-output relationship yield

$$\begin{Bmatrix} \mu_u(i^-) \\ \mu_d(i^+) \end{Bmatrix} = \begin{bmatrix} R(i^-,i^+) & T(i^+,i^-) \\ T(i^-,i^+) & R(i^+,i^-) \end{bmatrix} \begin{Bmatrix} \mu_d(i^-) \\ \mu_u(i^+) \end{Bmatrix} \tag{46}$$

where

$$R(i^-,i^+)=Q_{12}(i^+,i^-)[Q_{22}(i^+,i^-)]^{-1} \tag{47}$$

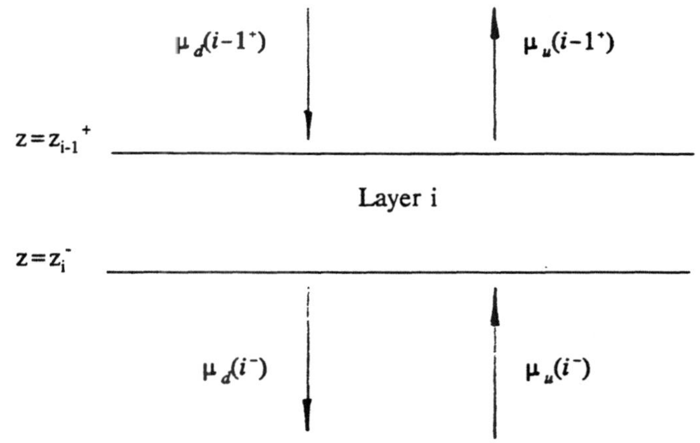

Fig. 3 Incoming and outgoing waves with respect to a layer

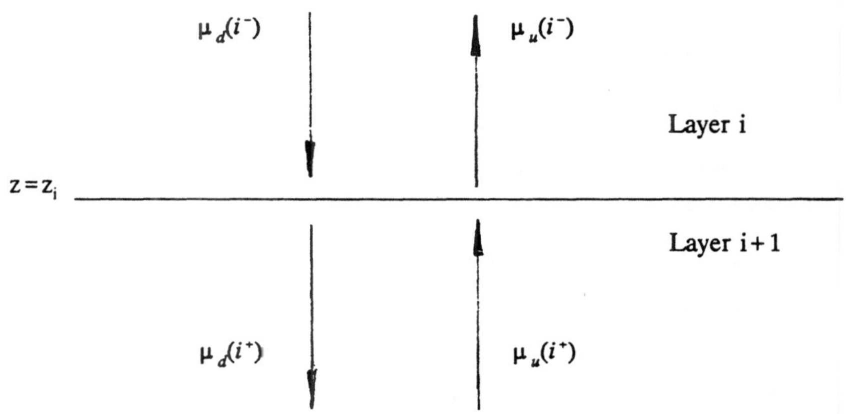

Fig. 4 Incoming and outgoing waves with respect to an interface between two layers

$$R(i^+,i^-)=-[Q_{22}(i^+,i^-)]^{-1}Q_{21}(i^+,i^-) \tag{48}$$

$$T(i^-,i^+)=[Q_{22}(i^+,i^-)]^{-1} \tag{49}$$

$$T(i^+,i^-)=Q_{11}(i^+,i^-)-Q_{12}(i^+,i^-)[Q_{22}(i^+,i^-)]^{-1}Q_{21}(i^+,i^-) \tag{50}$$

in which Q_{ij} (i,j=1,2) are the submatrices of $Q = [D_i]^{-1}[D_{i+1}]$. The square matrices on the right-hand-side of equations (43) and (46) are called scattering matrices, and their submatrices R and T are called the reflection and the transmission matrices, respectively. The physical meanings of R(j,k) and T(j,k) in these scattering matrices are clear; they represent, respectively, the reflection and transmission efficiencies of a waveguide from plane z_j to plane z_k. The arguments j and k for these matrices are $i-1^-$ and $i-1^+$ in equation (43), and i^- and i^+ in equation (46).

It is obvious that no wave reflection takes place in a uniform medium, therefore, the reflection matrices are null in equation (43). Furthermore, if the last layer, namely layer n+1, is assumed to be infinitely thick, then it is clear that

$$R(n^+,\infty)=0 \tag{51}$$

The boundary condition at the ground surface also can be represented by a reflection matrix. Write

$$\begin{Bmatrix} w_0 \\ f_0 \end{Bmatrix}=\begin{bmatrix} M_u(1) & M_d(1) \\ N_u(1) & N_d(1) \end{bmatrix}\begin{Bmatrix} \mu_u(0^+) \\ \mu_d(0^+) \end{Bmatrix} \tag{52}$$

where $M_u(1)$, $M_d(1)$, $N_u(1)$ and $N_d(1)$ are submatrices of D_1 which is given in equations (37)-(42). Since $f_0=\{0\}$ at the free boundary, it follows from equation (52) that

$$R(0^+,0)=-[N_d(1)]^{-1}N_u(1) \tag{53}$$

With the knowledge of the basic reflection and transmission matrices for a uniform layer, interface between two layers, and the ground surface, the reflection and transmission matrices characterizing a stratified medium between any two planes can be constructed according to the following composition rule [27]:

$$R(i,k)=R(i,j)+T(j,i)R(j,k)[I-R(j,i)R(j,k)]^{-1}T(i,j) \tag{54}$$

$$T(i,k)=T(j,k)[I-R(j,i)R(j,k)]^{-1}T(i,j) \tag{55}$$

where plane j is between planes i and k.

It is believed that earthquake is usually caused by dislocation of a fault. If typical dimension of a fault is short compared with its distance from the ground site of interest, then the effect of dislocation may be represented by equivalent jumps (e.g. Kennett [15]) in the state vector of displacements and forces in equation (24). For example, for a shear dislocation in the x-direction, occurring along a fault plane parallel to the ground surface, these jumps corresponding to a unit seismic moment are given by

$$[U]_-^+=[P]_-^+=[S]_-^+=[T]_-^+=0\ ,\ [V]_-^+=\pm\frac{1}{2}(\rho\beta^2)^{-1}\ ,\ [W]_-^+=-\frac{1}{2}i(\rho\beta^2)^{-1}\ ;\qquad m=\pm 1 \tag{56}$$

The magnitude of a seismic moment is obtained as (shear modulus) x (fault area) x (final slip). The reader is referred to Kennett [15] for other types of dislocations.

Let the equivalent jumps given by equation (56) be located at a source plane z_s within layer j, as shown in Fig. 1. It is convenient to treat these jumps as being produced by a source vector

$$\begin{Bmatrix} w_s \\ f_s \end{Bmatrix} = \begin{Bmatrix} w_{s^+} \\ f_{s^+} \end{Bmatrix} - \begin{Bmatrix} w_{s^-} \\ f_{s^-} \end{Bmatrix} \tag{57}$$

Substituting equation (36) into equation (57) and taking into consideration the reflection efficiencies above and below the source plane, we obtain

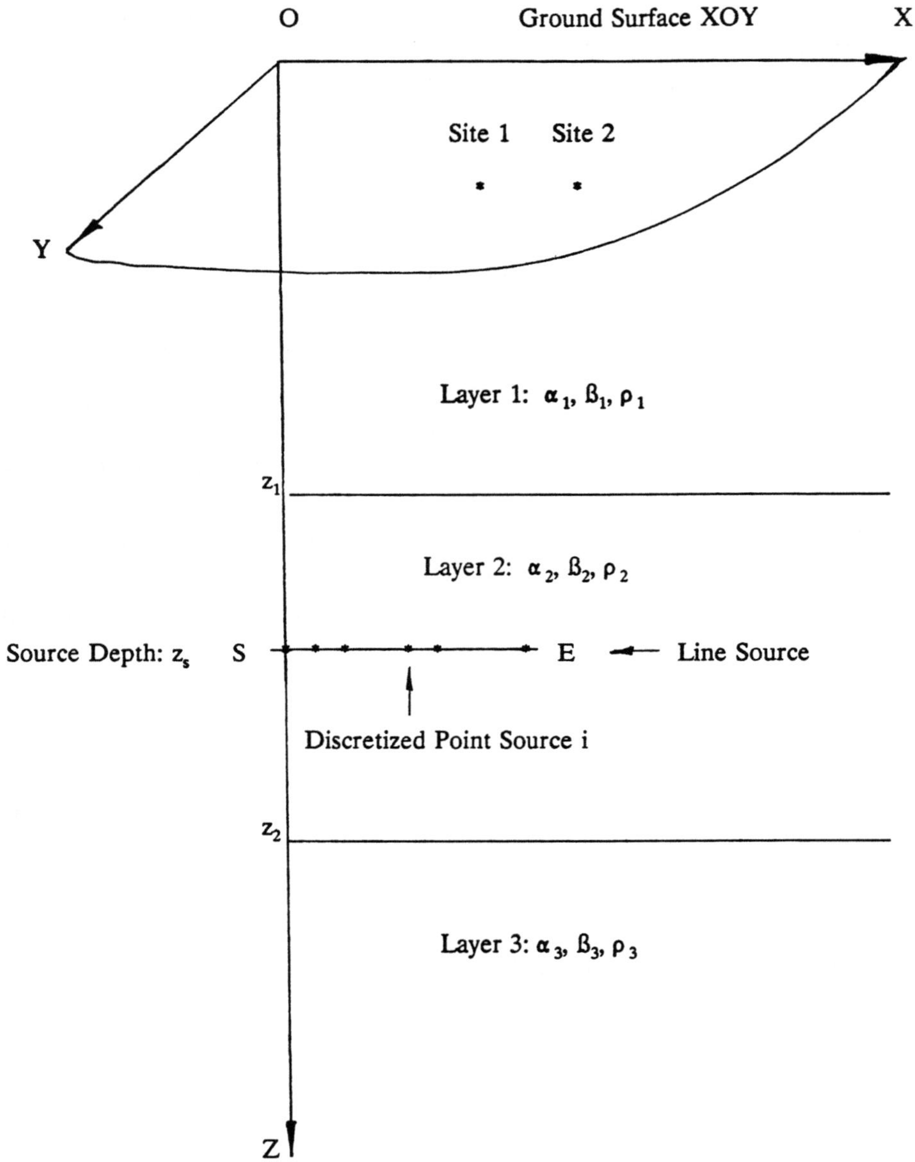

Fig. 5 Schematic representation of a numerical example

$$\begin{bmatrix} M_u(j) & M_d(j) \\ N_u(j) & N_d(j) \end{bmatrix} \begin{Bmatrix} R(s^+,n^+)\mu_d(s^+) \\ \mu_d(s^+) \end{Bmatrix} - \begin{bmatrix} M_u(j) & M_d(j) \\ N_u(j) & N_d(j) \end{bmatrix} \begin{Bmatrix} \mu_u(s^-) \\ R(s^-,0)\mu_u(s^-) \end{Bmatrix} = \begin{Bmatrix} w_s \\ f_s \end{Bmatrix} \tag{58}$$

where use has been made of $R(s^+,\infty)=R(s^+,n^+)$ which may be inferred from equation (51). Equation (58) can be used to solve for $\mu_u(s^-)$. Finally, the ground motion in the ω-k_n-m domain can be obtained as follows, using equation (36),

$$\{w_0\}=[M_u(1)+M_d(1)R(0^+,0)][I-R(0^+,s^-)R(0^+,0)]^{-1}T(s^-,0^+)\mu_u(s^-) \tag{59}$$

Numerical Example

Numerical calculations have been carried out for an example problem, in which the earth medium is modeled as being composed of three layers, and the line source is located in the second layer in the x direction, shown in Fig. 5. Physical parameters of the three layers are listed in Table I. The other parameters are $T_r=0.628$ s, $R=4.0\times10^5$ m, source depth $z_s = 6000$ m, fault length = 10000 m. The expected total seismic moment = 6.0×10^{17} N·m. The assumed expected rate $\lambda(t)$ at which the source points are activated is shown in Fig. 6. The selected sites for the ground motion investigation are (1) $x_1=2.0\times10^4$ m, $y_1=3.6\times10^4$ m, $z_1=0.0$ m, and (2) $x_2=2.6\times10^4$ m, $y_2=3.6\times10^4$ m, $z_2=0.0$ m.

Table I

	ρ (kg/m^3)	α (m/s)	β (m/s)	γ_α	γ_β	z (km)
Layer 1	1.67×10^3	3000	1730	0.08	0.04	0 - 0.5
Layer 2	2.28×10^3	5000	2890	0.04	0.02	0.5- 12
Layer 3	2.58×10^3	6000	3460	0.02	0.01	12 - ∞

To gain some insight into the basic characteristics of a Green's function computed from equation (29), let us examine first the ground motion due to a point source at $x_s=0.0$ m, $y_s=0.0$ m, and $z_s=6000$ m with a seismic moment = 2.0×10^{16} N·m, occurring at $\tau=0.0$

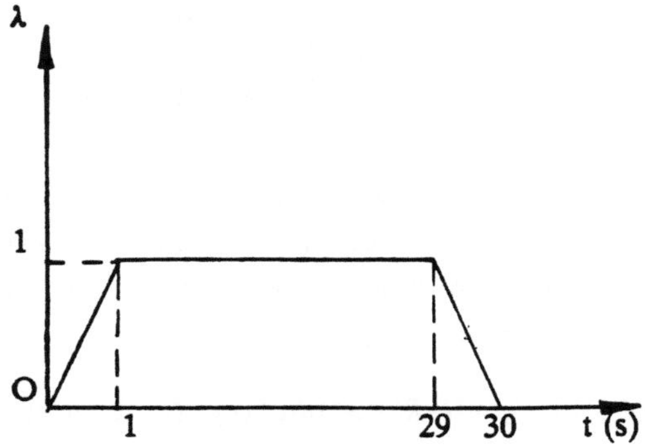

Fig. 6 Expected activation rate of point sources

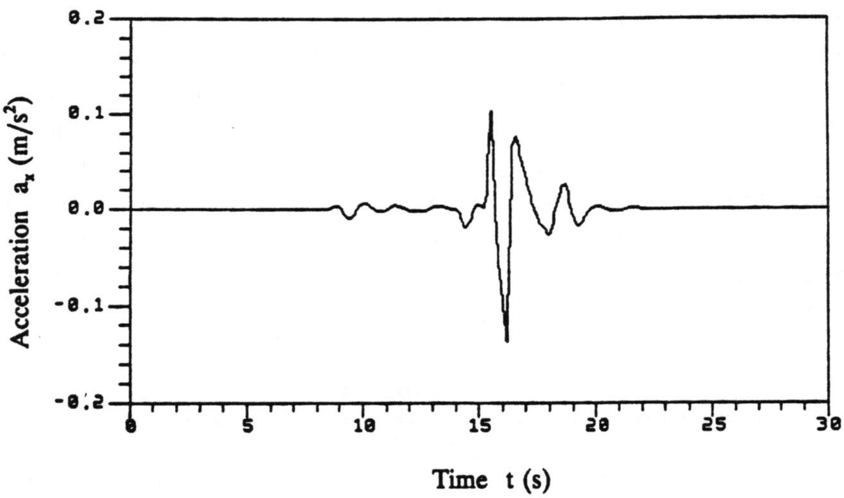

Fig. 7 Ground surface acceleration in the x-direction

s. Fig. 7 shows the ground acceleration in the x-direction at site 1. The response at the ground surface begins at the time when the first P wave signal arrives. It becomes much stronger when the S wave signals arrive several seconds later, and gradually dies down after the arrival of the last wave signal.

Fig. 8 shows the temporal covariance of the ground acceleration in the x-direction at the site 1 computed from equation (15) by setting $x_2=x_1$, $y_2=y_1$, and $z_2=z_1=0$. The mean square ground acceleration can be found along $t_1=t_2$. The arrivals of the first P- and S- waves, and eventual dying down of all signals can also be inferred from this figure. The profiles of the covariances away from $t_1=t_2$ appear zagged, but the basic trend of approaching zero at increasing $|t_1-t_2|$ is quite clear. The zagged appearance can be smoothed somewhat by taking smaller step size when carrying out numerical integration in equation (47), at the expense of greater computer cost. Fig. 9 represents the temporal cross-covariances of the ground acceleration in the x-direction at sites 1 and 2. The magnitudes along $t_1=t_2$ are still relatively large due to the relative proximity of the two sites. They decrease quickly as the time difference $|t_1-t_2|$ becomes large.

Fig. 10 is plot of the evolutionary spectral density of the ground acceleration in the x-direction at site 1. It shows the frequency distributions of earthquake accelerations at different times. The absence of the high frequency components is due to the filter effect of a layered medium. Finally, the evolutionary cross-spectral density of the ground acceleration in the x-direction at sites 1 and 2 is shown in Fig. 11. Its characteristics is similar to that shown in Fig. 10.

Concluding Remarks

In the proposed stochastic earthquake model the basic geophysical features of the earth medium and random nature of the seismic sources are taken into account. A numerically stable and accurate computational scheme is developed to obtain various statistical properties of seismic ground motion, including the covariance and cross-covariance, and evolutionary spectral

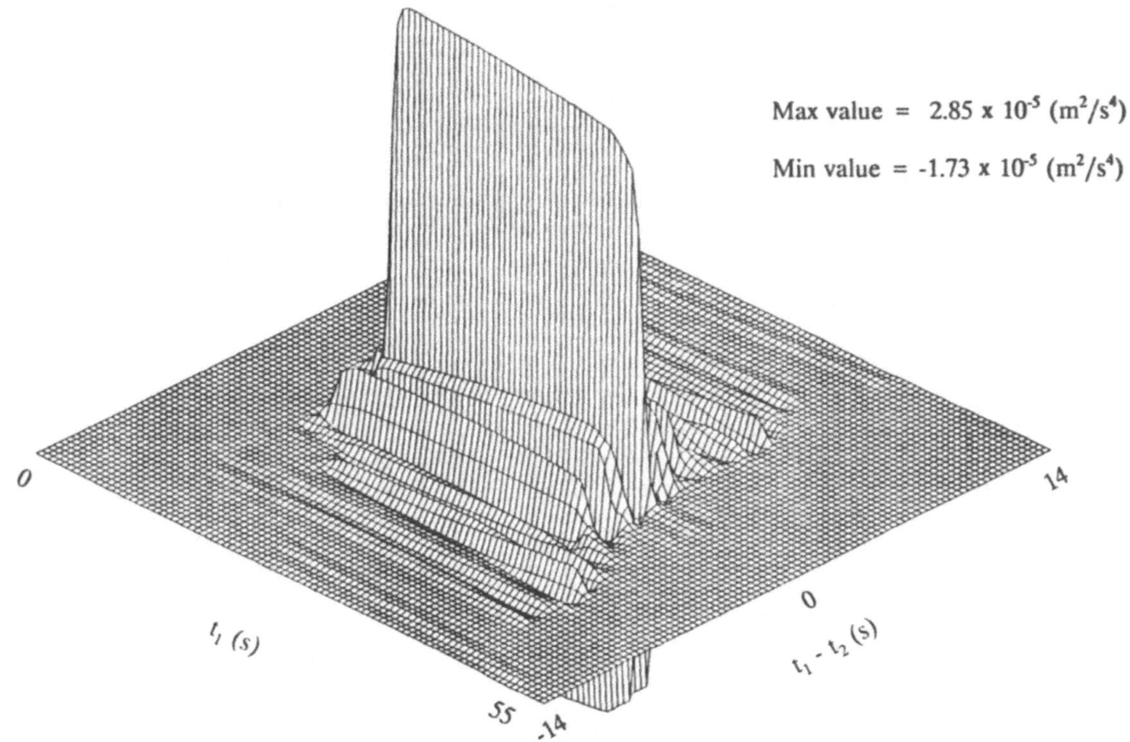

Fig. 8 Covariance function for ground acceleration in the x-direction at site 1

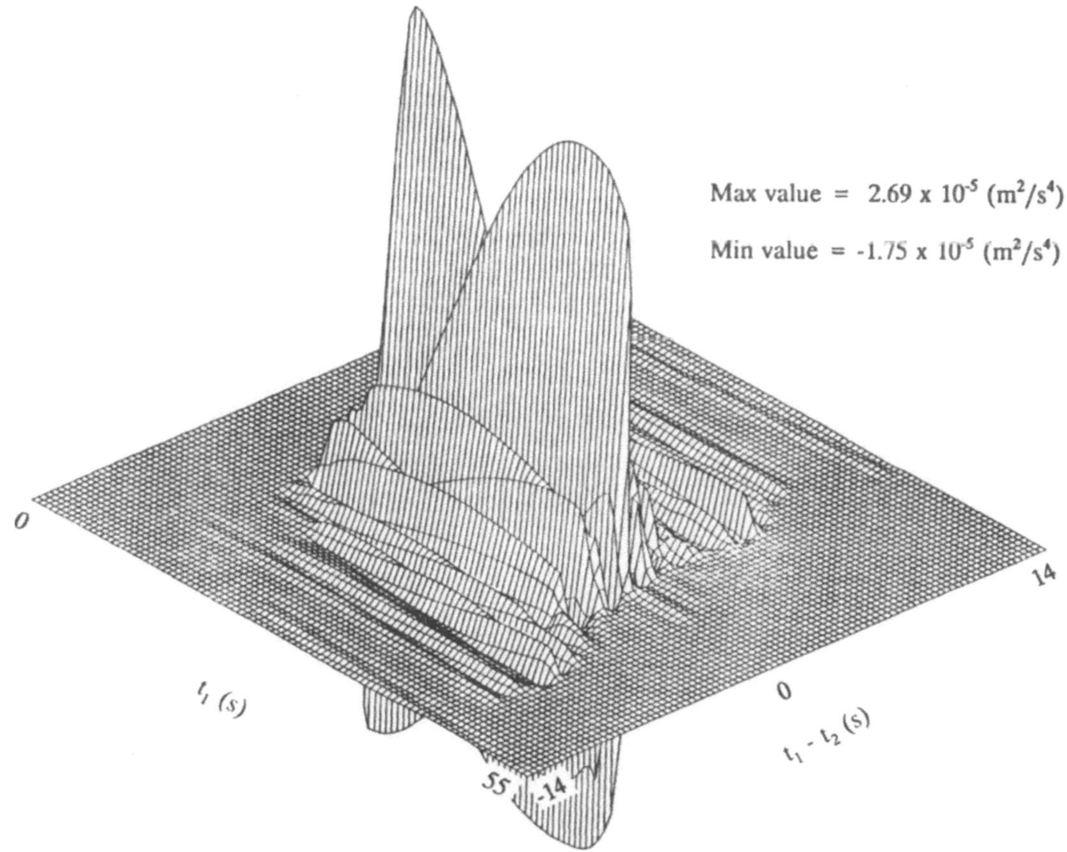

Fig. 9 Cross-covariance function for ground acceleration in the x-direction at sites 1 and 2

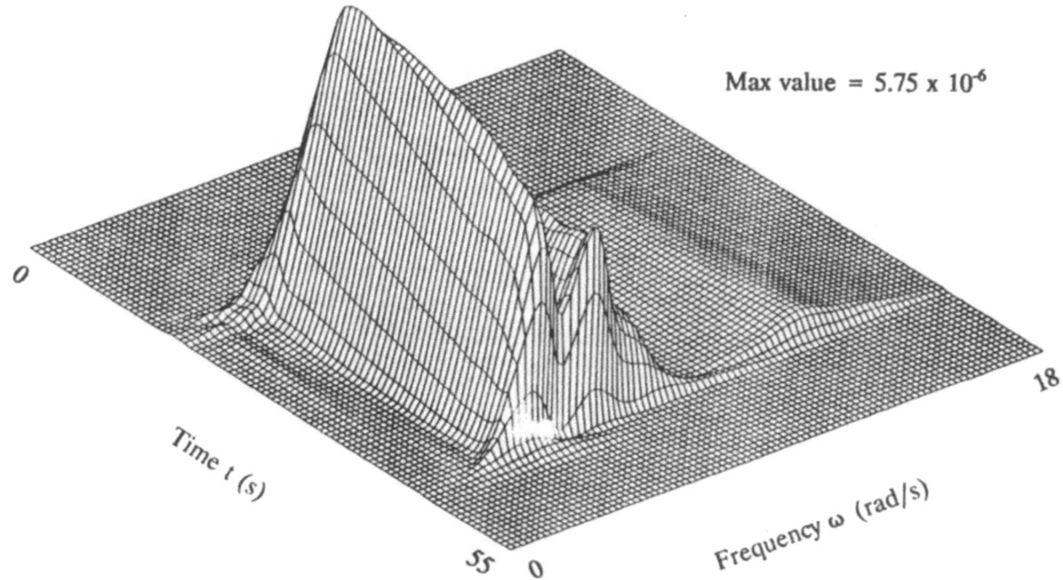

Fig. 10 Evolutionary spectrum of ground acceleration in the x-direction at site 1

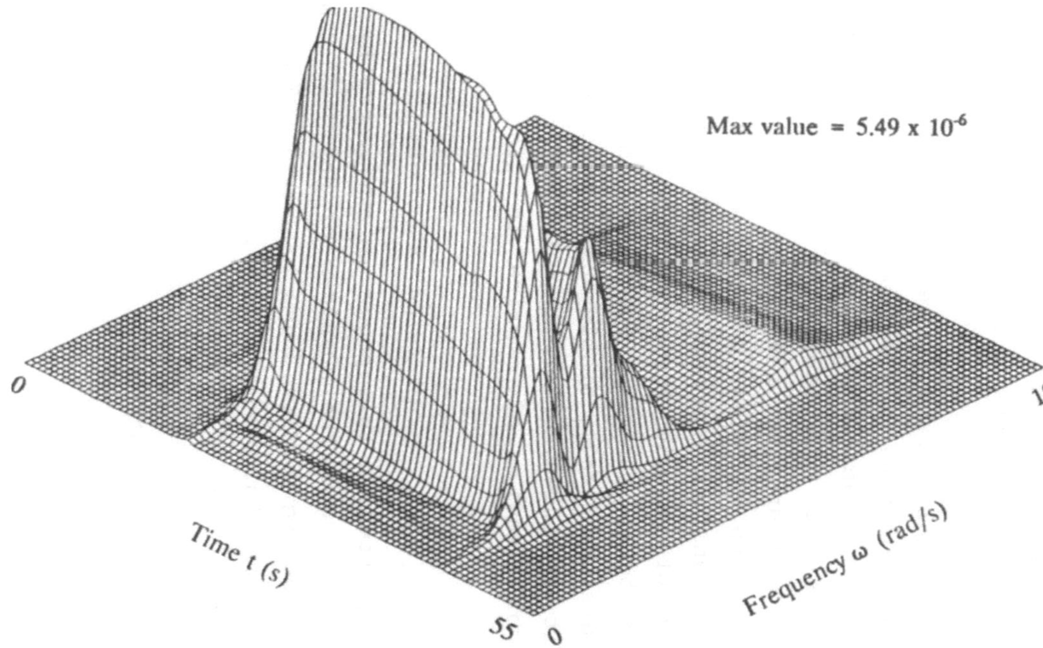

Fig. 11 **Evolutionary cross-spectrum of ground acceleration in the x-direction at sites 1 and 2**

density and evolutionary cross-spectral density, without recourse to Monte Carlo type simulations.

In solving the partial differential equations of motion for each earth layer, the Fourier finite-Hankel transform approach is used which results in an exact solution within the time window $t \leq T_w$ where T_w is the time lapsed for the fastest waves to propagate to a truncated radius R. The procedure is computationally more efficient compared with the conventional Fourier-Hankel transform approach. The problem of propagation and reflection of seismic waves in a multilayer medium is formulated in terms of wave scattering matrices so that calculations are numerically stable and accurate results can be obtained. It is shown in an example that a horizontally stratified earth medium gives rise to distinctive filter effects.

Acknowledgments

The research reported in this paper was supported by Contract No. NCEER 89-2004 under the auspices of National Center for Earthquake Engineering Research under NSF Grant ECE-86-07591 and by Florida State University through the allocation of supercomputer resources. Opinions, findings and conclusions expressed are those of the writers and do not necessarily reflect the views of our sponsors.

References

1. Aki, K.; Richards, P.G: Quantitative Seismology ---- Theory and Methods. W.H. Freeman and Company, New York, 1980.

2. Alekseev, A.S.; Mikhailenko, B.G.: The solution of dynamic problems of elastic wave propagation in inhomogeneous media by a combination of partial separation of variables and finite-difference methods. J. Geophys. 48 (1980) 167-172.

3. Apsel, R.J.; Luco, J.E.: On the Green's function for a layered half-space, part II, Bull. Seism. Soc. Am. 73 (1983) 931-951.

4. Bouchon, M.: Discrete wave number representation of elastic wave fields in three-space dimensions, J. Geophys. Res. Vol. 84, No. B7 (1979) 3609-3614.

5. Bouchon, M.: A simple method to calculate Green's functions for elastic layered media, Bull. Seism. Soc. Am. 71 (1981) 959-971.

6. Burridge, R.; Knopoff, L.: Body force equivalents for seismic dislocations, Bull. Seism. Soc. Am. Vol. 54, (1964) 1875-1888.

7. Chin, R.C.Y.; Hedstrom, G.; Thigpen, L.: Numerical methods in seismology, Journal of Computational Physics, Vol. 54 (1984) pp. 18-56.

8. Cornell, C.A.: Stochastic process models in structural engineering, Stanford Univ. Civil Eng. Dept. Tech. Rept. 34, 1964.

9. Deodatis G.; Shinozuka M.; Papageorgiou A.: Stochastic wave representation of seismic ground motion. I: F-K spectra, J. Engrg. Mech, Vol. 116, No. 11 (1990) pp. 2363-2379.

10. Deodatis G.; Shinozuka M.; Papageorgiou A.: Stochastic wave representation of seismic ground motion. II: simulation, J. Engrg. Mech, Vol. 116, No. 11 (1990) pp. 2381-2399.

11. Haskell, N.A.: Radiation pattern of surface waves from point sources in a multilayered medium, Bull. Seism. Soc. Am. 54 (1964) 54, 377-393.

12. Hudson, J.A.: A quantitative evaluation of seismic signals at teleseismic distances ---- I, Radiation from point sources, Geophys. J.R. Astr. Soc. 18 (1969) 233-249.

13. Kanamori, H.; Gordon, S.S.: Seismological aspects of the Guatemala earthquake of February 4, 1976, J. Geophys. Res. Vol 83, No. B7 (1978).

14. Kanamori, H.: Semi-empirical approach to prediction of long-period ground motions from great earthquakes, Bull. Seism. Soc. Am. Vol. 69 (1979) 1645-1670.

15. Kennett, B.L.N.: Seismic Wave Propagation in Stratified Media, Cambridge University Press, 1985.

16. Korn, M.: Computation of wavefields in vertically inhomogeneous media by a frequency domain finite-difference method and application to wave propagation in earth models with random velocity and density perturbations, Geophys.J.R.Astr.Soc. 88 (1987) 345-377.

17. Lin, Y.K.: Application of nonstationary shot noise in the study of system response to a class of nonstationary excitations, J. Applied Mechanics, Vol. 30, No. 4 (1963) 555-558.

18. Lin, Y.K.: On random pulse train and its evolutionary spectral representation, Probabilistic Engineering Mechanics, Vol. 1, No. 4 (1986) 219-223.

19. Lin, Y.K.; Yong, Y.: Evolutionary Kanai-Tajimi earthquake models, J. Engrg. Mech. Vol. 113, No. 8 (1987) 1119-1137.

20. Olson, A.H.; Orcutt, J.A.; Frazier, G.A.: The discrete wavenumber/finite element method for synthetic seismograms, Geophys. J.R. Astr. Soc. 77 (1984) 421-460.

21. Pestel, E.C.; Leckie, F.A.: Matrix Methods in Elastomechanics, McGraw-Hill, New York, 1963.

22. Priestley, M.B.: Evolutionary spectral and nonstationary process, J. Royal Statistical Society, B27 (1965) 204-228.

23. Sneddon, I.N.: Fourier Transforms, New York, Toronto, London, 1951.

24. Stratonovich, R.L.: Topics in the Theory of Random Noise, English translation by R.A. Silverman, Gordon and Breach, Science Publishers, New York, 1963.

25. Xu, P.-C.; Mal, A.K.: Calculation of the inplane Green's functions for a layered viscoelastic solid, Bull. Seism. Soc. Am. 77 (1987) 1823-1837.

26. Yong, Y: Stochastic earthquake modeling and dynamic response analysis, Ph.D. Thesis, Urbana, Illinois, 1987.

27. Yong, Y.; Lin, Y.K.: Propagation of decaying waves in periodic and piecewise periodic structures of finite length, J. Sound and Vibration 129(2) (1989) 99-118.

28. Zhang, R.; Yong, Y.; Lin, Y.K.: Earthquake ground motion modeling with stochastic line source, Part I, to appear in J. Engrg. Mech., (1991).

29. Zhang, R.; Yong, Y.; Lin, Y.K.: Earthquake ground motion modeling with stochastic line source, Part II, to appear in J. Engrg. Mech., (1991).

Analyzing Nonstationary Random Response to Evolutionary Random Excitation by Complex Modal Method

Zhang Tian-Shu and Fang Tong

Northwestern Polytechnical University
Xi'an, Shaanxi 710072, CHINA

Abstract

The time domain modal analysis of a nonstationary random response of a linear, time-invariant, MDOF system subjected to evolutionary random excitations is illustrated through some well known examples. Based upon the time-dependent correlation function of the response, a partial spectrum is defined in order to reveal the mean cross power distribution of the response in frequency domain.

By comparing the present results with those obtained by the frequency domain approach via the evolutionary spectrum, it is found that both methods give the same time-dependent mean square response, but the Fourier Transform of the evolutionary spectrum fails to predict the correct time-dependent correlation function.

Key Words: Random Vibration, Nonstationary Response, Structure Dynamics, Complex Modal Analysis.

1, Introduction

The analysis of nonstationary response of a structure subjected to a nonstationary random excitatiion is very important in many engineering fields, e.g. earthquake engineering, wind engineering and offshore engineering etc.. Since a general nonstationary random excitation is too difficult to deal with, attention has been focused on a kind of special nonstationary random excitation, namely the evolutionary random process. The evolutionary shot noise presented by Lin [1] belongs to this kind of nonstationary random excitation. Hammond [2] and Shinozuka [3] analyzed the nonstationary evolutionary responses of SDOF and MDOF systems by the method of evolutionary spectrum in frequency domain. However, such an approach is complicated and requires considerable computational effort.

Time domain analysis of nonstationary random response was first introduced

Y. K. Lin · I. Elishakoff (Eds.)
Stochastic Structural Dynamics 1
New Theoretical Developments
2nd International Conference on Stochastic Structural Dynamics
May 9-11, 1990, Boca Raton, Florida

by Caughey [4] in 1963. Some new developments on time domain modal analysis have been obtained in recent years [5,6,7], meanwhile Iwan and Hou suggested a time domain approach via the Transition Matrix with basically the same function [8,9]. In view of that most practicing engineers have been not aware of the time domain method [10] yet, in this paper we illustrate the method through some well known examples studied by Caughey and Stumpf [11], Amin and Ang [12], and Roberts [13]. Although we have considered here only some special kinds of modulated white noise excitations, the method can also apply to modulated filtered white noise excitations [5,6,7].

Based on the time dependent correlation function of a nonstationary random response, obtained by the time domain method, a partial spectrum of the response can be defined, which provides information about the mean cross power distribution of the response in frequency domain [14].

By comparing the partial spectrum and the evolutionary spectrum of the response to a suddenly applied white noise excitation, it is found that there doesn't exist such a simple relationship as Fourier Transform pairs between the time-dependent correlation function and the evolutionary spectrum of a nonstationary response.

2, Nonstationary Response of a SDOF Linear System

A SDOF damped linear system is considered at first. The differential equation of the system can be described as follows:

$$\ddot{x} + 2\zeta\omega_0\dot{x} + \omega_0^2 x = n(t) \tag{1}$$

where, ζ is the damping ratio of the system and ω_0 is the natural frequency of the undamped system; n(t) is evolutionary white noise with zero mean value, and its correlation function can be written as:

$$R_n(t_1,t_2) = I(t_1)\delta(t_2-t_1) \tag{2}$$

in which I(t) is a deterministic intensity function, and $\delta(t)$ is the Dirac delta function.

The damping of the system is assumed sub-critical, i.e. $\zeta<1$, then, the eigenvalues of the system occur in conjugate pairs, which can be put into

$$p_1 = \bar{p}_2 = -\zeta\omega_0 + j\omega_0\sqrt{1-\zeta^2} = \alpha + j\beta \tag{3}$$

Let

$$P = \begin{bmatrix} p_1 & \\ & p_2 \end{bmatrix}, \quad U = \begin{bmatrix} p_1 & p_2 \\ 1 & 1 \end{bmatrix}$$

and introduce the complex transform,

$$X = UY \tag{4}$$

where

$$X = \begin{bmatrix} \dot{x} \\ x \end{bmatrix}, \quad Y = \begin{bmatrix} y_1 \\ y_2 \end{bmatrix}$$

Eq.(1) can be reduced to

$$\dot{Y} - PY = F \tag{5}$$

where F is the generalized force in a vector form,

$$F = \begin{bmatrix} F_1 \\ F_2 \end{bmatrix} = \frac{j}{2\beta}\begin{bmatrix} -1 \\ 1 \end{bmatrix} n(t) \tag{6}$$

Accordingly, the correlation function matrix of the generalized force F can be expressed as

$$\langle F(t)\bar{F}^T(t+\tau)\rangle = GI(t)\delta(\tau) \tag{7}$$

where

$$G = \frac{1}{4\beta^2}\begin{bmatrix} 1 & -1 \\ -1 & 1 \end{bmatrix} \tag{8}$$

Eq.(5) can also be written in scalar form,

$$\dot{y}_i - p_i y_i = F_i, \qquad i = 1, 2 \tag{9}$$

The impulse response function of Eq.(9) is

$$h_i(t) = e^{p_i t}, \quad t \geq 0, \quad i = 1, 2 \tag{10}$$

For zero initial conditions, the solution of Eq.(9) can be formulated as:

$$y_i(t) = \int_0^t h_i(t) F_i(t-u)du \tag{11}$$

Then, the cross correlation function of the responses in Eq.(9) can be expressed as

$$\langle y_i(t)\bar{y}_s(t+\tau)\rangle = \int_0^t\int_0^{t+\tau} h_i(u)\langle F_i(t-u)\bar{F}_s(t+\tau-v)\rangle \bar{h}_s(v)dvdu$$

$$i, s = 1, 2 \tag{12}$$

From Eq.(7), we have

$$\langle F_i(t-u)\bar{F}_s(t+\tau-v)\rangle = G_{is}I(t-u)\delta(u+\tau-v) \tag{13}$$

where G_{is} is the element of the matrix G in Eq.(8).

By substituting Eq.(13) into Eq.(12) and making some reduction, the cross correlation function of the responses in Eq.(9) can be finally written as:

$$\langle y_i(t)\bar{y}_s(t+\tau)\rangle = G_{is} e^{\bar{p}_s \tau} e^{(p_i+\bar{p}_s)t} \int_0^t I(u) e^{-(p_i+\bar{p}_s)u} du$$

$$t,\ \tau \geq 0\ ,\quad i,\ s = 1,\ 2 \tag{14}$$

For typical modulating functions, the integration in Eq.(14) is readily obtained in analytical form, which is useful not only in SDOF system analysis, but in MDOF system analysis as well.

Now, we look for the correlation function of the nonstationary response of the system for the following three different kinds of I(t) respectively.

1) Case one

$$I(t) = U(t)$$

where U(t) is a unit step function.

2) Case two

$$I(t) = \begin{cases} U(t) I_0 \left(\frac{t}{t_1}\right)^2 & t \leq t_1 \\ I_0 & t_1 \leq t \leq t_2 \\ I_0 e^{-c(t-t_2)} & t \geq t_2 \end{cases}$$

where I_0, c, t_1, t_2 are constants. This kind of modulating function is often used in describing earthquake excitations.

3) Case three

$$I(t) = U(t)[c + d\cos(t)]$$

where c and d are constants.

By substituting the above mentioned three kinds of I(t) into Eq.(14) respectively, and after making some reduction, the correlation function of the nonstationary response of the system may be finally written as:

$$\begin{aligned} \langle x(t)x(t+\tau)\rangle &= \langle y_1(t)\bar{y}_1(t+\tau)\rangle + \langle y_1(t)\bar{y}_2(t+\tau)\rangle + \\ &\quad \langle y_2(t)\bar{y}_1(t+\tau)\rangle + \langle y_2(t)\bar{y}_2(t+\tau)\rangle \\ &= a(t) e^{\bar{p}_1 \tau} + b(t) e^{\bar{p}_2 \tau} \qquad \tau \geq 0 \end{aligned} \tag{15}$$

Formulae of a(t) and b(t) are given for three different kinds of I(t) in Appendix A.

It can be shown that the result of Case one is exactly the same as that obtained by Caughey and Stumpf [11], the result of the Case two is exactly the same as that obtained by Amin and Ang [12]. Our result of Case three contains the transient state and the steady state, while the result available in Ref.[13] is only a steady state one. However, both the steady state results do coincide. Some numerical results of Case two and Case three are illustrated in Fig.1 and Fig.2. In Fig.1, the mean square response curves are plotted for four different damping ratios and for $t_1=1.5$, $t_2= 15$, $c=0.18$, $\omega_0= 10\pi$ and $I_0= 1$. It can be seen that the mean square response curves have the similar shape to that of the I(t). The flat parts of the curves correspond to the steady state mean square responses to the stationary white noise excitation. For each response curve there is a transition right after $t=t_1$ until the mean square response comes up to the steady value. The smaller the damping ratio, the longer transition time it takes. Fig.2 is plotted for $c=d=1.0$ and $\omega_0= 10\pi$, and for four different damping ratios too. The influence of damping on the transient and steady state mean square responses can be readily seen in the figure.

3, Definition of Partial Spectrum [14]

In order to provide some information about power distribution in frequency domain of a nonstationary random response, based on the time dependent correlation function, we define a partial correlation function of a nonstationary random response. Let

$$R_x(t,\tau) = \begin{cases} \langle x(t)x(t+\tau)\rangle , & \tau\geq 0 \\ \langle x(t-\tau)x(t)\rangle , & \tau\leq 0 \end{cases} \tag{16}$$

The $R_x(t,\tau)$ is defined as the partial correlation function. Then, take the Fourier Transform of R (t,τ), i.e.

$$S_x(t,\omega) = \frac{1}{2\pi}\int_{-\infty}^{+\infty} R_x(t,\tau)e^{-i\omega\tau}d\tau \tag{16'}$$

The $S_x(t,\tau)$ is defined as the partial spectrum of a nonstationary response. The adjective "partial" implies that in Eqs.(16) and (16') t is taken as a fixed parameter.

In contrast to Påge's instantaneous power spectrum [15], which depends solely on the present and the past of the response, the partial spectrum depends solely on the present and the future of the response, so that it can provide reliable information at the right beginning of the response [14]. Besides, the partial spectrum is one to one correspondent to the time dependent correlation function.

For a SDOF damped linear system subjected to a suddenly applied white noise excitation, according to Eq.(15), Eq.(16) and Eq.(16'), the partial correlation function and the partial spectrum of the nonstationary response of the system may be written as follows:

$$R_x(t,\tau) = e^{\alpha|\tau|}[a(t)\cos\beta|\tau| + b(t)\sin\beta|\tau|] \tag{17}$$

where

$$a(t) = \frac{1}{4\alpha\beta^2(\alpha^2+\beta^2)}\{e^{2\alpha t}[2\alpha^2\sin^2(\beta t)-\alpha\beta\sin(2\beta t)+\beta^2]-\beta^2\}$$

$$b(t) = \frac{1}{4\beta^2(\alpha^2+\beta^2)}\{e^{2\alpha t}[\alpha\sin(2\beta t)-\beta\cos(2\beta t]+\beta\} \tag{18}$$

and

$$S_x(t,\omega) = \frac{1}{2\pi}\int_{-\infty}^{+\infty} R_x(t,\tau)e^{-i\omega\tau}\, d\tau$$

$$= \frac{1}{2\pi}|H(\omega)|^2\{-\frac{(\omega_0^2+\omega^2)}{2\beta^2(\alpha^2+\beta^2)}e^{2\alpha t}[2\alpha^2\sin^2(\beta t)-\alpha\beta\sin(2\beta t)+\beta^2]$$

$$+\frac{(\omega_0^2-\omega^2)}{2\beta(\alpha^2+\beta^2)}e^{2\alpha t}[\alpha\sin(2\beta t)-2\beta\cos^2(\beta t)+\beta]$$

$$+\frac{\omega_0^2}{(\alpha^2+\beta^2)}\} \tag{19}$$

where

$$|H(\omega)|^2 = \frac{1}{(\omega_0^2-\omega^2)^2+4\alpha^2\omega^2} \tag{20}$$

According to Eq.(3), we have

$$\alpha = -\zeta\omega_0 \quad , \quad \beta = \omega_0\sqrt{1-\zeta^2}$$

By substituting α and β into Eq.(17) and Eq.(19) and letting $\beta = \omega_d$, the partial spectrum $S_x(t,\omega)$ and the partial correlation function $R_x(t,\tau)$ of the nonstationary response of a SDOF linear system subjected to a suddenly applied white noise excitation can be put into another form. (see Eq.(B1), Eq.(B2) and Eq.(B3) in Appendix B.)

It is worthy of noting that for a nonstationary random process x(t), the two random variables, $x(t_1)$ and $x(t_2)$ for different t_1 and t_2, may be regarded as two different random variables with different characteristics.

Therefore, the time dependent partial correlation function has something common with a cross-correlation function, so has the partial spectrum with a cross-spectrum.

4, Comparison of Different Spectra

In section three, the partial correlation function and partial spectrum are defined. On the other hand, another kind of spectrum of a nonstationary response, i.e. the evolutionary spectrum, is defined in Ref.[16]. For a SDOF damped linear system subjected to a suddenly applied white noise excitation, the evolutionary spectrum of the nonstationary response has been obtained [2]. (see Eq.(B4) in Appendix B.)

From Eq.(B1) and Eq.(B4) in Appendix B, it is found that the evolutionary spectrum of the nonstationary response of the system, $S'_x(t,\omega)$, is different from the partial spectrum $S_x(t,\omega)$, though they yield the same time-dependent mean square response of the system, i.e.

$$\sigma_x^2(t) = \int_{-\infty}^{+\infty} S_x(t,\omega)d\omega = \int_{-\infty}^{+\infty} S'_x(t,\omega)d\omega \tag{21}$$

The inverse Fourier Transform of $S'_x(t,\omega)$, noted as $R'_x(t,\tau)$, can be obtained. (see Eq.(B5) in Appendix B.)

By comparing Eq.(B3) with Eq.(B5) in Appendix B, it is easy to find that $R'_x(t,\tau)$ is different from $R_x(t,\tau)$ greatly. However, the $R_x(t,\tau)$, obtained in section two, is an accurate representation of the time-dependent correlation function, while, the $R'_x(t,\tau)$, obtained by taking Fourier Transform of $S'_x(t,\omega)$, is not. In fact, the $S'_x(t,\omega)$ in Ref.[2] is obtained only in the sense that the integration of $S'_x(t,\omega)$ between the interval $(-\infty, +\infty)$ is equal to the mean square response of the system. It seems there is no such a simple relationship as Fourier Transform pairs between the evolutionary spectrum and the time-dependent correlation function. For comparison, some numerical results taken at three different instants are plotted in Fig.3-5 for $S_x(t,\omega)$ and $S'_x(t,\omega)$, and in Fig.6-8 for $R_x(t,\tau)$ and $R'_x(t,\tau)$. The differences between the two counterparts during the transitional phase are clearly shown. Since the system response goes finally to its steady state, the two counterparts coincide with each other then.

5, Nonstationary Response of a MDOF Linear System

The above method and discussions on a SDOF system can be readily extended to a MDOF system. Suppose the differential equation of a time invariant linear system can be written as

$$m\ddot{x} + c\dot{x} + kx = n(t) \tag{22}$$

where m, c and k are mass matrix, damping matrix, and stiffness matrix respectively; n(t) is evolutionary white noise process with zero mean value, and its correlation function can be written as

$$R_n(t_1,t_2) = DI(t_1)\delta(t_2-t_1)$$

in which, D is a real symmetrical matrix, I(t) is a deterministic intensity function, and $\delta(t)$ is the Dirac delta function.

So far as the system damping is below critical, the eigenvalues and associated eigenvectors of the system appear in conjugat% pairs. The eigenvalue matrix of the system can be written as

$$P = \mathrm{diag}[p_i] \quad , \quad i = 1, 2,\ldots,2n$$

where p_i's are eigenvalues of the system, with

$$p_{n+j} = \bar{p}_j \quad , \quad j = 1, 2,\ldots,n$$

Let the right eigenvectors of the system be noted as u_i and $\bar{u}_i$, and the left ones as v_i and $\bar{v}_i$. The right modal matrix and the left one can be put into

$$u = [u_1,u_2,\ldots,u_n, \bar{u}_1,\bar{u}_2,\ldots,\bar{u}_n] \qquad \text{(right)}$$
$$v = [v_1,v_2,\ldots,v_n, \bar{v}_1,\bar{v}_2,\ldots,\bar{v}_n] \qquad \text{(left)}$$

Introduce the following complex modal transform,

$$X = UY$$

where

$$U = \begin{bmatrix} uP \\ u \end{bmatrix} \quad , \quad X = \begin{bmatrix} \dot{x} \\ x \end{bmatrix}$$

Eq.(22) can be finally reduced to

$$\dot{Y} - PY = F \tag{23}$$

where

$$F = M^{-1}v^T n(t) = \{F_j\} \quad , \quad j = 1,2,\ldots,2n \tag{24}$$

$$M^{-1} = \begin{bmatrix} \mu & \\ & \bar{\mu} \end{bmatrix} \quad , \quad \mu = \mathrm{diag}[m_i^{-1}]$$

$$m_i = v_i^T\,[2p_i m + c]u_i \quad , \quad i = 1,2,\ldots,n$$

From Eq.(24), the correlation function matrix of the generalized force F can be expressed as

$$\langle F(t)\bar{F}^T(t+\tau)\rangle = GI(t)\delta(\tau)$$

with

$$G = M^{-1}v^T D\bar{v}\bar{M}^{-1}$$

Eq.(23) can also be written in scalar form

$$\dot{y}_i - p_i y_i = F_i \quad , \quad i = 1, 2,...,2n$$

which has the same form as Eq.(9). Hence, the cross correlation function of y_i and y_s can be put into the same formulation as Eq.(14), with i,s= 1,2,...,2n. Once the correlation function matrix of the complex modal response Y is obtained, the correlation function matrix of the system response x may be readily written as

$$\langle x(t)x^T(t+\tau)\rangle = u\langle Y(t)\bar{Y}^T(t+\tau)\rangle\bar{u}^T$$

6, Conclusions

(1) The complex modal method presented in this paper can apply to nonstationary random response problems of a time invariant linear system, which may be classically damped or not, and symmetrical or not. The method reduces the nonstationary response problem into algbraic operations only, and yields the characteristics in analytical form. As far as the time dependent mean square response to an evolutionary random excitation is concerned, the complex modal method is much more convenient than the frequency domain method.

(2) A partial spectrum of a nonstationary random response is defined, which has one to one correspondance with the time dependent correlation function of the response, and which reveals the mean cross power distribution for the present response with the future ones.

Though Priestley's evolutionary spectrum of a nonstationary random response can yield the correct time dependent mean square response, it fails to predict the correct time dependent correlation function simply via the Fourier Transformation.

Appendix A

In section two, the correlation function of nonstationary response of a SDOF damped linear system subjected to evolutionary white noise excitation can be written as

$$\langle x(t)x(t+\tau)\rangle = a(t)e^{\bar{p}_1\tau} + b(t)e^{\bar{p}_2\tau} \qquad \tau \ge 0$$

where , for Case one, we have

$$a(t) = -\frac{1}{4\beta^2}\left\{\frac{1}{2\alpha}[e^{2\alpha t}-1] - \frac{1}{2p_2}[e^{2p_2 t}-1]\right\}$$

$$b(t) = -\frac{1}{4\beta^2}\left\{\frac{1}{2\alpha}[e^{2\alpha t}-1] - \frac{1}{2p_1}[e^{2p_1 t}-1]\right\} \qquad \text{(A1)}$$

For Case two, a(t) and b(t) take the following different formulae respectively, for $t \le t_1$

$$a(t) = -\frac{1}{4\beta^2}\,[AA_{11}(t,t) - AA_{21}(t,t)]$$

$$b(t) = -\frac{1}{4\beta^2}\,[AA_{22}(t,t) - AA_{12}(t,t)] \qquad \text{(A2)}$$

for $t_1 \le t \le t_2$

$$a(t) = -\frac{1}{4\beta^2}\{[AA_{11}(t,t_1)+BB_{11}(t,t)]-[AA_{21}(t,t_1)+BB_{21}(t,t)]\}$$

$$b(t) = -\frac{1}{4\beta^2}\{[AA_{22}(t,t_1)+BB_{22}(t,t)]-[AA_{12}(t,t_1)+BB_{12}(t,t)]\} \qquad \text{(A3)}$$

and for $t \ge t_2$

$$a(t) = -\frac{1}{4\beta^2}\{[AA_{11}(t,t_1)+BB_{11}(t,t_2)+CC_{11}(t,t)]- [AA_{21}(t,t_1)+BB_{21}(t,t_2)+CC_{21}(t,t)]\}$$

$$b(t) = -\frac{1}{4\beta^2}\{[AA_{22}(t,t_1)+BB_{22}(t,t_2)+CC_{22}(t,t)]- [AA_{12}(t,t_1)+BB_{12}(t,t_2)+CC_{12}(t,t)]\} \qquad \text{(A4)}$$

where

$$AA_{is}(t,x) = \frac{I_0}{t_1^2(p_i+\bar{p}_s)^3}\{2e^{(p_i+\bar{p}_s)t} - e^{(p_i+\bar{p}_s)(t-x)}[(p_i+\bar{p}_s)^2x^2+2x(p_i+\bar{p}_s)+2]\} \qquad \text{(A5)}$$

$$BB_{is}(t,x) = -\frac{I_0}{p_i+\bar{p}_s}\left[e^{(p_i+\bar{p}_s)(t-x)} - e^{(p_i+\bar{p}_s)(t-t_1)}\right] \quad \text{(A6)}$$

$$CC_{is}(t,x) = -\frac{\bar{I}_0}{p_i-\bar{p}_s+c}e^{c(t_2-t)}\left[e^{(p_i+\bar{p}_s+c)(t-x)} - e^{(p_i+\bar{p}_s+c)(t-t_2)}\right] \quad \text{(A7)}$$

For Case three, we have

$$a(t) = a_1(t) + a_2(t)$$

$$b(t) = b_1(t) + b_2(t)$$

where

$$a_1(t) = \frac{d}{4\beta^2}\left\{\frac{1}{1+4\alpha^2}[2\alpha e^{2\alpha t} + \sin(t) - 2\alpha\cos(t)] - \frac{1}{1+4p_2^2}[2p_2e^{2p_2t} + \sin(t) - 2p_2\cos(t)]\right\}$$

$$b_1(t) = \frac{d}{4\beta^2}\left\{\frac{1}{1+4\alpha^2}[2\alpha e^{2\alpha t} + \sin(t) - 2\alpha\cos(t)] - \frac{1}{1+4p_1^2}[2p_1e^{2p_1t} + \sin(t) - 2p_1\cos(t)]\right\} \quad \text{(A8)}$$

$$a_2(t) = \frac{c}{4\beta^2}\left\{\frac{1}{2\alpha}(e^{2\alpha t}-1) - \frac{1}{2p_2}(e^{2p_2t}-1)\right\}$$

$$b_2(t) = \frac{c}{4\beta^2}\left\{\frac{1}{2\alpha}(e^{2\alpha t}-1) - \frac{1}{2p_1}(e^{2p_1t}-1)\right\} \quad \text{(A9)}$$

Appendix B

For a SDOF damped linear system subjected to a suddenly applied white noise excitation, the partial spectrum and the partial correlation function of nonstationary response of the system can be written as:

$$S_x(t,\omega) = \frac{1}{2\pi}|H(\omega)|^2\{1-e^{-2\zeta\omega_0t}[\frac{\zeta\omega_0}{\omega_d}\sin(2\omega_dt)+\cos^2(\omega_dt)+$$

$$\frac{(\omega^2+\zeta^2\omega_0^2)}{\omega_d^2}\sin^2(\omega_d t)]\} \tag{B1}$$

where

$$|H(\omega)|^2 = \frac{1}{(\omega_0^2-\omega^2)^2+4\zeta^2\omega_0^2\omega^2} \tag{B2}$$

$$\begin{aligned} R_x(t,\tau) = \frac{1}{4\zeta\omega_0^3} e^{-\zeta\omega_0|\tau|}\{[1-e^{-2\zeta\omega_0 t}(1+\frac{\zeta\omega_0}{\omega_d}\sin(2\omega_d t)+\frac{2\zeta^2\omega_0^2}{\omega_d^2} \\ \sin^2(\omega_d t))]\cos(\omega_d|\tau|)+\frac{\zeta\omega_0}{\omega_d}[1- \\ e^{-2\zeta\omega_0 t}(1+\frac{\zeta\omega_0}{\omega_d}\sin(2\omega_d t)-2\sin^2(\omega_d t))]\sin(\omega_d|\tau|)\} \end{aligned} \tag{B3}$$

The evolutionary spectrum and corresponding Fourier Transform of nonstationary response of the system can be written as:

$$\begin{aligned} S'_x(t,\omega) = \frac{1}{2\pi}|H(\omega)|^2\{1-2e^{-\zeta\omega_0 t}[\frac{\sin(\omega_d t)}{\omega_d}(\zeta\omega_0\cos(\omega t)+ \\ \omega\sin(\omega t))+\cos(\omega_d t)\cos(\omega t)]+e^{-2\zeta\omega_0 t}[\\ \frac{(\zeta^2\omega_0^2+\omega^2)}{\omega_d^2}\sin^2(\omega_d t)+\cos^2(\omega_d t)+\frac{\zeta\omega_0}{\omega_d}\sin(2\omega_d t)]\} \end{aligned} \tag{B4}$$

$$\begin{aligned} R'_x(t,\tau) = \frac{1}{4\zeta\omega_0^3}e^{-\zeta\omega_0|\tau|}\{[1+e^{-2\zeta\omega_0 t}(1+\frac{\zeta\omega_0}{\omega_d}\sin(2\omega_d t)+\frac{2\zeta^2\omega_0^2}{\omega_d^2}\sin^2(\omega_d t))] \\ \cos(\omega_d|\tau|)+\frac{\zeta\omega_0}{\omega_d}[1+e^{-2\zeta\omega_0 t}(1+\frac{\zeta\omega_0}{\omega_d}\sin(2\omega_d t)- \\ 2\sin^2(\omega_d t))]\sin(\omega_d|\tau|)\} \\ -\frac{1}{8\zeta\omega_0^3}e^{-\zeta\omega_0 t}\{e^{-\zeta\omega_0|t+\tau|}[(-\frac{2\zeta\omega_0}{\omega_d}\sin(\omega_d t)+2\cos(\omega_d t)) \\ (\cos(\omega_d|t+\tau|)+\frac{\zeta\omega_0}{\omega_d}\sin(\omega_d|t+\tau|)) \\ +\frac{2\omega_0^2}{\omega_d^2}\sin(\omega_d t)\sin(\omega_d(t+\tau))]+ \\ e^{-\zeta\omega_0|t-\tau|}[(-\frac{2\zeta\omega_0}{\omega_d}\sin(\omega_d t)+2\cos(\omega_d t))(\cos(\omega_d|t-\tau|)+ \\ \frac{\zeta\omega_0}{\omega_d}\sin(\omega_d|t-\tau|))+\frac{2\omega_0^2}{\omega_d^2}\sin(\omega_d t)\sin(\omega_d(t-\tau))]\} \end{aligned} \tag{B5}$$

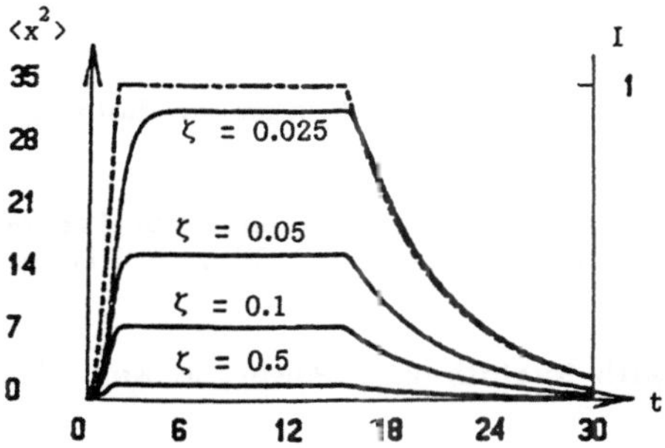

Fig.1

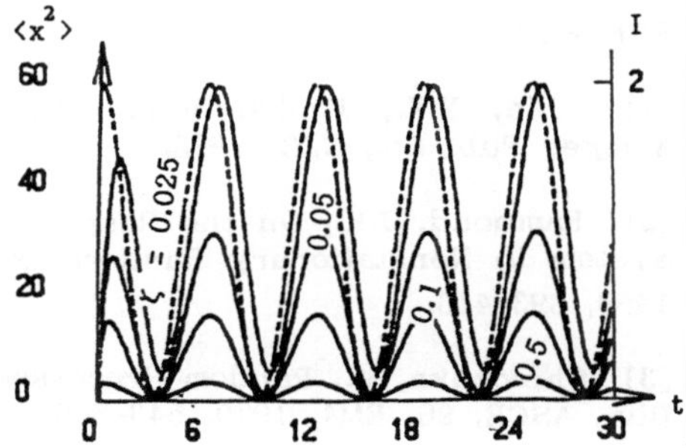

Fig.2

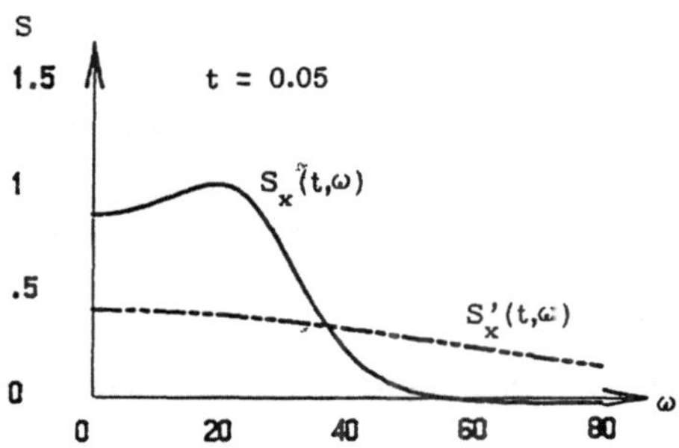

Fig.3

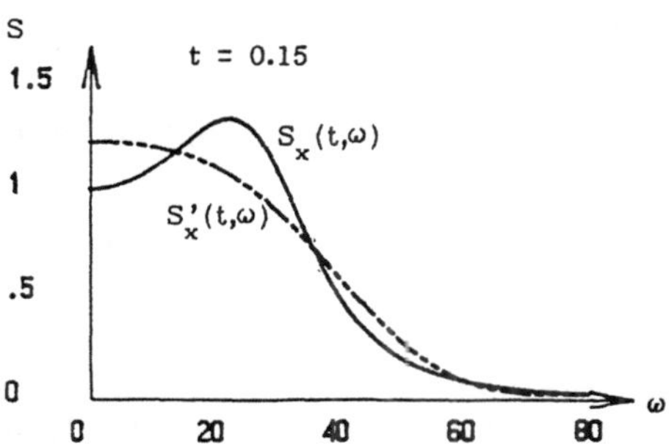

Fig.4

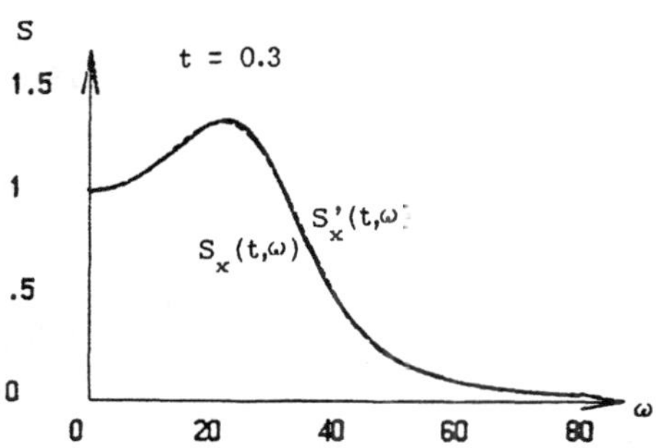

Fig.5

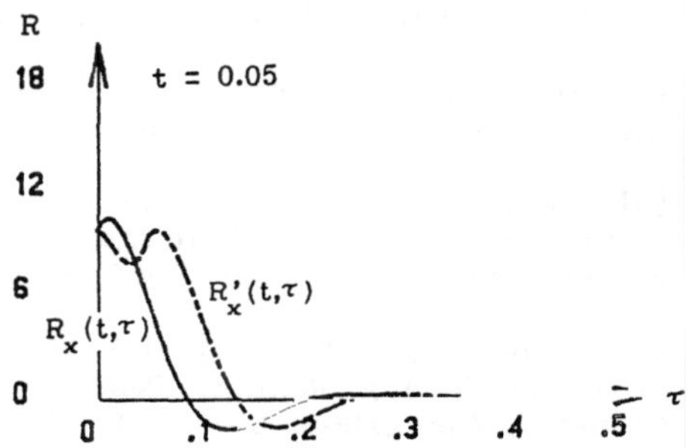

Fig.6

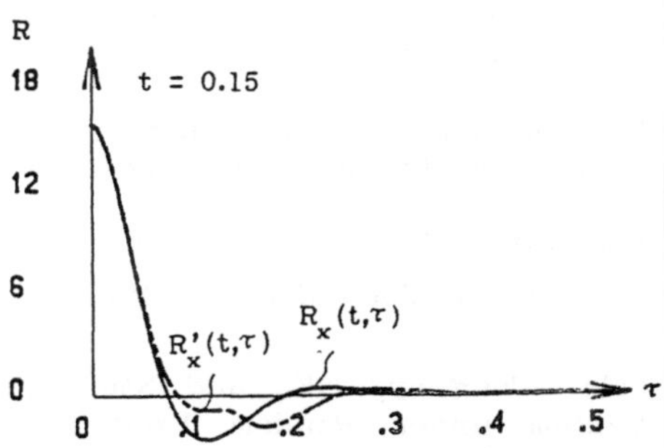

Fig.7

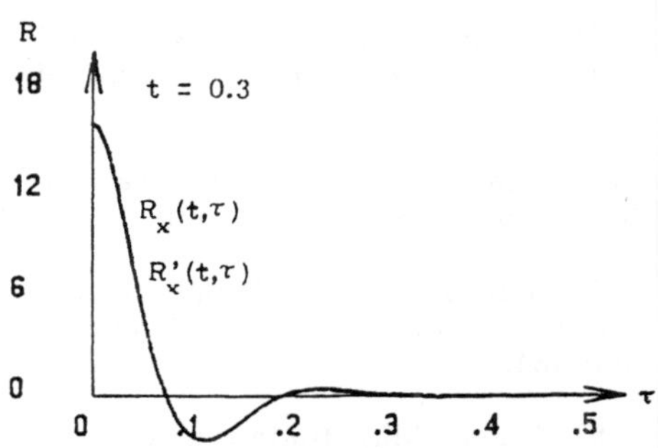

Fig.8

REFERENCES

[1] Lin, Y.K., Probabilistic Theory of Structural Dynamics, Robert E. Krieger Pub. Co., N.Y., 1976.

[2] Hammond, J.K., On the Response of Single and Multi-Degree-of-Freedom System to Nonstationary Random Excitation, J. Sound and Vib., Vol.7, No.3, 1968, 393-416.

[3] Shinozuka, M., Random Processes with Evolutionary Power, J. Eng. Mech. Div., ASCE, 96, EM4, 1970, 543-545.

[4] Caughey, T.K., Nonstationary Random Input and Responses, Chapter 3 of Random Vibration, Vol.2, ed., Crandall, S.H., MIT, Cambridge, 1963.

[5] Fang, T., and Wang, Z.N., New Developments in Time Domain Modal Analysis of Random Vibration, Proc. ASME Vib. Conf., DE-Vol.8, 1987, 37-41.

[6] Fang, T., and Zhang, T.S., Nonstationary Response due to Evolutionary Random Excitations, 17th ICTAM, Grenoble, France, Aug.21-27, 1988. See also J. Vib. Eng. (in Chinese), Vol.2, No.3, 1989, 36-41.

[7] Fang, T., and Zhang, T.S., Time Dependent Correlation Characteristics of Nonstationary Random Response, 4th National Conference on Vib. (in Chinses), Zhengzhou, China, May 5-8, 1990.

[8] Iwan, W.D., and Hou, Z.K., Explicit Solution for the Response of Simple System Subjected to Nonstationary Random Excitation, Structural Safety, Vol.6, 1989, 77-86.

[9] Hou, Z.K., Nonstationary Response of Structures and its Application to Earthquake Engineering, Report No. EERL 90-01, CIT, 1990.

[10] Lin, Y.K, and Schuëller, G.I., (Eds.) Stochastic Structural Mechanics, U.S.-Austria Joint Seminar, May 4-5, 1987, Boca Raton, Florida, U.S.A.

[11] Caughey, T.K., and Stumpf, H.J., Transient Response of a Dynamic System under Random Excitation, J. Appl. Mech., Vol.35, No.11, 1963, 1683-1692.

[12] Amin, M., and Ang, A.H.S., Nonstationary Stochastic Model of Earthquake Motions, J. Eng. Mech. Div., ASCE, Vol.94, No.EM2, 1968, 559-583.

[13] Roberts, J.B., On the Harmonic Analysis of Evolutionary Random Vibration, J. Sound and Vib., Vol.2, No.3, 1965, 336-351.

[14] Fang, T., and Zhang, T.S., On the Partial Spectrum of a Nonstationary Random Response, Procs. ICVPE, Wuhan-Chongqing, China, Jun.17-24, 1990, 460-464.

[15] Page, C.H., Instantaneous Power Spectra, J. Appl. Phy., Vol.23, No.1, 1952, 103-106.

[16] Priestley, M.B., Power Spectral Analysis of Nonstationary Random Processes, J. Sound and Vib., Vol.6, No.1, 1967, 86-97.

A Stochastic Linearization Technique Based on Minimum Mean Square Deviation of Potential Energies

Xiangting Zhang [1], Isaac Elishakoff[2] and Ruichong Zhang[2]

[1] Department of Engineering Mechanics, Tongji University, Shanghai, People's Republic of China.

[2] Center for Applied Stochastics Research and Department of Mechanical Engineering, Florida Atlantic University, Boca Raton, FL 33431-0991, U.S.A.

Summary

A new stochastic linearization technique is described. It is based on the requirement that mean square deviation of the potential energies of the original nonlinear system, and that of the equivalent linear one, be minimal. The accuracy of this version of stochastic linearization is checked by comparing the approximations resulting from the use of stochastic linearization method with some exact solutions available in the literature.

Introduction

Stochastic linearization techniques have been used widely in stochastic structural dynamics [1]. For a detailed review and the pertinent bibliography one may consult the papers by Sinitsin [2], Crandall [3] Spanos [4] as well as a recent monograph [5] by Roberts and Spanos. The central idea of the method is to replace the nonlinear term in the stochastic differential equation, by a linear term, which is equivalent to the original nonlinear one in some sense. For example, for the one-degree-of-freedom system

$$m\ddot{X} + c\dot{X} + g(X) = F(t) \tag{1}$$

the nonlinear term g(X) is replaced by a linear term $k_{eq}X$, so that the resulting equation is linear:

$$m\ddot{X} + c\dot{X} + k_{eq}^{(1)}X = F(t) \tag{2}$$

The error of linearization, a random process, reads

Y. K. Lin · I. Elishakoff (Eds.)
Stochastic Structural Dynamics 1
New Theoretical Developments
2nd International Conference on Stochastic Structural Dynamics
May 9-11, 1990, Boca Raton, Florida

$$\epsilon(t) = g(X) - k_{eq}^{(1)}X \tag{3}$$

which is the difference between the original nonlinear restoring force and its linearized counterpart. The usually employed method of attack is to make $E(\epsilon^2)$ as small as possible. This is accomplished by requiring that

$$\frac{\partial}{\partial k_{eq}^{(1)}} E\,(\epsilon^2) = 0 \tag{4}$$

and the result is

$$k_{eq}^{(1)} = \frac{E[Xg(X)]}{E(X^2)} \tag{5}$$

In this study another linearization technique is proposed; namely, the mean-square value of the difference of the potential energies, associated with the original nonlinear equation, and its equivalent linear counterpart, is to be minimized.

<u>Basic Equations</u>

In the proposed linearization scheme, we require that

$$E\,\{[U(X)-\frac{1}{2}k_{eq}^{(2)}X^2]^2\} = \min \tag{6}$$

This is accomplished by requiring

$$\frac{d}{dk_{eq}} E\,\{[\,U(X)\,-\frac{1}{2}k_{eq}^{(2)}X^2]^2\} = 0 \tag{7}$$

This results in the following expression for the equivalent spring stiffness

$$k_{eq}^{(2)} = \frac{2E\,[X^2U(X)]}{E(X^4)} \tag{8}$$

We will elucidate the accuracy of this linearization technique by comparing the computed mean square displacements from the use of statistical linearization techniques with some known exact solutions.

<u>Duffing's Oscillator</u>

The differential equation reads

$$m\ddot{X} + c\dot{X} + k_1X + k_2X^3 = F(t) \tag{9}$$

where m is the mass, c is the damping coefficient, and k_1 and k_2 are spring constants. The potential energies of the original and the substituting systems read

$$U(X) = \frac{1}{2}k_1X^2 + \frac{1}{4}k_2X^4 \tag{10}$$

and

$$U_{eq}(X) = \frac{1}{2}k_{eq}^{(2)}X^2 \tag{11}$$

where

$$k_{eq}^{(2)} = \frac{2E\,[X^2U(X)]}{E(X^4)} = k_1 + \frac{1}{2}k_2\frac{E(X^6)}{E(X^4)} \tag{12}$$

To calculated $E[X^4]$ and $E[X^6]$ we use the usual approximation that X has a normal probability density with zero mean:

$$p(x) = \frac{1}{\sqrt{2\pi}\,\sigma_x}\exp\left[-\frac{x^2}{2\sigma_x^2}\right] \tag{13}$$

Therefore

$$\begin{aligned} E(X^6) &= 15\sigma_x^6 \\ E(X^4) &= 3\sigma_x^4 \end{aligned} \tag{14}$$

Substitution of these into Eq (12), yields

$$k_{eq}^{(2)} = k_1 + 2.5k_2\sigma_x^2 \tag{15}$$

The spectral analysis results in the following mean square displacement

$$E(X^2) = \int_{-\infty}^{\infty} \frac{\Phi_F(\omega)d\omega}{(k_{eq} - m\omega^2)^2 + c^2\omega^2} \tag{16}$$

For an ideal white-noise excitation F(t), with mean zero and a constant spectral density s, we obtain

$$E(X^2) = \frac{\pi S}{ck_{eq}^{(2)}} = \frac{\pi S}{c(k_1 + 2.5k_2\sigma_x^2)} \tag{17}$$

This equation can be rewritten as

$$\sigma_x^2 = \frac{\sigma_{x_o}^2}{1 + 2.5(k_2/k_1)\sigma_x^2} \tag{18}$$

where $\sigma_{x_0}^2 = \pi s/ck_1$ is the mean square response for the linear system with zero k_2. For $k_2 \neq 0$, we obtain

$$\sigma_x^2 = \frac{\sqrt{1+10\sigma_{x_o}^2\, k_2/k_1} - 1}{5(k_2/k_1)} \tag{19}$$

If the following strong inequality holds

$$10\sigma_{x_o}^2 \frac{k_2}{k_1} \ll 1 \tag{20}$$

σ_x^2 can be approximated as

$$\sigma_x^2 = \sigma_{x_o}^2 (1 - 2.5\frac{k_2}{k_1}\sigma_{x_o}^2) \tag{21}$$

Now, the conventional stochastic linearization given by Eq. (5) yields

$$\begin{aligned} k_{eq}^{(1)} &= \frac{E[Xf(X)]}{E(x^2)} = \frac{E[k_1X^2 + k_2X^4]}{E(X^2)} \\ &= k_1 + k_2\,\frac{E(X^4)}{E(X^2)} \\ &= k_1 + 3k_2\sigma_x^2 \end{aligned} \tag{22}$$

Instead of Eq. (18) we obtain the following quadratic equation:

$$\sigma_x^2 = \frac{\sigma_{x_o}^2}{1+3(k_2/k_1)\sigma_x^2} \tag{23}$$

leading to

$$\sigma_x^2 = \frac{\sqrt{1+12\sigma_{x_o}{}^2\, k_2/k_1} \;\; -1}{6(k_2/k_1)} \tag{24}$$

Approximate expression analogous to Eq. (22) reads

$$\sigma_x^2 = \sigma_{x_o}{}^2 \left[1 - 3\frac{k_2}{k_1}\,\sigma_{x_o}{}^2\right] \tag{25}$$

In order to compare the results of these two stochastic linearization schemes with the exact solution, we recall that the exact expression for the probability density of X is [1,6,8]

$$p_X(x)=Cexp[-\frac{1}{\sigma_{x_o}^2}(\frac{1}{2}x^2+\frac{1}{4}\delta x^4)] \tag{26}$$

where

$$\delta=\frac{k_2}{k_1} \tag{27}$$

The constant C is found from the normalization condition

$$\int_{-\infty}^{\infty} p_X(x)dx = 1 \tag{28}$$

and equals

$$C = \left\{\frac{1}{\sqrt{2\delta}}\exp[\xi K_{1/4}(\xi)]\right\}^{-1} \tag{29}$$

with $K_{1/4}(\xi)$ representing the cylindrical function, and

$$\xi=\frac{1}{8\sigma_{x_o}{}^2\delta} \tag{30}$$

The exact mean-square displacement

$$\sigma_x^2 = \int_{-\infty}^{\infty} x^2 p_X(x)dx \tag{31}$$

can be found analytically or numerically. Fig. 1 portrays the mean-square displacement evaluated exactly, along with the results furnished by the two stochastic linearization techniques. Earlier comparison between the exact solution and the conventional stochastic linearization was performed by Lyon [9]. Fig. 1 displays, in addition, results of the new version of stochastic linearization. As is seen from the figure, for $\delta=1$, the proposed stochastic linearization yields results which are closer to the exact solution, than those calculated on the basis of conventional linearization method. Fig. 2 depicts relative errors

$$\eta_j = \frac{|E(X^2)_j - E(X^2)_{exact}|}{E(X^2)_{exact}} 100\% \tag{32}$$

associated with the conventional (j=1) and proposed (j=2) stochastic linearization versions. As is seen, the error produced by the new scheme is about one-fourth of the error associated with the conventional linearization. Remarkably, for $\sigma_{x_0}^2=0.7$ and $\delta=1$ the relative error η_2 vanishes, thus the proposed stochastic linearization technique yields the same results as the exact solutions in this particular case.

<u>Example 2</u>

Consider another oscillator, governed by the equation:

$$m\ddot{X} + c\dot{X} + k_1 X + k_2 X^5 = F(t) \tag{33}$$

where again F(t) is an ideal white-noise with zero mean. The exact solution reads [1,8,10]

$$p_X(x) = C_2 \exp[-\frac{1}{\sigma_{x_o}^2}(\frac{1}{2}x^2 + \frac{1}{6}\delta x^6)] \tag{34}$$

where constant C_2 is determined through the normalization requirement.

The conventional stochastic linearization yields

$$\sigma_x^2 = \sigma_{x_o}^2 \frac{E(X^2)}{E(X^2+\delta X^6)} = \frac{\sigma_{x_o}^2}{1+15\delta\sigma_x^4} \tag{35}$$

leading to a cubic equation for the mean-square displacement

$$15\delta(\sigma_x^2)^3 + \sigma_x^2 - \sigma_{x_o}^{\ 2} = 0 \tag{36}$$

Using the proposed linearization technique we obtain

$$\sigma_x^2 = \sigma_{x_o}^{\ 2} \frac{E(X^4)}{2E[(\frac{1}{2})X^2+(\frac{1}{6})\epsilon X^8\ ]} \tag{37}$$

which results in the following algebraic equation

$$\frac{35}{3}\delta(\sigma_x^2)^3 + \sigma_x^2 - \sigma_{x_o}^{\ 2} = 0 \tag{38}$$

Fig. 3 contrasts the two linearization techniques with the exact solution. Curve 1 is associated with the exact solution, curve 2-with the proposed technique, whereas curve 3 depicts the conventional linearization method. Here too, as in the case of Duffing's oscillator, the proposed method turns out to be superior than the conventional one. The relative errors are shown in Fig. 4. The present method shows error of about one half of the conventional version of the linearization.

In the above two cases the oscillators exhibited a hardening-type behavior. We will contrast these two methods for the case of a softening-type behavior.

Example 3

Consider the following one-degree-of-freedom system

$$m\ddot{X} + c\dot{X} + k_1X + k_2\tanh(\alpha X) = F(t) \tag{39}$$

In Ref. 10 it is shown that the stationary probability density of X(t) is

$$p_X(x) = C_3[\cosh(\alpha x)]^{\gamma} \tag{40}$$

where

$$\gamma=\frac{1}{\sigma_{x_o}^{\ 2}\alpha^2} \tag{41}$$

From normalization condition we arrive at C_3

$$C_3 = \{\sqrt{\gamma}\sigma_{x_o}B(\frac{1}{2},\frac{\gamma}{2})\}^{-1} \tag{42}$$

where B(x,y) is an Euler's integral of the first type.
The exact solution for σ_x^2 as well as its approximations through two stochastic linearization techniques are portrayed in Fig. 5. Here the proposed method yields an error which is about twice as much as that of the conventional linearization procedure (Fig. 6). It should be noted that in Ref. 6 it was suggested to use some weighing function for softening oscillators; namely, instead of Eq. 8 the suggestion was made to use the following formula:

$$k_{eq}^{(2)} \fallingdotseq \frac{2E[w(X)X^2U(X)]}{E[w(X)X^4]} \tag{43}$$

with

$$w(X) = 1+U^2(X) \tag{44}$$

In examples considered in Ref. 6 introduction of the weighing function turned out to lead to reducing the errors.

<u>Conclusion</u>

The paper contrasts the new stochastic linearization technique with the conventional one. To gain some insight, three examples are considered: namely, (1) the Duffing oscillator; (2) an oscillator with the nonlinear part of the restoring force proportional to the fifth power of the displacement and, finally; (3) an oscillator, whose restoring force is a hyperbolic tangence function of the displacement. Whereas the first two oscillators are of the hardening type, the third exhibits a softening behavior. The proposed method turns out to be superior than conventional stochastic linearization, for hardening oscillators described in this Note. However, for the softening oscillator, the conventional linearization technique yields more accurate results. More study appears to needed to be better define the range of applicability of the proposed method.

References

1. Lin, Y. K. Probabilistic Theory of Structural Dynamics, McGraw Hill, New York 1967 (second edition, Robert Krieger Company, Malabar, FL, 1976).
2. Sinitsin, I.N.: Methods of Statistical Linearization (Survey), Automation and Remote Control 35 (1974) 765-776.
3. Crandall, S.H.: On Statistical Linearization for Nonlinear Oscillators, in "Problems of the Asymptotic Theory of Nonlinear Oscillations", Academy of Sciences of the Ukrainian SSR, Naukova Dumka, 1977, 115-122. Reprinted in "Nonlinear System Analysis and Synthesis", Vol. 2, Techniques and Applications, eds. R.V. Ramnath, J.K. Hedrick, H.M. Paynter, ASME, New York, 1980, 199-209.
4. Spanos, P. D. Stochastic Linearization in Structural Dynamics, Applied Mechanics Reviews, (1981) 1-8.
5. Roberts, J. B.; Spanos, P. D.: Random Vibration and Statistical Linearization, Wiley, Chichester: 1990.
6. Wang, C.; Zhang, X. T.: An Improved Equivalent Linearization Technique in Nonlinear Random Vibration, Proceedings of the International Conference on Nonlinear Mechanics. (1985) 959-964.
7. Zhang, X. T.: Equivalent Potential Technique for Deterministic and Random Response Analysis of Nonlinear Systems, Applied Mechanics, IAP, 1989 (in Chinese).
8. Bolotin, V. V.: Random Vibrations of Elastic Systems, The Hague: Martinus Nijhoff Publishers 1984.
9. Lyon, R. H.: On the Vibration Statistics of Randomly Excited Hard-spring Oscillator, J. Acoustical Society of America, 32, (1960) 716-719.
10. Piszczek K; Nizoil, J.: Random Vibration of Mechanical Systems, New York: Ellis Horwood Limited, 1986.
11. Roberts, J. B.: Response of Nonlinear Mechanical Systems to Random Excitation - Part 2: Equivalent Linearization and Other Methods, Shock and Vibration Digest, 13(15), (1972) 15-19.

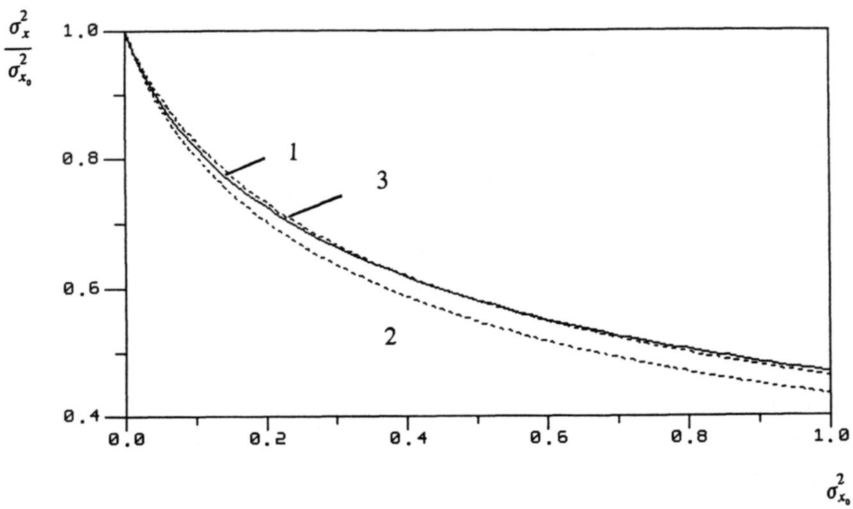

Fig. 1. Comparison of mean square displacements in the Duffing's oscillator via stochastic linearization with exact solution (curve 1 -- exact, curve 2 -- conventional linearization, curve 3 -- proposed linearization)

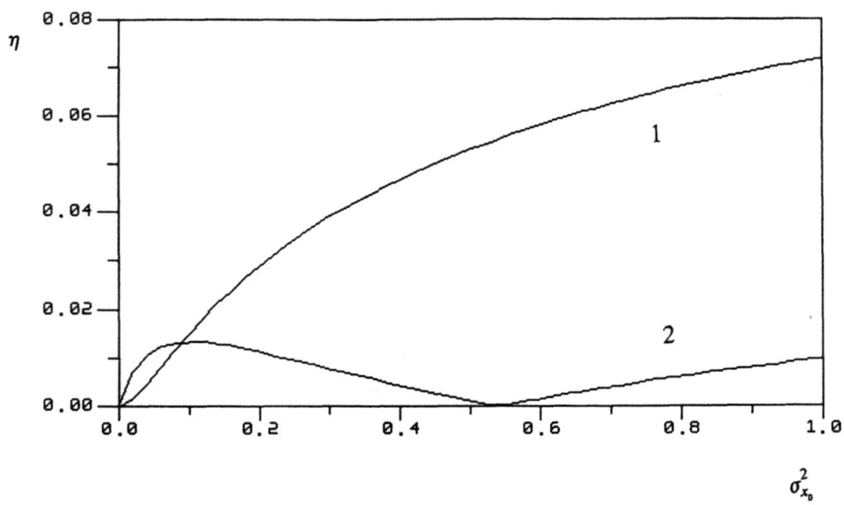

Fig. 2. Percentagewise errors in the Duffing's oscillator associated with the linearization (curve 1 -- conventional linearization, curve 2 -- proposed linearization)

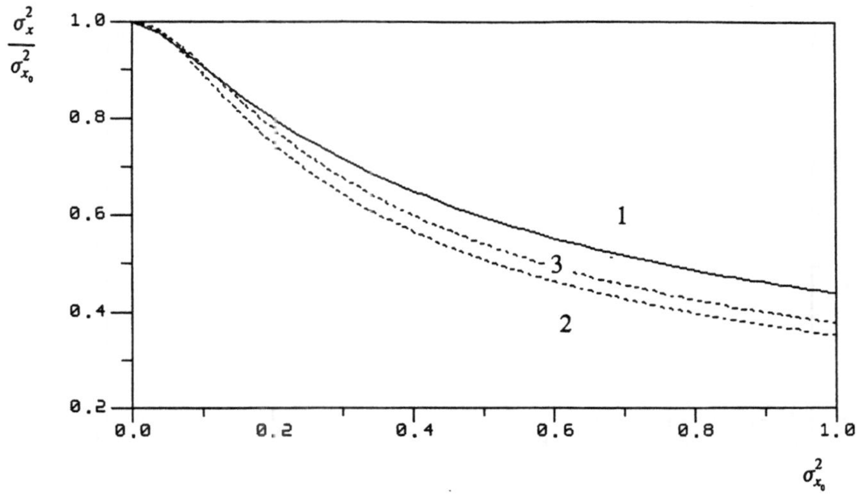

Fig. 3. Comparison of mean square displacements in the quintic oscillator via stochastic linearization with exact solution (curve 1 -- exact, curve 2 -- conventional linearization, curve 3 -- proposed linearization)

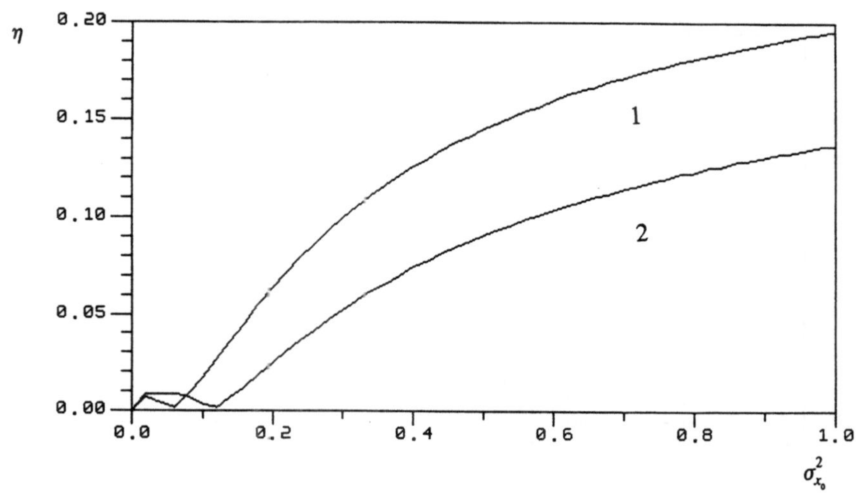

Fig. 4. Percentagewise errors in quintic oscillator associated with the linearization (curve 1 -- conventional linearization, curve 2 -- proposed linearization)

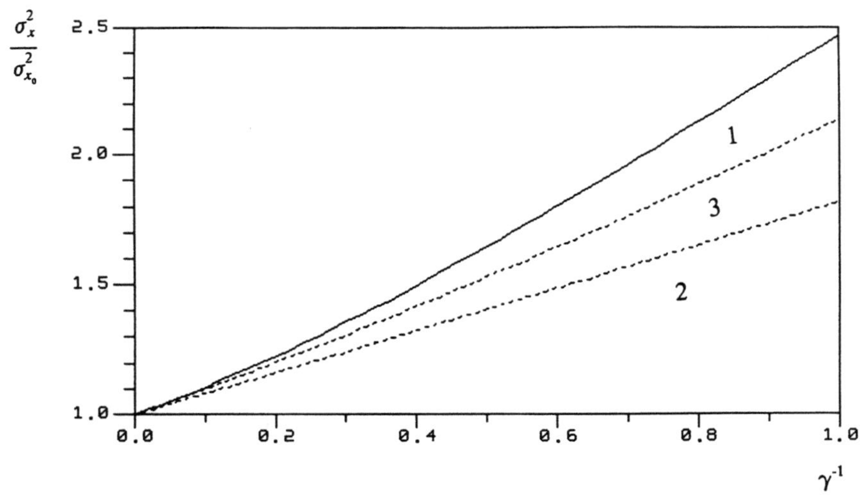

Fig. 5. Comparison of mean square displacements in an oscillator with restoring force in the form of hyperbolic tangence function via stochastic linearization with exact solution (curve 1 -- exact, curve 2 -- conventional linearization, curve 3 -- proposed linearization)

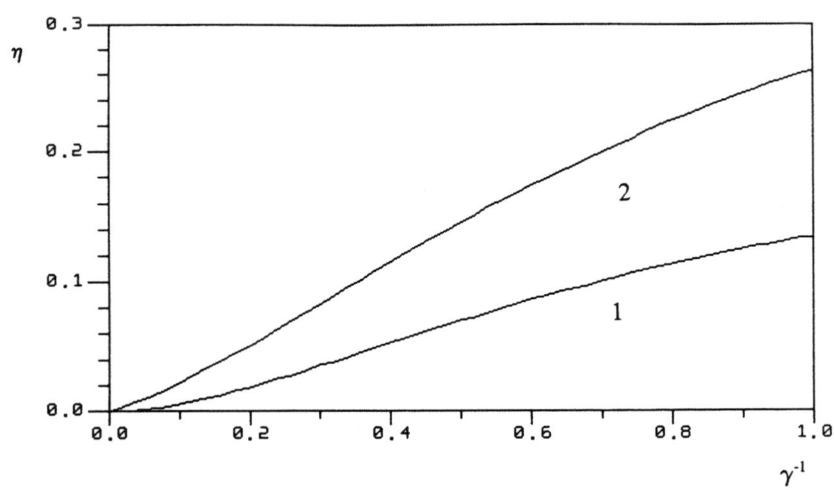

Fig. 6. Percentagewise errors in an oscillator with restoring force in the form of hyperbolic tangence function associated with the linearization (curve 1 -- conventional linearization, curve 2 -- proposed linearization)

A Stochastic Theory of Cumulative Fatigue Damage

W. Q. ZHU[1] and Y. LEI[2]

[1]Department of Mechanics, Zhejiang University, Hangzhou, China

[2]Department of Architectural Engineering, Shantou University, Shantou, Guangdong, China

Summary

A stochastic theory for the cumulative fatigue damage of structural component with random fatigue strength under random loading is proposed on the basis of Stratonovich - Khasminskii theorem. The analytical solutions for the probability densities of the cumulative fatigue damage and fatigue life and for the reliability function are given for steel and reinforced concrete components with constant fatigue strength subject to narrow band stationary Gaussian stress process with zero mean. The results agree very well with those of digital simulation. It is noted that the theory can be applied, in principle, to metallic and some non-metallic materials, to both narrow band and wide band stress process, and can be adapted to a sequence of n stationary stress processes or quasi-stationary stress process. The scatter and degradation of fatigue strength can also be incorporated into the theory.

Introduction

Failure of a large number of mechanical/structural components is a result of accumulation of fatigue damage. Because of the inherent variability in fatigue strength of materials as well as statistical nature of the service loads experienced by the structures, the stochastic approach is important to the analysis and design of fatigue - critical components.

The cumulative damage stems from material behavior at the atomic or molecular level. To date, it appears that material behavior at this level is not sufficiently well understood. Therefore, current models of cumulative damage must be phenomenological in nature and based upon experimental data. Most available fatigue data to date are obtained from constant amplitude tests. The problem is how to predict the time to

Y. K. Lin · I. Elishakoff (Eds.)
Stochastic Structural Dynamics 1
New Theoretical Developments
2nd International Conference on Stochastic Structural Dynamics
May 9-11, 1990, Boca Raton, Florida

reach a prescribed level of fatigue damage for structural components under random loading. When the stress is modelled by a random process, the central limit theorem is usually invoked to obtain the probability density for the cumulatative fatigue damage [1-4]. Obviously, this is correct only asymptotically.

In the present paper a stochastic theory for cumulative fatigue damage of structural components with random fatigue strength under random loading is proposed on the basis of Stratonovich - Khasminskii limit theorem [5,6]. With the assumption that the damage accumulation is a sufficiently slow process as compared with the stress process (this is the case for high - cycle fatigue), the cumulative fatigue damage of a component with constant fatigue strength is approximately a diffusion process. The conditional probability densities for the cumulative fatigue damage and fatigue life and the conditional reliability function can be obtained by using Stratonovich's stochastic averaging method. The unconditional probability densities for the cumulative fatigue damage and fatigue life and the unconditional reliability function can then be obtained on the basis of the total probability theorem [7]. For steel and reinforced concrete components subject to narrow band stationary Gaussian stress process with zero mean, the analytical solutions for the conditional probability densities of the cumulative fatigue damage and fatigue life and the conditional reliability function are obtained. Comparisons of the theoretical results and those of digital simulation show that the agreement is excellent. It is noted that the theory is applicable in many practical cases.

2. Stochastic Theory of Cumulative Fatigue Damage

In a phenomenological model of damage accumulation, the measure of fatigue damage is usually introduced axiomatically. For a structural component it is prescribed by means of a non-negative, non-decreasing time function $D(t)$. The time rate of increment in $D(t)$ depends upon the levels of the damage and the stress at given time, i.e.,

$$\frac{dD}{dt} = f(D, X, r) \tag{1}$$

where D=D(t) is cumulative fatigue damage; X=X(t) a function of time characterizing stress; r a parameter characterizing fatigue strength. The form of function f is determined on the basis of tests or on a model characterized by a S - N relationship. Generally, X=(t) is a random process while r is a random variable with a distribution p(r) independent of X(t). So the cumulative fatigue damage is also a random process. In addition, if X(t) is stationary, then so is the cumulative fatigue damage.

Assume that fatigue damage accumulation is a sufficiently slow process as compared with the stress process X(t), that is, the correlation time of X(t) is small as compared to that of D(t). It is concluded from Stratonovich - Khasminskii limit theorem [5,6] that the cumulative fatigue damage D(t) governed by equation (1) with constant r is approximately a one - dimensional homogeneous diffusion process, which is described by the following Fokker-Planck-Kolmogrov (FPK) equation

$$\frac{\partial p}{\partial t} = - m(D,r) \frac{\partial p}{\partial D} + \frac{1}{2} \frac{\partial^2}{\partial D^2}[\sigma^2(D,r)p] \tag{2}$$

where $p = p(D,t \mid D_0;r)$ is the conditional probability density of the cumulative fatigue damage; m(D,r) and $\sigma^2(D,r)$ are the drift and diffusion coefficients, respectively. They are determined by using Stratonovich stochastic averaging method [5] as follows:

$$m(D,r) = E[f] + \int_{-\infty}^{0} K[\frac{\partial f}{\partial D} \mid t, f_{t+\tau}]d\tau \tag{3}$$

$$\sigma^2(D,r) = \int_{-\infty}^{\infty} K[f_t, f_{t+\tau}]d\tau \tag{4}$$

where $f_t = f[D,X(t),r]$, $f_{t+\tau} = f[D,X(t+\tau),r]$, and $K[f_1,f_2]$ is the cross-covariance of f_1 and f_2. The conditional probability density and moments of the cumulative fatigue damage are obtained by solving FPK equation (2).

If the initial and critical values of the fatigue damage are D_0 and D_{cr}, respectively, the conditional reliability function is determined by the integration of p, i. e.,

$$R(t \mid D_0;r) = \int_{D_0}^{D_{cr}} p(u, t \mid D_0;r)du \tag{5}$$

The conditional probability density of fatigue life is then obtained as the derivative of R with respect to time, i.e.,

$$p(T \mid D_0;r) = -\frac{\partial R}{\partial t}\Big|_{t = T} \tag{6}$$

The conditional mean and variance of the fatigue life are obtained from $p(T \mid D_0;r)$ as follows:

$$m_T(D_0;r) = \int_0^\infty Tp(T \mid D_0;r)dT \tag{7}$$

$$\sigma_T^2(D_0;r) = \int_0^\infty [T - m_T]^2 p(T \mid D_0;r)dT \tag{8}$$

Finally, the unconditional statistics of the cumulative fatigue damage of the fatigue life and the unconditional reliability function are determined on the basis of the total probability theorem [7].

For a structure having n fatigue - critical locations or a structure composed of n components, a cumulative fatigue damage vector $\mathbf{D}(t) = [D_1(t), D_2(t), \ldots, D_n(t)]^T$ can be introduced to describe the fatigue damage of the structure. In this case, equation (1) should be modified as follows:

$$\frac{d\mathbf{D}}{dt} = \mathbf{f}(\mathbf{D},\mathbf{X},\mathbf{r}) \tag{9}$$

where $\mathbf{X} = \mathbf{X}(t)$ is a stress vector process and $\mathbf{r}$ a fatigue strength vector. $\mathbf{D}(t)$ is approximately an n-dimensional homogeneous diffusion process provided that fatigue damage accumulation is slow compared with the stress processes. Then the conditional statistics of the cumulative fatigue damage and of the fatigue life of the structure can be obtained by using the stochastic averaging method.

If a component or a structure is subjected to a sequence of n successive stationary stress processes, then an equation of the form (1) or (9) can be set up for each of n stationary stress processes. The transition probability density of the cumulative fatigue damage between any two instants of time can be obtained by integration of the product of $k(\leq n)$ successive transition probabilities. In the case of quasi-stationary stress process, the statistics of cumulative fatigue damage and fatigue life can be obtained by using the method described in Reference 7.

To improve structural safety and reliability, periodic inspection maintenance is often introduced during the service of a structure. Once the structure passes an inspection, the statistical distribution of fatigue damage may be modified significantly. This modification of fatigue damage can be easily incorporated as an initial condition in the theory.

So far, the analytical solution to equation (1) can be obtained only for steel and reinforced concrete components subject to narrow band stationary Gaussian process. However, the theory can be adapted to wide band stress processes by introducing a modification factor which depends on a bandwidth parameter obtained using the rain flow counting method [8].

3. Conditional Statistics of Fatigue Damage and Fatigue Life

Most fatigue data are obtained from constant amplitude tests. Such tests are repeated at different stress levels to establish the familiar S - N curve relating the stress amplitude to the number of cycles to failure. For high - cycle fatigue of steel, the S - N relationship is characterized by the classical model

$$N_S = \begin{cases} N_{S_e}(S_e/S)^{\beta}, & s \geq s_e \\ \infty, & s < s_e \end{cases} \tag{10}$$

if there exists an endurance limit, or

$$N_S = \left(\frac{S_u}{S} \right)^{\beta} \tag{11}$$

for the case without an identifiable endurance limit. In equations (10) and (11), N_S is the number of cycles to failure at a stress amplitude s; s_e the endurance limit; N_{S_e} the number of cycles to failure at endurance limit; S_u ultimate strength; β the slope parameter for the S - N curve.

Fatigue data are usually widely scattered especially for high - cycle fatigue. Equations (10) and (11) describe the medium curves of the scattered data. The spread of the data can be represented by lettering S_u or S_e be a random variable characterized by a probability distribution such as a two - parameter Weibull or lognormal distribution.

Now suppose that a steel component is subjected to a random stress process. The time rate of increase in the fatigue damage is [2]

$$\frac{dD}{dt} = \frac{1}{N_S} E[M_T(t)] \tag{12}$$

where $E[M_T(t)]$ is the expected total number of stress peaks per unit time. For a stationary stress process, $E[M_T(t)]$ is a constant. Note that no restriction is imposed here concerning the bandwidth of the stress process.

To obtain an analytical solution to equation (12), assume that the stress is a narrow-band stationary Gaussian random process with zero mean and the S - N relationship is represented by equation (10). In this case, $E[M_T(t)] = f_e$, where f_e is the expected frequency of the stress process, and equation (12) becomes

$$\frac{dD}{dt} = \frac{f_e}{N_{S_e}} \left(\frac{S}{S_e}\right)^{\beta} \tag{13}$$

For high - cycle fatigue, the quantity on the right hand side of equation (13) is small. Thus, D(t) is a slowly varying process and the stochastic averaging method can be applied to equation (13). D(t) is then approximately a one - dimensional homogeneous diffusion process with the transition probability density $p = p(D,t \mid D_0;S_e)$ governed by the following FPK equation

$$\frac{\partial p}{\partial t} = - m(s_e) \frac{\partial p}{\partial D} + \frac{1}{2} \sigma^2(s_e) \frac{\partial^2 p}{\partial D^2} \tag{14}$$

where

$$m(s_e) = \frac{f_e}{N_{s_e} s_e^{\beta}} E[s^{\beta}] \tag{15}$$

$$\sigma^2(s_e) = \left(\frac{f_e}{N_{s_e} S_e^{\beta}}\right)^2 \int_{-\infty}^{\infty} C_{s^{\beta}}(\tau) d\tau \tag{16}$$

and $C_{s^{\beta}}(\tau)$ is the covariance function of $s^{\beta}(t)$.

For a narrow-band stationary Gaussian stress process X(t) with a zero mean, the probability density of the peaks is

$$p(s) = \frac{s}{\sigma_X^2} \exp\left(- \frac{s^2}{2\sigma_X^2} \right) \tag{17}$$

where σ^2_X is the variance of the stress process X(t). Substituting equation (17) into equation (15), one obtains

$$m(s_e) = \frac{2^{\beta/2}\Gamma(1+\beta/2, s_e/2\sigma_X^2)}{T_0} \left(\frac{\sigma_X}{s_e}\right)^{\beta} \tag{18}$$

where $T_0 = N_{s_e}/f_e$ and $\Gamma(m,n)$ is an incomplete gamma function.

To obtain $C_{s^{\beta}}(\tau)$ the second - order probability density of the stress peak is required . Unfortunately, it remains to be found. In the case of a narrow band stationary Gaussian stress process, however, it is reasonable to approximate this second order probability density by that of the envelope of the stress process, which is

$$p(s_1,s_2;\tau) = \frac{S_1S_2}{\sigma_X^4[1-\rho^2(\tau)]} \exp\{-\frac{S_1^2+S_2^2}{2\sigma_X^2[1-\rho^2(\tau)]}\} I_0\{\frac{S_1S_2\rho(\tau)}{\sigma_X^2[1-\rho^2(\tau)]}\} \tag{19}$$

where $S_1=S(t_1)$, $S_2=S(t_2)$, $t_2-t_1 = \tau$; $\rho^2(\tau) = (4/\sigma_X^4)\{[\int_0^\infty S_X(\omega)\cos(\omega-\omega_0)\tau d\omega]^2 + [\int_0^\infty S_X(\omega)\sin(\omega-\omega_0)\tau d\omega]^2\}$; I_0 is the Bessel function of the first kind with imaginary arguments. Using the relationship [9]

$$\frac{(uvw)^{-\alpha/2}}{(1-w)} \exp(-w\frac{(u+V)}{(1-w)}) I_\alpha(\frac{2\sqrt{uvw}}{(1-w)}) = \sum_{n=0}^{\infty} n! \frac{L_n^{\alpha}(u)L_n^{\alpha}(v)w^n}{\Gamma(n+\alpha+1)} \tag{20}$$

$p(S_1,S_2;\tau)$ can be expanded in the series of Laguerre polynomials $L_n^{(0)}$ of power n and zero order

$$p(S_1,S_2;\tau) = \frac{S_1S_2}{\sigma_X^4} \exp(-\frac{S_1^2+S_2^2}{2\sigma_X^2}) \sum_{n=0}^{\infty} \rho^{2n}(\tau)L_n^{(0)}(\frac{S_1}{2\sigma_X^2})L_n^{(0)}(\frac{S_2}{2\sigma_X^2}) \tag{21}$$

Using equation (21), one finds that

$$C_{S\beta}(\tau) = \sigma_X^{2\beta} \sum_{n=1}^{\infty} C_n^2 \rho^{2n}(\tau) \tag{22}$$

where

$$C_n = 2^{\beta/2} \sum_{k=0}^{n} \frac{(-1)^k n!}{(k!)^2(n-k)!} \Gamma(1+k+\beta/2, S_e^2/2\sigma_X^2) \tag{23}$$

Substituting equation (22) into equation (16), one has

$$\sigma^2(S_e) = \frac{1}{T_0^2} \left(\frac{\sigma_X}{S_e} \right)^{2\beta} \sum_{n=1}^{\infty} c_n^2 \tau_n \tag{24}$$

where

$$\tau_n = 2 \int_0^{\infty} \rho^{2n}(\tau) d\tau \tag{25}$$

The solution of FPK equation (14) subject to the initial condition

$$p(D,t \mid D_0; S_e) = \delta(D - D_0), \; t = 0 \tag{26}$$

is

$$p(D, t| D_0;S_e) = \frac{1}{\sqrt{2\pi t}\, \sigma(S_e)} \exp\{- \frac{[D-D_0-m(S_e)t]^2}{2\sigma^2(S_e)t}\}, \; -\infty < D < \infty \tag{27}$$

It should be noted that equation (27) admits those D values which are smaller than D_0 and physically unreasonable. The error which arises from not restricting $D > D_0$ is great when t is small, and decreases as t increases. For high - cycle fatigue, the error should be negligible. Furthermore, this error can be compensated by renormalizing the total probability at any instant of time, i.e., by changing equation (27) to [10]

$$p(D,t| D_0;S_e) = \frac{1}{\sqrt{2\pi t}\, \sigma(S_e)\, \Phi[\frac{m(S_e)\sqrt{t}}{\sigma(S_e)}]} \exp\{- \frac{[D-D_0- m(S_e)t]^2}{2\sigma^2(S_e)t}\},$$

$$D > D_0 \tag{28}$$

where Φ is the standard Gaussian probability distribution function, which approaches 1 when $m(S_e)\sqrt{t} >> \sigma(S_e)$. Although equation (28) is not the solution of FPK equation (14), it agrees very well with results of digital simulation as shown below. Therefore, equation (28) will be used as a conditional probability density of the cumulative fatigue damage.

The conditional mean and variance of the cumulative fatigue damage are obtained from equation (28) as follows:

$$m_D(D_0;S_e) = D_0 + m(S_e)t + \frac{\sigma(S_e)\sqrt{t}}{\sqrt{2\pi}\,\Phi[\frac{m(S_e)\sqrt{t}}{\sigma(S_e)}]} \exp\{-\frac{[D_0+m(S_e)t]^2}{2\sigma^2(S_e)t}\} \tag{29}$$

$$\sigma_D^2(D_0;S_e) = \sigma^2(S_e)t + \frac{[D_0+m(S_e)t]\sigma(S_e)\sqrt{t}}{\sqrt{2\pi}\,\Phi[\frac{m(S_e)\sqrt{t}}{\sigma(S_e)}]} \exp\{-\frac{[D_0+m(S_e)t]^2}{2\sigma^2(S_e)t}\} \tag{30}$$

Assume that failure occurs when the cumulative damage reaches critical value D_{cr}. Then the conditional reliability function is

$$R(t|\,D_0;S_e) = 1 - \Phi[\frac{m(S_e)\,t + D_0 - D_{cr}}{\sigma(S_e)\sqrt{t}}]/\Phi[\frac{m(S_e)\sqrt{t}}{\sigma(S_e)}] \tag{31}$$

The conditional probability density of the fatigue life is then

$$p(T\,|\,D_0;S_e) = \frac{D_{cr} - D_0 + m(S_e)T}{2T\sqrt{2\pi T}\,\sigma(S_e)\,\Phi[\frac{m(S_e)\sqrt{T}}{\sigma(S_e)}]}$$

$$\exp\{-\frac{[m(S_e)T + D_0 - D_{cr}]^2}{2\sigma^2(S_e)T}\}$$

$$-\frac{m(S_e)\Phi[\frac{m(S_e)T + D_0 - D_{cr}}{\sigma(S_e)\sqrt{T}}]}{2\sqrt{2\pi T}\,\sigma(S_e)\Phi^2[\frac{m(S_e)\sqrt{T}}{\sigma(S_e)}]} \exp\{-\frac{m^2(S_e)T}{2\sigma^2(S_e)}\} \tag{32}$$

The conditional mean and variance of the fatigue life can be obtained from equation (32) according to equations (7) and (8), respectively.

Following the tradition, let $D_0 = 0$ and $D_{cr} = 1$. When $m(S_e)\sqrt{t} >> \sigma(S_e)$, equations (28) ~ (32) become

$$p(D,t|S_e) \approx \frac{1}{\sqrt{2\pi t}\,\sigma(S_e)} \exp\{-\frac{[D-m(S_e)t]^2}{2\sigma^2(S_e)t}\} \tag{33}$$

$$m_D(S_e) \approx m(S_e)t \tag{34}$$

$$\sigma_D^2(S_e) \approx \sigma^2(S_e)t \tag{35}$$

$$R(t|S_e) \approx 1 - \Phi[(m(S_e)t-1)/\sigma(S_e)t] \tag{36}$$

$$p(T|S_e) \approx \frac{1 + m(S_e)T}{2T\sqrt{2\pi T}\,\sigma(S_e)} \exp\{-\frac{[m(S_e)T - 1]^2}{2\sigma^2(S_e)T}\} \tag{37}$$

Accordingly, the conditional mean and variance of the fatigue life are

$$m_T(S_e) \approx \frac{1}{m(S_e)}[1 + \frac{\sigma^2(S_e)}{2m(S_e)}] \tag{38}$$

and

$$\sigma^2(S_e) \approx \frac{1}{m^2(S_e)}[\frac{\sigma^2(S_e)}{m(S_e)} + \frac{5}{4}\frac{\sigma^4(S_e)}{m^2(S_e)}] \tag{39}$$

respectively. Equations (33) ~ (39) are essentially the same as those obtained by Bolotin [3] on the basis of the central limit theorem.

For the fatigue model equation (11), equations (27) ~ (39) remain valid provided that $m(S_e)$ and $\sigma^2(S_e)$ are replaced, respectively, by

$$m(S_u) = 2^{\beta/2} f_e\Gamma(1 + \beta/2)(\frac{\sigma_x}{S_u})^{\beta} \tag{40}$$

and

$$\sigma^2(S_u) = 2^{\beta} f_e^2 \left(\frac{\sigma_X}{S_u}\right)^{2\beta} \sum_{n=1}^{\infty} c_n'^2 \tau_n \tag{41}$$

where

$$c_n' = \sum_{k=0}^{n} \frac{(-1)^k k!}{(n-k)!(k!)^2} \Gamma(1 + k + \beta/2) \tag{42}$$

For reinforced concrete, the time rate of fatigue damage increment is [4]

$$\frac{dD}{dt} = f_e e^{-\gamma(s-s_0)} \tag{43}$$

where γ and S_0 are material constants. Following the same procedure as that used earlier for steel, one obtains once again equations (27) ~ (39), with $m(S_e)$ and $\sigma^2(S_e)$ replaced by

$$m(S_0) = \frac{f_e}{2} e^{(\gamma s_0+\gamma^2\sigma_X^2/2)}\left[e^{\gamma^2\sigma_X^2/2} - \frac{\gamma\sigma_X}{\sqrt{2}} \Gamma(1/2,\gamma^2\sigma_X^2/2)\right] \tag{44}$$

$$\sigma^2(S_0) = f_e^2 e^{(2\gamma S_0+\gamma^2\sigma_X^2)} \sum_{n=1}^{\infty} C_n''^2 \tau_n \tag{45}$$

where

$$c_n'' = \sum_{k=0}^{n} \frac{(-1)^k n!}{(n-k)!(k!)^2} d_k \tag{46}$$

$$d_k = \sum_{i=0}^{2k+1} (-1)^i \binom{2k+1}{i} \left(\frac{\gamma\sigma_X}{\sqrt{2}}\right)^i \Gamma(1+k - i/2, \gamma^2\sigma_X^2/2) \tag{47}$$

4. Comparisons With Digital Simulation and Other Theoretical Results

Comparison was made between the results of the present theory and digital simulations for the example described in section 3.6 of Reference [1]. The conditional reliability function, the conditional probability density of the fatigue life, the conditional mean and variation coefficient of the fatigue life and the mean and standard deviation of the cumulative damage for damping ratio ζ = 0.01 and 0.005 are shown in Fig. 1 ~ 5. The Solid liens represent the results of the present theory and the crosses those of digital simulations. It is seen that the two results agree very well.

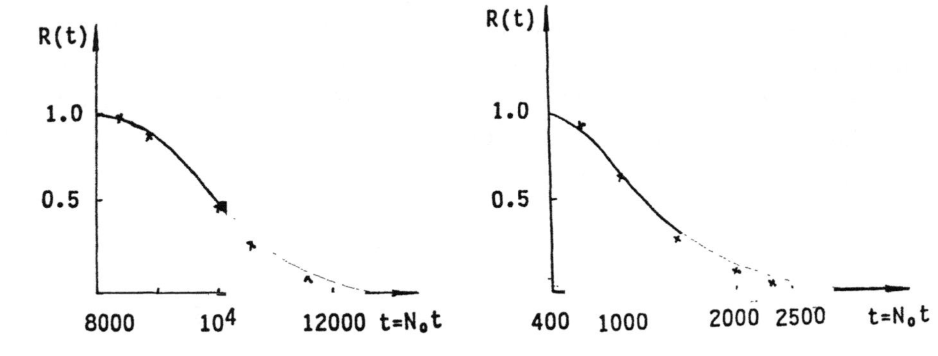

ζ = 0.01 ζ = 0.005

Fig. 1. Reliability function

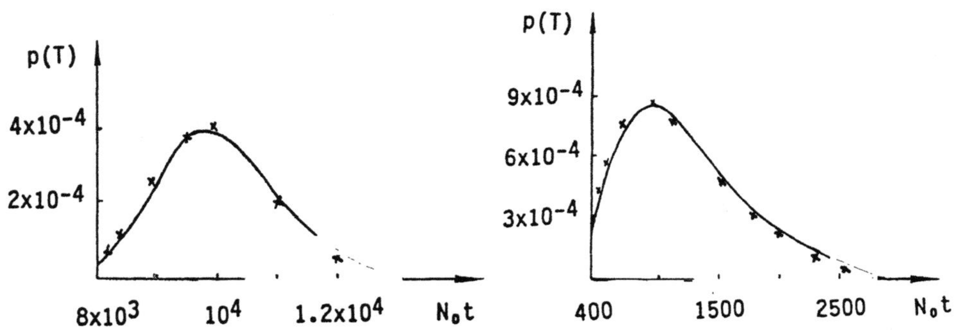

ζ = 0.01 ζ = 0.005

Fig. 2. Probability density of fatigue life

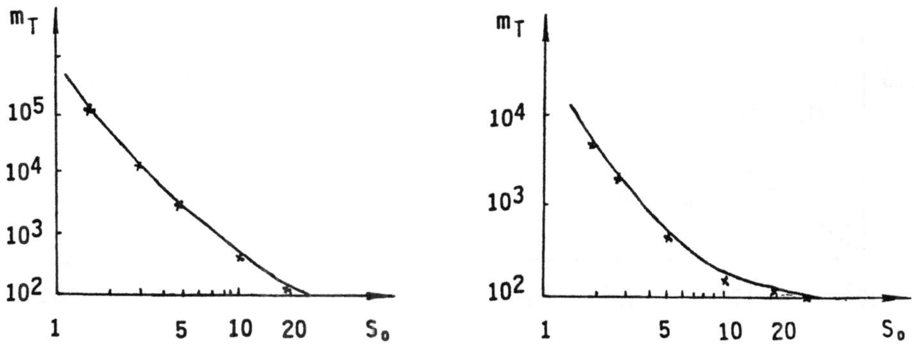

$\zeta = 0.01$ $\qquad\qquad$ $\zeta = 0.005$

Fig. 3. Mean fatigue life

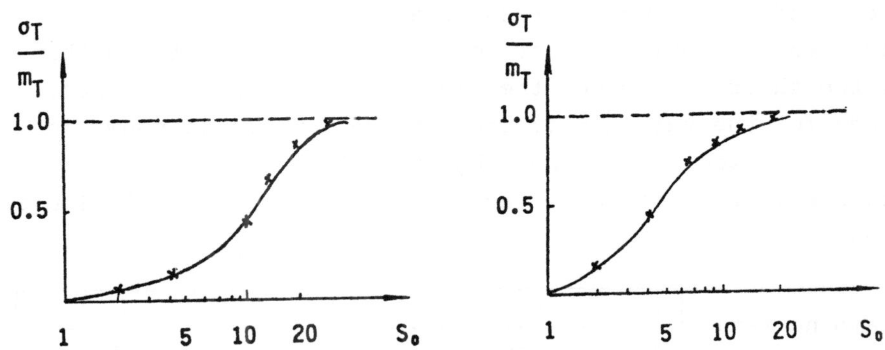

$\zeta = 0.01$ $\qquad\qquad$ $\zeta = 0.005$

Fig. 4. Variation coefficient of fatigue life

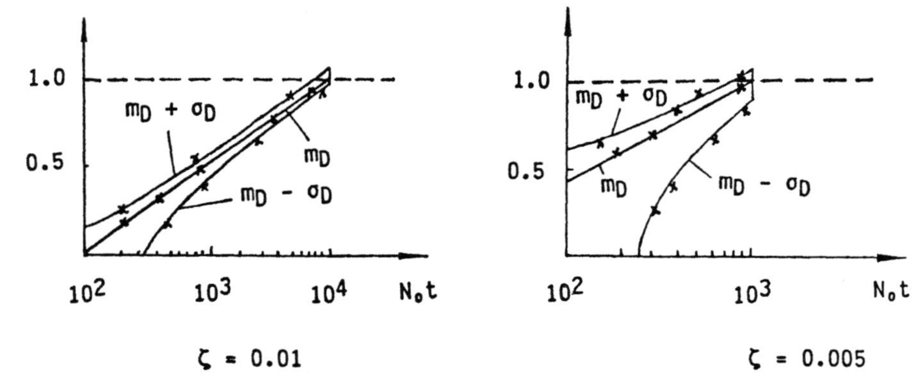

Fig. 5. Mean and standard deviation of cumulative damage

Comparison was also made between the results of the present theory and those due to Crandall and Mark[1] for the example described in section 3.6 of Reference 1. It was found that the two theories yielded the same mean value but different values for the variance of the fatigue damage, as shown in Tab. 1. It would seem that the present theory is more accurate in view of better agreement with the digital simulations.

Damping ratio ζ	σ_D	
	The present theory	Reference [1]
0.005	$0.108\sqrt{t}$	$0.15\sqrt{t}$
0.01	$0.84 \times 10^{-2}\sqrt{t}$	$0.13 \times 10^{-1}\sqrt{t}$

Table 1. Standard deviation of fatigue damage.

5. Concluding Remarks

A stochastic theory is developed for the analysis of cumulative fatigue damage on the basis of the Stratonovich - Khasminskii limit theorem. The analytical expressions for the conditional probability densities of the cumulative fatigue damage and fatigue life and for the conditional reliability function have been obtained for steel and reinforced concrete components subject to a narrow-band stationary Gaussian stress process with zero mean. The theoretical results agree very well with those of digital simulations. Although the theoretical results due to Bolotin on the basis of central limit theorem agree with the asymptotic values of our solutions, the present theory is mathematically more rigorous.

The theory proposed has many advantages. It is applicable to high - cycle fatigue problem for many kinds of materials. The stress can be a stationary process, a sequence of n stationary processes and a quasi - stationary process. The scatter and degradation of fatigue strength can be taken into account.

Acknowledgements

This research has been supported by the National Natural Science Foundation of China. The authors wish to express their gratitude to Professor Y. K. Lin for his reading of the manuscript and helpful comments.

References

1. Crandall, S. H.; Mark, M.D.: Random vibration in mechanical systems, Academic Press, New York, 1963.

2. Lin, Y. K.: Probabilistic theory of structural dynamics, Krieger Publishing Co., Melbourne, FL, 1976.

3. Bolotin, V. V.: Random vibration of elastic systems, Martinus Nijhoff Publishers, The Hague, 1984.

4. Schuëller, G. I.; Bucher, C. G.: Nonlinear damping and its effect on the reliability estimates of structures, in Elishakoff, I. and Lyon, R. H. (Eds), Random Vibration - Status and Recent Developments, Elsevier Science, Amsterdam, 1986, 389-402.

5. Stratonovich, R. L.: Topics in the theory of random noise, Vol. 1, Gordon & Breach, New York, 1963.

6. Khasminskii, R. Z.: A limit theorem for the solutions of differential equations with random right - hand sides, Theory of Probability and Application, 11 (1966) 390-405.

7. Zhu, W. Q.: A Method for estimating reliability of structures subject to Quasi - stationary and /or Quasi - homogeneous Gaussian random excitations, in Structural Safety and Reliability, Vol. 1, Proceedings, ICOSSAR'85, Kobe, Japan, 1985, 219-228.

8. Wisching, P. H.; Light, M. C.: Probability based fatigue design criteria for ocean structures, PRAC Project 15, Final Report, American Petroleum Institute 1979.

9. Gradshteyn, I. S.; Ryzhik, I. M.: Table of integrals, series, and products, Academic Press, New York, 1980.

10. Lin, Y. K.; Yang, J. N.: A stochastic theory of fatigue crack propagation, AIAA Journal, 23 (1985) 117-124.

MIX
Papier aus verantwortungsvollen Quellen
Paper from responsible sources
FSC® C105338

If you have any concerns about our products,
you can contact us on
ProductSafety@springernature.com

In case Publisher is established outside the EU,
the EU authorized representative is:
Springer Nature Customer Service Center GmbH
Europaplatz 3, 69115 Heidelberg, Germany

Printed by Libri Plureos GmbH
in Hamburg, Germany